GLOBAL MIGRATION, GENDER, AND HEALTH PROFESSIONAL CREDENTIALS

Edited by Margaret Walton-Roberts

Bringing together diverse approaches and case studies of international health worker migration, *Global Migration, Gender, and Health Professional Credentials* critically reimagines how we conceptualize the transfer of value embodied in internationally educated health professionals (IEHPs).

This volume provides key insights into the economistic and feminist concepts of global value transmission, the complexity of health worker migration, and the gendered and intersectional intricacies involved in the workplace integration of immigrant health care workers. The contributions to this edited collection uncover the multitude of actors who play a role in creating, transmitting, transforming, and utilizing the value embedded in international health migrants.

MARGARET WALTON-ROBERTS is a professor in the Department of Geography and Environmental Studies at Wilfrid Laurier University and the Balsillie School of International Affairs.

Global Migration, Gender, and Health Professional Credentials

Transnational Value Transfers and Losses

EDITED BY MARGARET WALTON-ROBERTS

University of Toronto Press
Toronto Buffalo London

Toronto Buffalo London
utorontopress.com
Printed in the U.S.A.

ISBN 978-1-4875-0520-2 (cloth) ISBN 978-1-4875-3175-1 (EPUB)
ISBN 978-1-4875-2373-2 (paper) ISBN 978-1-4875-3174-4 (PDF)

Library and Archives Canada Cataloguing in Publication

Title: Global migration, gender, and health professional credentials : transnational value transfers and losses / edited by Margaret Walton-Roberts.
Names: Walton-Roberts, Margaret, 1968– editor.
Description: Includes bibliographical references and index.
Identifiers: Canadiana (print) 20220139903 | Canadiana (ebook) 20220139997 | ISBN 9781487505202 (cloth) | ISBN 9781487523732 (paper) | ISBN 9781487531751 (EPUB) | ISBN 9781487531744 (PDF)
Subjects: LCSH: Medical personnel – Employment – Foreign countries. | LCSH: Medical personnel – Supply and demand – Foreign countries. | LCSH: Emigration and immigration. | LCSH: Labor mobility. | LCSH: Globalization – Economic aspects. | LCSH: Feminism.
Classification: LCC RA410.6.G56 2022 | DDC 331.12/79108861–dc23

This book has been published with the help of a grant from the Federation for the Humanities and Social Sciences, through the Awards to Scholarly Publications Program, using funds provided by the Social Sciences and Humanities Research Council of Canada.

We wish to acknowledge the land on which the University of Toronto Press operates. This land is the traditional territory of the Wendat, the Anishnaabeg, the Haudenosaunee, the Métis, and the Mississaugas of the Credit First Nation.

University of Toronto Press acknowledges the financial support of the Government of Canada, the Canada Council for the Arts, and the Ontario Arts Council, an agency of the Government of Ontario, for its publishing activities.

Canada Council for the Arts Conseil des Arts du Canada

Funded by the Government of Canada Financé par le gouvernement du Canada

Contents

Section 5. Recasting Brain Drain and Global Circulation

Tables and Figures

Tables

Figures

Acknowledgments

This edited collection emerges from a SSHRC funded international workshop held in May 2017 at the Balsillie School of International Affairs, Waterloo, Ontario Canada. Financial support for the development of the collection was gratefully received from the Social Sciences and Humanities Research Council of Canada, Wilfrid Laurier University Research Support Fund, and the Balsillie School of International Affairs. I would like to acknowledge the wonderful assistance of Busra Hacioglu and Marina Ghosh in planning and running the workshop, and Carla Angulo-Pasel and Zabeen Khamisa for their assistance in copyediting earlier drafts of the collection, and Hari Kc for indexing. I am grateful to all the workshop participants, and those who submitted their chapters for this collection. The comments of two reviewers were helpful in improving the focus of the collection, and the team at the University of Toronto Press, especially Jodi Lewchuk, were enormously supportive of the project. I would also like to acknowledge the scholarly, community, and personal support of the International Migration Research Center (IMRC) at Wilfrid Laurier University, including past directors Jenna Hennebry, Kim Rygiel, and Alison Mountz, as well as the many students, staff, faculty, and community members who have made the IMRC a splendid example of collaboration and research synergy.

GLOBAL MIGRATION, GENDER, AND HEALTH PROFESSIONAL CREDENTIALS

Introduction. Global Migration, Gender, and Health Professional Credentials: Transnational Value Transfers and Losses

MARGARET WALTON-ROBERTS

1 Origins, Inspirations, and Contribution of This Collection

This collection brings together diverse approaches and international case studies of international health worker migration to critically reimagine how we conceptualize the transfer of value embodied in internationally educated health professionals (IEHPs). The key objective is to offer a new analytical framework for interdisciplinary scholarship on health worker migration using the lens of embodied value and its transfer in the process of international migration. The core conceptual framing of the collection attempts to bring global value and commodity chain approaches from the development and economic globalization literature into dialogue with feminist transnational perspectives on global care chains and health care worker migration.

Such analysis is responsive to the feminization and increasing economic significance of the health sector and care work, especially in light of the growing demands for elder care; one of the most preeminent social challenges currently facing many higher-income nations.

The health care services industry is one of the most significant and growing parts of the global economy. The United Nations International Standard Industrial Classification (ISIC) categorizes the health care industry as hospital activities; medical and dental practice activities; and other human health activities that occur under the supervision health professions in various areas. In terms of value, a 2019 health care outlook report by Deloitte indicated that health care spending represents 10.2 per cent of global GDP (over $8 trillion) and is expected to stay at this level till 2023, with expenditure on global geriatric care likely exceeding $1.4 trillion by 2023 (Deloitte, 2019). On a per capita basis, health expenditures vary from $12,262 in the United States, to a mere $45 in Pakistan, reflecting the geographically uneven health landscapes through which international health worker migration takes place. As with other sectors, the role of labour in the health industry has been subject to division, standardization, and deskilling together with the rise of managerial superstructures (Rastegar, 2004). The health sector has seen widespread but differentiated engagement with New Public Management approaches to increase efficiency and enhance innovation (Pollitt & Bouckaert, 2017).

Analysis of international health professional migration brings into focus the evolving nature of global economic relations, since the supply of trained health professionals is increasingly sourced from lower- and middle-income nations such as India, the Philippines, Sub-Saharan African nations, and small Island States in the Caribbean and South Pacific. The role of the state in the production of this embodied value is wide-ranging and diverse. States provide the structures for education and professional testing, and create the taxation and investment contexts that support or deter private actors from engaging in these fields. The effective organization of these systems is central to generating and capturing more value, since the state can play an important role in facilitating skill upgrading, and can also develop new innovative ways to benefit from the global deployment of such labour. Health sector development and integration is also increasingly driven by global service orientation (Collyer & White, 2011). The globalization of health care systems currently underway is spurred on by what has been characterized as a "paradigm shift" in our understanding of the delivery of health services; namely, trade in health-related services (Lunt et al., 2011, p. 6). Interaction between states and global markets is essential to assess since, "state action and inaction creates the enabling conditions that shape whether and how firms, regions and nations are able to engage with global markets and their capacities to upgrade these engagements" (Neilson et al., 2014, p. 8).

The chapters in this collection explore the roles played by an array of state and market actors – education providers, migration intermediaries, accreditation agencies, medical tourism agencies, and formal and informal networks active in the production and valuation of IEHPs' professional credentials. The contributions uncover the multitude of actors, beyond the sending and receiving countries and migrants themselves, who play a role in creating, transmitting, transforming, and utilizing the value embodied in international health worker migrants.

While there are many excellent books on health care worker migration, especially nurses (for example, Choy, 2003; Flynn, 2011; George, 2005; Kingma, 2006; Reddy, 2015), they tend to be more micro ethnographic in their scale and approach. Policy and technical reports by international health and development agencies also examine health care worker migration, but typically structure debate in terms of bilateral transfers from one state to another. They also tend to examine the issue in terms of health worker to population ratios and health service delivery in national contexts rather than as global processes of value creation and circulation that entangle multiple actors and interests. There is an important and related field that explores medical training and competence (for example, see Hodges & Lingard, 2012), but this also tends to focus on *national* systems of education, and does not necessarily engage with the complexities of international credential transfer and the creation of value that takes place through transnational circulations and global amalgamations of expertise. There is a need for analysis of inter-state and multi-stakeholder interaction, but the multiscalar complexity of professional migration besets policy and academic research, which is constrained even at the national scale (Riley et al., 2012).

This collection addresses this complex realm of transnational value transfer in the field of the health professions, and offers a transnational and global orientation that analyses how the value embodied in these workers is generated, transferred, and reintegrated across a global health service market. To this end, the collection is highly indebted to the work of Nicola Yeates (2004, 2009, 2014) whose research extends the literature on global care chains to that of global nurse care chains, as well as offering analysis of the global governance of social policy, especially with regard to health worker migration and recruitment (Yeates & Pillinger, 2019). Yeates has highlighted how the networks and governance of health professional migration are well established, yet she argues that they do not speak to each other. Yeates suggests that in the past, we had to choose between one framework or the other, but now she asks if there is a way to strike up a dialogue between network and governance approaches. This collection makes an attempt at creating that dialogue between network (production) and governance (power asymmetry) approaches.

The book was in its final stages of completion as the global coronavirus pandemic occurred and health systems buckled under the weight of managing COVID-19 cases. The unprecedented global shutdown due to the pandemic exposed the weaknesses of public health systems globally. Furthermore the pandemic exposed how health care workers are made vulnerable in part through policy decisions regarding the amount of personal protective equipment available, the level of infection control in place, staffing ratios and how these factors interact with and intensify already marginal conditions of work. For example, in many countries, the spread of COVID-19 was most acute in long-term care homes, where agency or part-time staffing is common and workers often have to seek employment across different facilities to make a living wage; workers thus become vectors of the disease as they move between work sites. This situation played itself out in the first US COVID-19 outbreak at a Life Care Center nursing home in Seattle, Washington (Davis, 2020). Underinvestment in health and long-term care effectively undermines infection control and exposes health workers and patients to increased risk.

The experience in Toronto, Ontario, of the severe acute respiratory syndrome (SARS) epidemic in 2003 provided similar examples (Keil & Ali, 2008). The damage committed to health care workers through austerity and profit-seeking can be seen as a form of structural violence. This is arguably the case in the

United States where it has been argued that burnout among physicians and other health care workers is more appropriately viewed as "moral injury," since these workers struggle to provide patient care in a national health care system that is broken by profit-driven business models (Talbot & Dean, 2018). These factors contribute further to the need for more careful analysis of how the value embodied in health worker migrants is realized, distributed, and exploited. What also became apparent during the coronavirus pandemic was the international composition of the health care workforce in several member nations of the Organisation for Economic Co-operation and Development (OECD), revealing the truth that not only was the pandemic global, but so too was the supply of health workers.

2 Overarching Themes

The chapters included in this collection offer immense diversity regarding our understanding of health worker migration. This includes geographical, theoretical, disciplinary, and empirical variety. As a whole, the contributions highlight five important themes and new directions for research on the global migration of health care workers. These include the need (1) to forge dialogue between economistic and feminist concepts of global value transmission; (2) to theorize from different locations and escape methodological nationalism; (3) to recognize the increasing dynamism and geographical complexity of health worker migration; (4) attend to the embodied and embedded nature of value transfer and stratification; and (5) to escape methodological sexism and analyse intersectional complexity in the mobility and integration of health care and allied workers.

2.1 Forge Dialogue between Economistic and Feminist Concepts of Global Value Transmission

The critical synthesis offered by the chapters in this collection draw upon concepts of chains and networks in global value chains (GVCs), but enhance it through a deeper consideration of sex (gender and the household) and feminist analysis of care and global care chains (GCCs). Global economic value transfers include concepts such as commodity chains and global production networks (GPNs) (Coe & Yeung, 2015). Ravenhill (2014) considers these labels as largely interchangeable, and argues that chains are about actors and power – the asymmetry between actors, and how power is generated and distributed among actors. Coe and Yeung (2015) assert that "we see GPN analysis as something that can and should be applied to all industries within the global economy" (p. 28). Despite this claim, GPN research has not fully explored gender, services, or the role of labour as a value creator in GPNs (Christian, 2016; Flecker et al., 2013; Taylor et al., 2013). Both GPNs and GVCs have been critiqued for being overly productivist in their conceptual framing and application (Neilson et al., 2014). This is echoed by Dunaway (2014), whose explicitly gendered analysis of commodity chains critiques the overly productivist and gender-blind analysis of contemporary GVC work on the global economy. Application of these approaches to services industries is still relatively underdeveloped, but Coe et al. (2010) have made an important argument for the application of GPN approaches to services. More recently, Coe and Hess (2013) have urged the development of a transnational approach that highlights the role of global health service centres, global staffing agencies, professional associations, and global production networks in general to understanding the externalization of labour processes. Some have portrayed these externalizing practices optimistically, in terms of externalization creating new opportunities for skill development by various actors, but there is also pessimism, in terms of viewing the chains or networks as "risk-and-flexibility transfer schemes," which spread precarious employment and distribute risk to others in the chain (Frade & Darmon, 2005). This is clearly developed by a number of authors in this collection, including Crystal Ennis (chapter 7), Héctor Goldar Perrote and Margaret Walton-Roberts (chapter 13), Caitlin Henry (chapter 6), and Maddy Thompson (chapter 9). The role of labour is particularly underplayed in GPN approaches, since processes are viewed through the role of lead firms, and labour is only understood as either disrupting or claiming a share of value creation through labour agency (Barratt et al., 2019). What is not assessed is how labour is embedded in certain places and regulated territorially; how social reproduction sustains firms and workers, and the factors that shape how workers and firms interact in these activities. Highlighting the role of labour in the health sector is a particularly complex issue and one where GPN or GVC analysis has largely been absent. Nevertheless, it is apparent that "health worker migrants as skilled labour constitute an important commodity with exchange value in a global market" (Parry et al., 2015, p. 6). Forging a conversation between economistic and feminist concepts of global value transmission and terms of transfer must be developed in relation to health care workers.

Antecedents of such conversations already exist. Wilma Dunaway's (2014) edited collection, *Gendered Commodity Chains*, explored how early global systems thinking in commodity chains identified gender and the household as an essential element of the surplus value process in global production, which has been lost in subsequent economic analyses. Dunaway (2014) challenges literature on service and production systems in the globalizing economy by making an explicitly gendered and social reproduction (care services) intervention, which builds upon and advances earlier work in the field of gender, care, and the global economy.

Industrial systems externalize labour process costs – the social reproduction of the workforce – and women overwhelmingly carry that burden. The "commodity chain structures the maximal exploitation of underpaid and unpaid labor," and this is manifest in the unequal exchange "embedded in the gendered relations of the household" (Dunaway, 2014, as cited in Yeates, 2014, p. 177). Nicola Yeates's (2004, 2009, 2014) work on global care chains and global nurse care chains and her analysis of the global governance of social policy are also germane to this effort. Yeates (2014) elaborates on the commodity chain approach of the world systems theory where chains are of varying length and are differentially networked between and across multiple geographic units that are "tethered to geographic hierarchies" (p. 177). Feminist-inspired research on services in the care economy (child, dependent and elder care, health and personal services) is evident in literature on care chains (Yeates, 2009), but this approach has not been incorporated into broader literature on global value chains and production networks. This deficiency must be addressed, since the corporatization, commoditization, and privatization of health and personal care services mean these sectors are an increasingly central feature of the productive economy. The time is ripe to apply the critical aspects of both the care chain and value chain approaches to the global mobility of health workers, the interaction of which has not been widely engaged with by either health or economic researchers (Connell & Walton-Roberts, 2016). By prioritizing more diverse approaches to registering value transfer in the case of international health migration, this collect attends to this effort to bring these two approaches into dialogue.

2.2 Theorize from Different Locations and Escape Methodological Nationalism

The second theme is the importance of theorizing from different locations and escaping methodological nationalism. For example, educational systems are not only oriented to national demand; the international employment context infiltrates and reshapes the conditions of training, and this has consequences regardless of whether an individual worker engages in international migration (Ortiga, 2014). This is clear in migrant sending regions such as the Philippines (see Thompson, chapter 9), but also in mostly migrant receiving countries, where training systems and health worker accreditation processes are increasingly attuned to internationally educated migrants. This is clear in practices such as bridging courses (see Bourgeault, Atanackovic, and Neiterman, chapter 5), changes in labour market and immigration regulations (see Sweetman, chapter 3; Jafri, chapter 11), and in the modes of regulatory structures employed to govern, control, and regulate foreign trained health workers (see van Riemsdijk, chapter 10; Ennis, chapter 7). Theorizing from different locations attunes us to the work of intermediaries (public and private) that are increasingly responsive to regulatory changes occurring at the national and global level of governance (see Chikanda, chapter 8), as well as understanding the spatial and temporal complexity of the migration pathways emerging (see Crush, chapter 16). Furthermore, by theorizing from the perspective of the institutions and societies that receive the value embodied in workers but integrate them into workplace contexts where their contribution is marginalized or devalued, we develop radical new readings of how these workers generate and contribute immense amounts of knowledge relevant to their societies over the short and long term (see Raghuram, Bornat, and Henry, chapter 17). All these examples make it clear that analysis must escape the closure of national borders when accounting for the value and contributions that states gain from the migration of health workers.

2.3 Recognize the Increasing Dynamism and Geographical Complexity of Health Worker Migration

Third, the collection highlights the need to recognize the increasing dynamism and geographical complexity of health labour flows, in terms of both their spatial and sectoral characteristics. In terms of the geographical complexity of migration, chapters include the dominant processes of South-to-North migration (see Adekola, chapter 15; Chikanda, chapter 8; Thompson, chapter 9; Connell and Negin, chapter 12; Henry, chapter 6), South-to-South movement (see Kaspar, chapter 2; Ennis, chapter 7), and east to west migration (see van Riemsdijk, chapter 10). In terms of sectoral complexity, the collection makes plain the hybrid nature of health care (public and private and combinations of this), how globalization processes are fundamental to this and related phenomena (peripatetic professions, global corporations, eHealth, trade in health services, increased use of technology in care, the privatization and outsourcing of care). Likewise, the human needs related to elder care are challenging social welfare systems, which in turn have come to depend upon integrating international workers into formal and informal care systems. This can happen through secondary processes of migrant labour integration (see Connell and Negin, chapter 12), as well as through new forms of care delivered through human-robot interaction, adding ethical and legal complexity to our analysis of this phenomena (see Goldar Perrote and Walton-Roberts, chapter 13). Chapters also engage with the changing nature of systems and flows, seeing migrants embedded in new types of networks that allow for different forms of circular migration to occur across diverse temporal frames (see Crush, chapter 16). Revealing these processes, the collection illustrates the importance of geography and place to how the global movement of health care workers is regulated. There are regional systems of organization structured through colonial and postcolonial networks, as well as constantly shifting engagements with new sources for migrant workers based on chains and networks created through economic globalization and regional pressures of integration/disintegration. Regionalization and economic processes can lead to integrative pressure in training and assessment processes, but the variability in training, assessment, and workplace relations and expectations (both formal and informal) must be understood as rooted in specific histories and geographies of health care professional development (see Raghuram, Bornat, and Henry, chapter 17; van Riemsdijk, chapter 10; Ennis, chapter 7). The puzzle of fitting those systems together is manifest in the assessment of each foreign-trained professional, and such processes account for how and if the value embodied in the worker is revalued in different jurisdictional contexts.

2.4 Embodied and Embedded Nature of Value Transfer and Stratification

Fourth, the chapters in this collection critically embrace the idea of embodiment and embeddedness. The reorganization of employment and labour processes that occurs within the GVCs and GPNs must be assessed from the perspective of major labour market divisions, including but not limited to those of gender, race, and nationality. These contexts are relevant to the increasingly globalized circulation of health care workers, since the initial training of workers and the competencies that workers develop and represent in an increasingly global market normally begin with domestic training in their home context, where intersectional identities such as gender, class, region, religion, and sexuality determine the educational decisions made, the dominant training models in place, and the employment and migration opportunities accessed. This recognition of the importance of place and identity is a key feature of critical geographies of migration that are "increasingly addressing the relationship between migrant status, race and sexuality, adding to a longer-standing interest in migration and gender" (Gilmartin & Kuusisto-Arponen, 2019, p. 18). Gilmartin and Kuusisto-Arponen (2019) maintain that a focus on precarious care work and labour migration productively details embodied geographies and "articulates both global political and power structures and also multiple embodiments of human and institutional settings, which are part of the wider processes of social reproduction" (p. 24).

Chapters in this collection highlight the international value transfer that accompanies migrant health workers' mobility though chain and network dynamics. This includes detailed analysis of the historical stratification of racialized workers into institutional settings (see Peralta, chapter 14; Raghuram, Bornat, and Henry, chapter 17) and current manifestations of new forms and features of such stratification (see Goldar Perrote and Walton-Roberts, chapter 13; Connell and Negin, chapter 12). Chapters have also addressed how this stratification and devaluation of migrant workers might be resisted or overcome in terms of engaging with ongoing education (see Adekola, chapter 15), creating spaces to salvage identity and reassert value (see Peralta, chapter 14; Raghuram, Bornat, and Henry, chapter 17), and generating new models to promote effective integration of international health workers (see Bourgeault, Atanackovic, and Neiterman, chapter 5; Bauman, Crea-Arsenio and Antonipillai, chapter 4). There are also chapters that relate the type of policies and processes that can ameliorate aspects of this activity (see Sweetman, chapter 3; Jafri, chapter 11). By understanding the messy reality of the embodied nature of such movement, we can incorporate the complexities recounted by Yeates (2014) and others in terms of conceptualizing global care chains and revisions to it (see Henry, chapter 6), and move beyond current interpretations of the nature of work and workplace relations (Goldar Perrote and Walton-Roberts, chapter 13). Chapters in this collection have also employed approaches that resist atomizing the migrant; rather, they understand how migrants are embedded in households, communities, institutions, networks, and chains, how these can structure the migration process, and offer new interpretations of workers' embeddedness as members of a global intimate workforce (see Henry, chapter 6). There is also some recognition of how migrants are embedded in new relations across the human-robot frontier (see Goldar Perrote and Walton-Roberts, chapter 13). The multiscalar analysis of the transmission of care, the messy fleshy reality of this process (separated families, intimate patient care, racialized hierarchies), together with the political, policy, and corporate structuring of health care systems reveals a dynamic landscape deserving of far deeper analytical engagement from those fascinated with the current structure of the global economy.

2.5 Escape Methodological Sexism and Analyse Intersectional Complexity in the Mobility and Integration of Health Care and Allied Workers

Fifth, we need to escape methodological sexism by describing "how care can be gained, learned and improved, including in migration contexts" (Dumitru 2014, p. 210). Dumitru (2014) offers a critique of the global "care drain" discourse, which she deems as "methodological sexism" because it builds on sexist stereotypes, it devalues and misrepresents care work, and "it misses the opportunity for a theoretical change about how skills in migration contexts can be understood" (p. 203). This collection takes Dumitru's directive to heart, especially the call to develop "a more dynamic account of skills" by paying closer attention to international skilled migrants and their position in the increasingly important area of health services. One way to do this is to explore the health sector, including the role of care workers within it, not from the position of naturalized and devalued feminine traits, but as codified, regulated, and authorized forms of skill and competency that add value to patients' lives, workers' incomes, employment sectors, corporate health organizations, and national economic accounts. This collection examines the transfer of value through GVCs and GPNS, but rarely has the focus of these approaches been on services and labour, and even less on the gendered dimensions of these processes. The gender implications of health worker migration are particularly significant, since women account for almost half of today's 191 million international migrants, and female migrants are overrepresented in the health and personal care sectors of the global economy, which are often undervalued in terms of income security and status (Valiani, 2011). While the specific policy context of IEHPs' integration into different national labour markets varies (Picot & Sweetman, 2011), there is a structural trajectory of convergence of IEHPs being incorporated into national health systems facing restructured or diminished state spending (Williams, 2012; Yeates, 2009). In light of these structural transformations, we need to see how different types of gender hierarchies intersect in ways that might under value or devalue the work delivered. In terms of migration, sending states might exploit their patriarchal systems to create competitive conditions for the "export" of labour or goods, and the receiving states of such goods or labour can likewise find opportunities in their own systems of gendered hierarchy to facilitate cost savings. In the case of migration, we can see this in terms of labour market and migration policies that undervalue women's skills:

> As immigration states become more competitive in the race for talent and as selecting nations place greater emphasis on human capital credentials, language abilities, vocational skills and work experience, the importance of gender is amplified. In short, the global race for talent is gendered, with significant implications for the skill accreditation, labour market outcomes, rights of stay, gendered family dynamics, including freedom from domestic violence and financial independence, of female immigrants. (Boucher, 2016, p. 3)

Deeper analysis of skills development, sex or gender considerations, and their role in the expanding health service sector enhances our understanding of contemporary global economic processes and can provide openings for new policy approaches and demands to fully account for the value embedded in health worker migration.

3 Overview of the Book

3.1 Key Debates

This collection's opening section includes two chapters, one by political scientist John Ravenhill on GVCs and one by geographer Heidi Kaspar, who examines transfers of care in transnational medical travel. These two chapters illustrate the use of two different conceptual approaches: productivist GVCs on the one hand, and critical feminist theorizing of the value of care on the other.

In chapter 1, John Ravenhill provides an overview of the GVC literature and its limitations with regard to illustrating the political economy of development in the contemporary global economy. Those limitations include a paucity of theory, whether or not the terminology of GVCs captures the complexity of production, and the relative lack of critical enquiry into the distributional aspects of GVCs. Ravenhill highlights two additional critiques that are most relevant to the issues discussed here: the lack of focus on services and labour. The focus on GVCs attempt to identify the source of value that is added into the manufacturing or service process; this is something traditional trade statistics do not tell us. Services are significant in terms of value added, and as Ravenhill indicates, services bring in contributions from the higher value added component of the "smiling curve." Deeper analysis is needed to understand how lower-income nations might benefit (or lose out) from the services provided by their citizens when they engage in international migration. Ravenhill makes the point that the issues of gender, household labour, paid and unpaid labour, and the complexity of how to value care due to its intangible features, provide challenges to the GVC formulations. Care itself is an immensely challenging "product" to assess because of its non-economic nature (Folbre, 2006). Despite these challenges, as Ravenhill writes in chapter 1, thinking about how to value care offers an "important application and extension of some of the core ideas of the value chain literature."

In chapter 2, Heidi Kaspar examines these intangible transfers of care within transnational medical travel. She offers an entirely different reading of value and what kind of values are transferred in global health circulations. She challenges us to think about the kind of values we measure and those that we imagine are important to people's experiences of health care. Medical travel represents situations where those in need of care become mobile to access needed specialist care, often at an affordable cost. Kaspar argues that we need to broaden GCC research in order to elaborate on the inverse mobility of patients and care recipients (generally moving from higher- to lower-income nations). The content, direction, and nature of these transfers are increasingly heterogeneous and this diversity is important to consider in terms of the resulting distribution of resources and power hierarchies formed. Kaspar focuses on embodied care work, which is typically an underexplored feature of transnational health care. The quality, nature, and extent of actual care, not just the medical procedures on offer, are an important reason why patients travel for medical treatment to certain places. Her research suggests that patients may be escaping the "rationed" care delivered in austerity-driven health systems in the Global North, to seek care, in all of its intangible dimensions, in other locations where care is more plentiful, and produces a better context for the provision of health services. The quality of care is indeed a complex feature of the asymmetries involved in health and other care services, and Kaspar details the relational complexities involved and how they are spatialized. Medical travel involves multiple actors not just the medical and care staff, it includes family caregivers, translators, and facilitators. A whole industry has emerged around this medical travel process, revealing another new dimension and sector of health care–related employment for middle- and lower-income nations. Kaspar examines how to coordinate these transfers of care – as a gift and/or commodity – and gives us detailed insight into the role of facilitators who fashion their work as a gift rather than as a commodity for exchange. Does this "love labour" indicate surreptitiously commodified ties, or does it reveal how health care itself loses value if the deeply humanizing traits of care, love, and reciprocity are rationed, measured, and ultimately converted into a commodity?

3.2 Conceptualizing Workplace Integration and Stratification: Immigration Policy, International Credentials, and Intersectional Disadvantage

Section two examines the transfer of skills and credentials in practice, paying attention to immigration and workplace policy, credential recognition barriers and solutions, and the importance of intersectional disadvantage and how it shapes migrant health workers' experiences. The chapters in this section illustrate some of the complexities of applying a GVC approach to health services workers, because the "product" or worker is not homogenized nor are the workplaces. This adds important complexity to how value is generated, recognized, measured, transferred, and rewarded. The chapters in this section highlight various institutional mechanisms that accompany the integration of workers, and how immigration regulation is part of the skills development/devaluation pathway that transforms alongside the international mobility of health care workers.

In chapter 3, Arthur Sweetman details one of the key problems in the area of health worker migration: the fact we have a series of odd imbalances in terms of facing a global health care worker shortage/crisis that has resulted in policies to limit the damage done by health care worker migration, (e.g., World Health Organization code on recruitment), yet in Canada we see a surplus in many health professions. In early 2020, HealthForceOntario (an agency of the Ontario government) announced that in Ontario alone there were 13,000 foreign-educated physicians and 6,000 foreign-educated nurses not working in their trained field (immigrants as well as Canadians who studied abroad). Comprehensive numbers for other provinces are not readily available, and even in Ontario only those who voluntarily step forward are counted by HealthForceOntario. Although only reflecting a portion of the national situation, these point to the magnitude of the problem. Sweetman considers that most of these immigrant health professionals will never be licensed, so these people will end up working in alternative careers. Sweetman argues this reflects a policy failure – a lack of coordination between federal and provincial governments – as well as, historically, a failure to communicate and put in the effort to understand the regulatory process framing immigration and licensure on both immigrants' and policymakers' sides. Sweetman details the complexity of health professional labour markets and the added complexity of integrating immigrant professionals. He suggests there are policy options to address this, he references the Ontario Fairness Commissioner and Fair Access to Trades legislation (see Jafri, chapter 11), but this will not solve the problem of their being too large a number of potential workers. Rather, he argues that we need to address the "front door," in that regulated health professionals need better pre-migration information. He also suggests that regulated health professionals be taken out of the federal immigration pathways, with the provinces (which actually manage the health care system) taking the lead on the recruitment of health workers.

Sweetman's suggested immigration policy change is one means by which employers (in this case the provincial government) can be drawn more fully into workforce planning (something Bourgeault and her colleagues in chapter 5 also encourage). Sweetman's chapter and his earlier collaborative work (Olaizola & Sweetman, 2019; Sweetman et al., 2015) provide a valuable overview of how economists interpret the issue of professional credentials, skilled immigration, and labour market outcomes. The earlier work of Sweetman and his colleagues (2015) indicates there is surprisingly little research on credential issues, especially outside of the United States, but nurses have been one of the occupational groups identified early on by economists interested in these issues (for example, see White, 1987).

Andrea Baumann, Mary Crea-Arsenio, and Valentina Antonipillai in chapter 4 examine Canadian workforce issues in terms of the integration of IEHPs. They highlight the increased rate of international migrants in the health sector in OECD nations. Looking at the Canadian case, they highlight the importance of developing health workforces that reflect national demographic complexity (for purposes of appropriate and safe treatment of the population). Similar to Sweetman (in chapter 3), Baumann, Crea-Arsenio, and Antonipillai identify how the orthodox supply/demand model does not work well in health care and consider the failure of past efforts to forecast labour force issues. The complexity of labour market planning is heightened by including international migration and the vagaries of immigration policy. Policy formulation and change now occur more rapidly, with some immigration policies lasting only a few years. As the inclusion of IEHPs becomes more important, as their numbers and the demands for culturally appropriate health services increase, policy makers and health force planners need to focus immigration policy to align with labour market needs, rather than immigration and credential barriers driving changes in labour market distribution. The authors highlight the labour market complexity that IEHPs, especially internationally educated nurses (IENs), are facing.

There are spatial restrictions at play here as well, for example, in Ontario the majority of employers are around the Greater Toronto Area (GTA), but outside of the GTA employers face challenges in filling jobs and matching candidates to the available employment. IENS also have to manage the specific culture of hiring and the nature of work across different health institutions. The authors recommend that employers develop best practices, mandates, and a vision that includes diversity, all of which is needed to serve an increasingly diversified community.

Ivy Bourgeault, Jelena Atanackovic, and Elena Neiterman in chapter 5 present a framework that explains professional migration and integration in the context of multiple differences that accompany the international mobility of health workers, particularly gender, racialization, and class. Their framework engages with the idea of glocalization, where local integration processes and adaptations and global forces intersect. The authors examine the integration process for IEHPs, highlighting how intersectional factors, especially gender, connect with workplace, family, life cycle, and other contextual factors. These intersections can produce isolation and precarity, even when workers are integrated into the labour market. An important element of the workplace integration process is bridging programs, which are developed to assist IEHPs to bridge competency gaps and enter the professions for which they have trained. While bridging programs tend to focus on individual needs and identified or assumed competency "gaps," the authors also examine systemic issues in terms of changing the outlook and orientation of higher education institutions towards understanding the needs of different populations, including IEHPs. The authors argue for a tighter focus on health workforce planning and how it intersects with the diverse pathways IEHPs utilize. This includes the need for more research on the rising importance of international study-work pathways, and more sophisticated assessments of how gender, race, and class inform worker integration into the elder care sector.

In chapter 6, Caitlin Henry further develops an intersectional analysis of migrant health care workers' experiences by providing a careful examination of IENs' integration experiences in stratified workplaces in the United States. Henry conceptualizes the idea of the global intimate workforce, a concept that echoes Yeates's (2009) idea of the global nurse care chain. Henry argues that research on IENs under examines the relevance of scale and intersectoral processes. Based on interviews with IENs in the United States, the author shows how GCC literature is too linear and fails to capture processes that social policy and care scholars such as Fiona Williams (2012) have identified. For example, too strong a focus on the household can be problematic, since it does not capture structural factors such as the care surplus in the Global North and the resulting unevenness that contributes to the global care deficit. Henry builds on the idea of the global intimate (following Pratt & Rosner, 2012) to overcome the idea of spatial binaries such as local/global. These terms are not defined against one another; rather, they are elliptically related domains. Intimacy expands far beyond the household or private sphere and reveals the mutual constitution of global and local and their recurring patterns of intimacy. When applied to nursing, Henry shows the multiple ways subjects manage intimacy at work and home. Communities create forms of social reproduction through collective care work that deviates from the GCC norm, for example husbands who are "astronauts" (leaving the home for an extended period of overseas work) play a role in care as do visiting relatives who "drop in" and provide care. Some Global North communities face care deficits, which are presented as a technical spatial issue rather than as a structural reality of a system that produces poor, racialized hospitals. These areas are served by nurses with special IEN visas, which has the effect of devaluing the workers who serve devalued populations (also see Raghuram, Bornat, and Henry, chapter 17). In effect, this case reveals how the United States maps its own interior national spatial inequalities to global inequalities through the channelling of IENs to work in poorly served community hospitals. The politics of the provision of health care workers in this case is not a zero sum game of workforce planning - some here, some there- but invokes deeply ethical and material decisions and consequences for patients and workers alike.

3.3 Transnational Health Mobilities: Networks, Regulation, and Intermediaries

This section examines the web of actors and regulations structuring the transnational dimension of value transfer that accompanies the movement of health care workers. These three chapters illustrate the geographical dynamism and regulatory complexity shaping the (im)mobility of health care workers and their workplace experiences, and how international demand infiltrates into national contexts and reshapes the identities and experiences of health workers.

In chapter 7, Crystal Ennis examines the plight of IENs in the Arabian state of Oman, who she interprets as being in an in-between profession, nestled between the lower skilled construction and domestic migrant workers and "ex-patriate" higher skilled workers. Positioned in limbo in this space, Ennis reviews how this labour is managed through what she terms a multiplex migration governance system shaped by Oman's *kafala* system. Ennis focuses on the networking that occurs through multiple levels of interactions embedded in these labour market processes. Gulf Cooperation Council (GCC) countries, especially Saudi Arabia and United Arab Emirates, represent an important South-to- West migration corridor that has substantial economic importance. The money that migrants send home in this corridor represents over one sixth of global remittances. Researchers have noted that there is "nothing more permanent in the Gulf than temporary workers." Labour market segmentations in

Gulf labour markets are formed through the multiple axes of identity, nationality, gender, skill level, and location in the public/private sector. Gulf migration is mainly focused on recruiting lower skilled workers, but many migrants are higher skilled, especially in health and education. *Kafala* is a labour sponsorship or guest worker model that applies to temporary workers when they are resident in Oman for a fixed period of time. Oman's labour market is marked by attempts to increase labour flexibility while facing relatively high levels of domestic unemployment. Within Oman, nationals and non-nationals are not competing for the same jobs, since the cost of their labour is vastly different. Global policy interaction with the *kafala* system is a useful indicator as to how the labour market is transnationally constructed and how value is created within this policy framework. National labour within Oman is overvalued, making the continued use of migrant workers relatively cheap, flexible, and productive, which makes it challenging for employers to move away from their reliance on migrant workers. The Oman case demonstrates how regulatory norms have embedded the use of migrant health care workers into regional systems, how dependent health care employers have become on this flexible workforce, and how nationalization efforts on the part of national governments cannot easily substitute national workers for international migrants.

Abel Chikanda in chapter 8 asks how migration governance has changed the way that recruiters work in the health field, especially in light of global codes on the recruitment of migrant health workers. He explores the various types of intermediaries involved in health care worker migration. These intermediaries can be state led or they can represent a placement model or agent hire (body shopping) approach. Evidence suggests a decline in the number of recruiters in the United States. Issues of ethical recruitment may have influenced this process, but there are questions about the ethical right of those workers who are located in "crisis countries" to emigrate in order to improve their standard of living. Chikanda comments on how difficult it is to get a handle on the scale of recruitment in any systematic way, but he focuses on key destination markets to illustrate recent trajectories in the hiring of IEHPs. His observations indicate the importance of understanding the waxing and waning of flows between major supply and demand nations, and the kinds of policy instruments that curtail (for example, visa retrogression in the US) and facilitate (Australian rural workforce planning) these movements. Part of the mandate of ethical recruitment is to reduce the problematic recruitment of health professionals from regions that are facing a crisis in terms of the ratio of health care workers to patients. This approach may actually cause significant problems in light of the growing practice of overproducing health workers for export and not investing in hiring them at home, which is certainly evident in India and the Philippines. Recruitment intermediaries have reorganized in light of the global codes of practice, consolidating in some cases and changing their methods in others in order to bypass some of the restrictions imposed by the code, while still appearing to be in accordance with the demands of employers. Chikanda's chapter helps us understand how intermediaries are responding to governance changes, and how their role in transmitting the value of health care workers between different jurisdictions remains vital, even as the nature of their work alters in response to governance frameworks that attempt to minimize the damage of global health worker mobility.

In chapter 9, Maddy Thompson provides an analysis of how international recruitment reshapes domestic conditions of employment for nurses in the Philippines. Over 20,000 Philippine nurses go overseas each year, yet 200,000 to 300,000 nurses remain unemployed at home. The Philippines as a state has to adapt to changing international demand and it must maintain its "brand" for its migrant nurses. International mobility is endemic in terms of how it shapes the conditions of employment for nurses in the Philippines, and this reality has come to shape and deform how value is created within the national system of training and employment (Ortiga, 2014). Thompson provides a review of the different actors involved in creating and maintaining what is effectively a system of learned subordination, which, from a state perspective, contributes to a national comparative advantage when it comes to training nurses. Current developments in the Philippines service international demand, which is reforming the very nature of nurses' education, training, and working conditions in the Philippines, regardless of whether they actually venture overseas. Thompson highlights the corrosive nature of hospital volunteerism demanded of recently graduated nurses in the Philippines, and the degree to which state and non-state actors are complicit in this exploitation. This represents a transnational interpenetration of global exploitation in occupationally specific ways. The state has become a key actor in producing migrant nurses, but even before nurses become "migrants," their subjectivity is manipulated through the intersection of domestic nursing with international structures (see also Peralta, chapter 14). This provides an example of how nurses' agency is constrained even before they have left the Philippines. This doubling back and recirculation of exploitative relations in the sending nation raises questions about the specific levers and enforcers of processes of devaluation, and signals the need to expose and address multiple transnational channels of exploitation.

3.4 Domestic Policies in Receiving Countries: Value Transfer, Integration, and Regulation

This section looks more deeply at some examples of how receiving countries value, integrate, and regulate IEHPs. Here we can see how chains and networks are operating to position and regulate worker mobility and professional integration, the asymmetrical nature of the power of the "multiplex" actors in this process, and policy differences in the approaches used to translate and translocate workers' skills into specific health care systems.

In chapter 10, Micheline van Riemsdijk explores how nursing training systems makes the placement of IENs in different national contexts a regime of complexity. Valuation and transfer of credentials is subject to geographic variability. From an ethical perspective, the nurse-to-patient ratio is an indicator of the surplus of health care workers that exists in places like Norway and Switzerland. Most IENs in Norway are from the EU with a relatively small number from the Philippines, and most (70 per cent) of these workers are employed in elder care. Of those with labour market authorizations, the largest countries of origin are Denmark and Sweden, whose mobility issupported by Nordic mutual recognition agreements and European Union directives on mutual recognition for nurses. Language cannot be a reason for denial of authorization by the state, but it can be for the employer (for example, safe handling of medicine does require certain language tests in Norwegian). Other factors of identity obviously intersect with this, and in some cases can act as "shorthand" for the administrators in terms of their assessment of the applicant and the quality of their education. Van Riemsdijk's chapter reveals interesting cases of "the Swedish trick" where internationally trained nurses "game" or arbitrage national differences encoded in Norway's credential assessment system in order to benefit from mutual recognition agreements.

In chapter 11, Nuzhat Jafri discusses the role of an innovative regulatory body in Ontario Canada, the Ontario Fairness Commissioner (OFC). The OFC overseas the work of forty-two regulatory bodies and twenty-two trades in Ontario. The OFC assesses the processes used to evaluate the credentials of international applicants applying to join professional associations. The OFC stipulates that testing should be fair, transparent, and objective, and that fees charged to applicants must only cover cost recovery. In 2016, 1,153,273 people in Ontario were eligible to work in a regulated profession, and of these, 128,431 were internationally trained. India, the United States, and the Philippines are the top source countries for internationally trained applicants in Ontario. In many cases, the work of the OFC has resulted in novel approaches to improving the credential assessment process, and in response to the concerns of the OFC, and with federal support, many health professional associations have redesigned their assessment processes in order to improve credential recognition and transfer processes. There remain some key challenges, but the OFC is an example of state policy intervention that can continuously work to improve the processes that frame the international mobility of health workers and improve the efficiency of transferring the skills and credentials of IEHPs.

John Connell and Joel Negin in chapter 12 review the changing demand for "aged care workers" employed in aged care facilities in Australia; the largest single occupational group employed in such facilities. Connell and Negin review the regulatory conditions that govern work in this sector and the skills of migrants working in it. As of 2016, there were more than 160,000 non-nurse aged care workers in Australia, some with international nursing credentials. Those entering the sector include migrant workers who have arrived in Australia through other migration pathways (since there is no direct route for the low skilled occupation of "aged care worker"). Many of these workers do have previous skills or qualifications relevant to this field, but Connell and Negin found extensive language limitations evident in the workforce. This reflects a notable shift in the nationality of those working in the aged care sector between 2006 and 2016 away from the UK and European origin migrants, to those from Asia and Africa. The largest increase in foreign workers in the care workforce has been those from Nepal followed by Nigeria, Pakistan, and India. These new migrant workers tend to cluster in urban cores, which suggest limited access to services for the elderly in rural and more remote regions of Australia. This case reflects how the consequences of international migration interact with the prevailing elder care context in Australia, a case where social policy needs have not been meaningfully planned. Connell and Negin's chapter highlights the concern that Australia's elder care policy is effectively relying on an approach that devalues international care workers by integrating them into a feminized and racialized system that provides needed, necessary, and important care, but care that is symbolically, structurally, and materially devalued.

In chapter 13, Héctor Goldar Perrote and Margaret Walton-Roberts examine the super-aging society of Japan and how care demands are being addressed through a combination of human and robot labour. The chapter reflects on the practical, technical, and ethical challenges of generating and realizing value in human-robot interaction in elder care in Japan. Skill mismatches in the labour market, women's incorporation into the workforce, and the pressures derived from an aging society are forcing Japan to develop new approaches to utilize foreign labour for the provision of elder care. Since 2008, Japan has recruited more migrant nurses and caregivers and has created more visa categories to attract care workers, some of whom are eligible for long-term settlement. The nature of these policies is symbolic of the more liberal tendencies that are reshaping Japan's generally restrictive immigration laws. Rather than simply import more labour to solve its elder care crisis, one solution pursued is to invest in robotics. This chapter explores some of the technical, social, and ethical challenges linked to human-robot interaction in contexts where migrant labourers might enjoy fewer rights than robots. The case of Japan is an important one to reflect upon as we consider how the increasingly significant challenge of elder care will be addressed globally.

3.5 Recasting Brain Drain and Global Circulation

The final section of this collection focuses on the workplace as a convergence of the multiple socio-spatial processes examined thus far. The chapters in this section layer the complex geographies, histories, and intersectional and structural inequalities that comprise the conditions of international value transfer for health and care workers. The result is a call for renewed analysis of what international migration means for the political economy of health workers (costs of initial and ongoing training, regulation, conditions of employment, etc.), who benefits, how, and by what means? The chapters also reveal the significance of layering scales of analysis from the global to the intimate scale of workplaces and even individual biographies of pioneering migrants to reveal how systems and structures emerge, how they operate, and how they are reproduced.

Colonialism, imperialism, and globalization processes have connected places, but those places still exert influence over the specific nature of their assemblage and the manner of integration between occupational value systems. Asymmetrical power relations clearly inform how and under what conditions this happens, with evidence of persistent rooting in geographical difference clearly apparent and surprisingly resilient.

Christine Peralta in chapter 14 explores the historical formation of racialization and segmentation in the nursing profession in the United States and Canada using the experience of Filipino nursing pioneer Julita Sotejo. Examining the biography of such nurses at key transition points for the Philippines (between colonialism and American imperialism) reasserts the idea of agency for the migrant but also provides intimate portrayals of the stratification systems that emerged to differentially incorporate these internationally educated nurses. This historical case study demonstrates the importance of locating the health professions and their development in specific (work) place contexts. Sotejo is a product of the structural processes of remaking nursing in the Philippines, she was a graduate of the Rockefeller Foundation's Nurse Fellowship Program and spent time working in the United States and Canada in the 1940s as part of a wider neo-imperial development agenda. Peralta's chapter contributes to understanding the legacy and resilience of exclusionary racial formations in terms of nursing, education, professional politics, and its continued international orientation. The story of Julita Sotejo also reminds us of the importance of attending to stories of individual agency, no matter the constraining contexts that may accompany its exercise.

In chapter 15, Sheri Adekola examines the contemporary period in terms of increased migration of skilled African workers to Canada, especially in the health professions. Using microanalysis, she explores the experiences of fifty-nine Nigerian-educated nurses and other migrant health workers in Canada by focusing on their education and training before and after their migration.

Her research findings show that these migrants perceive themselves as having had positive experiences in Canada, despite their settlement and labour market integration challenges. The immigrants in Adekola's research perceive international migration not as a singular event but as integrated into longer-term educational and career development goals. She argues that their experiences most closely align with positive discourses of international migration such as brain (re)train, where migrants accumulate more training and education throughout their migration journey. Even if their location in the labour market appears to demonstrate deskilling (for example, moving from a registered nursing position pre-migration to a personal support worker post-migration), migrants themselves interpret their labour market position in relation to a larger set of factors, including an enhanced quality of life. In this case, the value of the training they embody and accumulate is centrally important to rebuilding their lives after migration. For the immigrants in Adekola's research their experience extends beyond the workplace, and incorporates a much broader set of concerns about identity, family, and quality of life.

Jonathan Crush in chapter 16 explores reversing brain flight through the case of South African health workers. South African brain drain to the United States, Canada, Australia, and the United Kingdom is calculated to be worth US$1.4 billion. Surveys have indicated a low level of interest in return among South African health care workers overseas, but data show increases in the per cent of health workers living in South Africa who have previously worked overseas, so returnee numbers appear to be increasing. Crush's survey of health care workers in South Africa shows that at some point over 40 per cent had worked overseas for less than three years. Crush considers these returnees as a form of "peripatetic physician" and identifies five types: "Transnational" are professionals who move between locations on an annual recurring basis. The "underqualified" are physicians who want to increase their skills and use international fellowships to achieve this. "Locum" physicians engage in short-term vacation-like travel. "Private" physicians are those who work in the 230 private hospitals in the UK and for intermediaries who organize contracts for these hospitals. There are also the "Can't stay away" physicians who are those who emigrated but came back to South Africa after a ten-year absence. Crush considers how far the licensure process and health care system restrictions in the receiving countries frame the return and circular mobility of these South African health professionals.

In chaper 17, Parvati Raghuram, Joanna Borat, and Leroi Henry focus on knowledge and the effects of its transfer on health systems through the case of South Asian medical and allied health professional migrants and their role in the development of geriatrics in the United Kingdom. This focus on knowledge moves the lens away from migrant bodies and focuses on understanding the systems involved in the transmission, generation, and incorporation of medical knowledge; a significant form of value. The chapter also explores structural transformations that occur because of professional mobility, and how health systems become hubs of innovation due to the very marginalization and devaluation experienced by racialized workers within these systems. The authors illustrate the importance of the long-term consequences of connecting people to places. South Asian migrant doctors were "ghettoed" into gerontology in the UK, but then become embedded into "place-based practices of innovation" in what has become an emerging field. In their chapter, they draw attention to the "stickiness of knowledge to places and how passing through particular places shapes migrants' medical knowledge"(348). Various types of skills are outlined, including embrained, encoded, encultured, embodied, and embedded. Different levels of transferability, translatability, and transmitability of knowledge are detailed. Regulation of this knowledge transfer is important to problematize: What is of value in these contexts and how is this value transferred? Space and place play an important role in these transmission processes, since there is a spatial hierarchy of nations at work here. In the case of South Asia and the UK, the postcolonial is deeply implicated in relations between places. The specialty and its professional status also play a role in the process of transfer. In the case of geriatrics, immigration regulations, professional accreditation processes, and social networks (especially the old boys club of the UK medical profession at that time), conspired to channel South Asian medical professionals into this "devalued" medical field.

Raghuram, Borat, and Henry then move beyond framing this process as one of deficits (brain drain and segmented labour markets) to one of integration and innovation. South Asian doctors (together with many white Jewish doctors) played a central role in the development of geriatrics in the UK. In addition to social marginality, the specialty also developed because of spatial marginality, the specialty of geriatrics developed in small regional hospitals; effectively blank canvases in a way that large metropolitan hospitals of the day were not. The chapter demonstrates the production of knowledge in place, and how the value of the skills transferred between locations became manifest in the development of new medical specialties.

Conclusion

Those who focus on global economic transfers through the lens of GPNs, GVCs, and the like, and who have tended to see health and related care work as tangential to global economic exchange, must take on board the work of feminist scholars in understanding the increasing economic significance of the health sector and related care work, and the experiences of those who provide it. This collection critically examines one key aspect of the industrialization of global health care – transnational labour migration – and it does so through a novel conceptual approach, that of global value transfer where value incorporates the investments made in workers in terms of training and development and their capacity to provide various forms of care. A value chain approach to the transition of human skills in services must be attentive to how skills embodied in people are subjected to losses due to multiple intersectional factors (identity as well as regulatory). Transitions between national or regional labour markets are marked by regulatory barriers that exact costs, especially foreign credential devaluation. Scholars have begun to examine this regulatory devaluation of knowledge transfer as evidence of an emerging form of unequal exchange, but closer analysis can also reveal how the agency and embodied knowledge of such workers contributes to forms of medical and educational innovation that benefit immigrant-receiving nations as well as sending nations.

The sensitivity of GPNs to the value of labour input can be particularly useful for the analysis of the migration of health workers, because workforce hierarchies distribute power unequally within workplaces. Across locations of work, the quality of migrant labour can add value to the national reputation of a workforce that is globally orientated – concomitantly, poor quality can diminish national reputation – and international migration can spatially and socially distort the return on the value added in the labour process. Workforce hierarchies are also imbued with cultural and politically charged issues of identity, difference, and inequality. The messy and fleshy relevance of these distinctions in the field of health work, which represents deeply intimate bodywork, is immense. When it comes to people in GVCs, the "product" is not homogenized, and this diversity adds important complexities to how value is generated and transferred along the global value/care chain.

REFERENCES

Barratt, T., McGrath-Champ, S., & Bailey, A. (2019, April). *Labour and global production: The (mis)combination of workers and theories of global production networks.* [Conference presentation]. American Association of Geographers Meeting, Washington, DC, United States.

Boucher, A. (2016). *Gender, migration and the global race for talent.* Oxford University Press.

Choy, C.C. (2003). *Empire of care: Nursing and migration in Filipino American history.* Duke University Press.

Christian, M. (2016). Kenya's tourist industry and global production networks: Gender, race and inequality. *Global Networks, 16*(1), 25–44. https://doi.org/10.1111/glob.12094.

Coe, N.M., & Hess, M. (2013). Global production networks, labour and development. *Geoforum, 44*, 4–9. https://doi.org/10.1016/j.geoforum.2012.08.003.

Coe, N., & Yeung, H. (2015). *Global production networks: Theorising economic development in an interconnected world.* Oxford University Press.

Collyer, F., & White, K. (2011). The privatisation of medicare and the National Health Service, and the global marketisation of healthcare systems. *Health Sociology Review, 20*(3), 238–44. https://doi.org/10.1080/14461242.2011.11003086.

Connell, J., & Walton-Roberts, M. (2016). What about the workers? The missing geographies of health care. *Progress in Human Geography, 40*(2), 158–76. https://doi.org/10.1177/0309132515570513.

Davis, M. (2020, March 20). The coronavirus crisis is a monster fueled by capitalism. *In These Times*. https://inthesetimes.com/article/coronavirus-crisis-capitalism-covid-19-monster-mike-davis.

Deloitte. (2019). 2020 Global health care outlook: Laying a foundation for the future. https://www2.deloitte.com/content/dam/insights/us/articles/GLOB22843-Global-HC-Outlook/DI-Global-HC-Outlook-Report.pdf.

Dumitru, S. (2014). From "brain drain" to "care drain": Women's labor migration and methodological sexism. *Women's Studies International Forum, 47* (Part B), 203–12. https://doi.org/10.1016/j.wsif.2014.06.006.

Dunaway, W. (Ed.). (2014). *Gendered commodity chains: Seeing women's work and households in global production*. Stanford University Press.

Flecker, J., Haidinger, B., & Schönauer, A. (2013). Divide and serve: The labour process in service value chains and networks. *Competition & Change, 17*(1), 6–23. https://doi.org/10.1179/1024529412Z.00000000022.

Flynn, K.C. (2011). *Moving beyond borders: A history of Black Canadian and Caribbean women in the diaspora* (Vol. 37). University of Toronto Press.

Folbre, N. (2006). Measuring care: Gender, empowerment, and the care economy. *Journal of Human Development, 7*(2), 183–99. https://doi.org/10.1080/14649880600768512.

Frade, C., & Darmon, I. (2005). New modes of business organization and precarious employment: Towards the recommodification of labour? *Journal of European Social Policy, 15*(2), 107–21. https://doi.org/10.1177/0958928705051509.

George, S. (2005). *When women come first: Gender and class in transnational migration*. University of California Press.

Gilmartin, M., & Kuusisto-Arponen, A.K. (2019). Borders and bodies: Siting critical geographies of migration. In K. Mitchell, R. Jones, & J.L. Fluri (Eds.), *Handbook on critical geographies of migration* (pp. 18–29). Edward Elgar Publishing.

Hodges, B.D., & Lingard, L. (Eds.). (2012). *The question of competence: Medical education in the twenty-first century*. Cornell University Press.

Keil, R., & Ali, S.H. (2008). SARS and the restructuring of health governance in Toronto." In S.H. Ali & R. Keil (Eds.), *Networked disease: Emerging infections in the global city* (pp. 55–69). John Wiley & Sons.

Kingma, M. (2006). *Nurses on the move: Migration and the global health care economy*. Cornell University Press.

Lunt, N., Smith, R., Exworthy, M., Green, S.T., Horsfall, D., & Mannion, R. (2011). *Medical tourism: Treatments, markets and health system implications: A scoping review*. OECD Directorate for Employment, Labour and Social Affairs. http://www.oecd.org/els/health-systems/48723982.pdf.

Neilson, J., Pritchard, B., & Yeung, H.W.C. (2014). Global value chains and global production networks in the changing international political economy: An introduction. *Review of International Political Economy*, *21*(1), 1–8. https://doi.org/10.1080/09692290.2013.873369.

Olaizola, A., & Sweetman, A. (2019). Brain gain and waste in Canada: Physicians and nurses by place of birth and training. In *Recent trends in international migration of doctors, nurses and medical students* (pp. 123–39). Paris: OECD Press. https://doi.org/10.1787/5571ef48-en.

Ortiga, Y.Y. (2014). Professional problems: The burden of producing the "global" Filipino nurse. *Social Science & Medicine*, 115, 64–71. https://doi.org/10.1016/j.socscimed.2014.06.012.

Parry, B., Greenhough, B., & Dyck, I. (Eds.). (2015). *Bodies across borders: The global circulation of body parts, medical tourists and professionals*. Ashgate.

Picot, G., & Sweetman, A. (2011). Canadian immigration policy and immigrant economic outcomes: Why the differences in outcomes between Sweden and Canada? IZA Policy Paper No. 25. http://ftp.iza.org/pp25.pdf.

Pollitt, C., & Bouckaert, G. (2017). *Public management reform: A comparative analysis – new public management, governance, and the neo-weberian state* (4th ed.). Oxford University Press.

Pratt, G., & Rosner, V. (Eds.). (2012). *The global and the intimate: Feminism in our time*. Columbia University Press.

Rastegar, D.A. (2004). Health care becomes an industry. *The Annals of Family Medicine*, *2*(1), 79–83. https://doi.org/10.1370/afm.18.

Ravenhill, J. (2014). Global value chains and development. *Review of International Political Economy*, *21*(1), 264–74. https://doi.org/10.1080/09692290.2013.858366.

Reddy, S.K. (2015). *Nursing and empire: Gendered labor and migration from India to the United States*. UNC Press Books.

Riley, P.L., Zuber, A., Vindigni, S.M., Gupta, N., Verani, A.R., Sunderland, N.L., Friedman, M., Zurn, P., Okoro, C., Patrick, H., & Campbell, J. (2012). Information systems on human resources for health: A global review. *Human Resources for Health*, *10*(7). https://doi.org/10.1186/1478-4491-10-7.

Sweetman, A., McDonald, J.T., & Hawthorne, L. (2015). Occupational regulation and foreign qualification recognition: An overview. *Canadian Public Policy*, *41* (Supplement 1), S1–S13. https://doi.org/10.3138/cpp.41.s1.s1.

Talbot, S.G., & Dean, W. (2018, July 26). Physicians aren't "burning out." They're suffering from moral injury. STAT. https://www.statnews.com/2018/07/26/physicians-not-burning-out-they-are-suffering-moral-injury/.

Taylor, P., Newsome, K., & Rainnie, A. (2013). "Putting labour in its place": Global value chains and labour process analysis. *Competition & Change*, *17*(1), 1–5. https://doi.org/10.1179/1024529412Z.00000000028.

Valiani, S. (2011). *Rethinking unequal exchange: The global integration of nursing labour markets*. University of Toronto Press.

White, W. D. (1987). The introduction of professional regulation and labor market conditions. *Policy Sciences, 20*(1), 27–51. https://doi.org/10.1007/BF00137048.

Williams, F. (2012). Converging variations in migrant care work in Europe. *Journal of European Social Policy, 22*(4), 363–76. https://doi.org/10.1177/0958928712449771.

Yeates, N. (2004). Global care chains. *International Feminist Journal of Politics, 6*(3), 369–91. https://doi.org/10.1080/1461674042000235573.

Yeates, N. (2014). Global care chains: Bringing in transnational reproductive laborer households. In W. Dunaway (Ed.), *Gendered commodity chains: Seeing women's work and households in global production* (pp. 175–89). Stanford University Press.

Yeates, N., & Pillinger, J. (2019). *International health worker migration and recruitment: Global governance, politics and policy*. Routledge.

SECTION 1

Health Worker Migration and Global Value Transfer: New Approaches and Challenges

1 The Study of Global Value Chains: Bringing Services and People In

JOHN RAVENHILL

The global value chain (GVC) became the dominant metaphor in the study of globalization and the political economy of development around the turn of the twenty-first century. The concept had its origins in the focus on commodity chains in world systems theory approaches developed in the 1970s and 1980s (Hopkins & Wallerstein, 1977). In the 1990s, it was taken up by sociologists, political scientists, students of business organization, and eventually by economists and became a mainstay of reports from multilateral economic institutions. In the process, not only did the terminology change – from commodity to value chains, in recognition that the latter embraced a broader range of activities and end-products (Gereffi et al., 2001) – but so too did the unit of analysis and the questions that scholars were posing.

World systems theory saw commodity chains as a key component of the unique transformation of the global economy brought about by capitalism. Commodity chains were the mechanism through which unequal exchange was conducted between core and periphery. The primary focus, therefore, was on the role of commodity chains in structuring and reproducing a hierarchical system. World systems theorists denied that national economies were a relevant unit of analysis, although, through their concern with how countries might move from core to periphery, national economies inevitably came into focus. In contrast, the value chain approach emerged from different theoretical perspectives (Bair, 2005). It was regarded as a novel form of industrial organization brought about by globalization in the last quarter of the twentieth century. The unit of analysis was typically neither the world system nor national economies but specific industries (although, there is a frequent and sometime problematic elision between industry and national economy in much of the literature).

The literature on value chains owes much to studies of industrial organization, particularly the transaction costs approach to economics pioneered by Coase (1937) and Williamson (1975). Here, the central question is why firms choose to undertake certain tasks in-house (a non-market environment) rather than purchase them from others in the marketplace. In the 1980s, some firms adopted a radically different strategy in which a much larger number of tasks were delegated to other companies. Nike and Apple were pioneers in outsourcing much of the manufacturing of their products while taking their profits from their control of core competencies – research and design, brand name, distribution channels, and core (patented) technologies (Quinn & Hilmer, 1994). Other companies quickly followed suit. Borrus and Zysman (1997), for instance, referred to the strategies of Intel and Microsoft in making their profits from control of the core technologies of Windows-based personal computers as "Wintelism."

The starting point of value chain analysis was the assumption that globalization had entered a fundamentally different phase: whereas in the past, companies took responsibility for the whole process of design, manufacturing, and marketing, they now frequently manufactured to a blueprint provided by others. A simple dichotomy between market and hierarchy no longer captured the dynamics and complexities of relationships within and between firms in the contemporary global economy. Trust was an important dimension of many relationships, for instance, where fabless semiconductor companies provided blueprints containing critical intellectual property to semiconductor foundries. Writers on GVCs who had a background in sociology accordingly drew on the work pioneered by Granovetter (1985), which saw economic activity as embedded in social relations, and interfirm relationships as sustained by trust and good will.

The progressive "mainstreaming" of the GVC concept since the mid-1990s led to a loss of its original critical orientation (Klein, 2011a). The central focus of much of the work in sociology and business economics was on the organization of value chains among firms. This embraced both a mapping exercise – identifying which firms from which countries were involved in various activities in the value chain – and an exploration of questions of governance and power within value chains. The latter revolved around the characteristics of lead firms (Gereffi's [1994, 2001] original distinction between buyer-led and producer-led chains and their different sources of power) and the institutional context in which they were embedded. Regulatory standards, for instance, could make it more difficult for producers from low-income economies to export to industrialized economies, enhancing the power of lead firms on which others depended for technical assistance and/or distribution channels. The importance not just of national but also of international regulations to relations within the value chain was illustrated by Ponte's (2002) study of the international coffee industry.

Discussion of governance relations and power distribution naturally led to questions relating to the distribution of gains from participation in GVCs, moving the discussion to the study of who gets what, when, and how – Lasswell's (1936) definition of politics. Most of the analysis in the academic literature revolved around questions of how *firms* might capture more value added through upgrading, building on the typology first developed by Humphrey and Schmitz (2002) that distinguishes product, process, intersectoral (applying existing competencies to new sectors), and functional (undertaking different tasks) upgrading. But with growing interest among multilateral economic institutions in the concept, the unit of analysis frequently morphed into national economies, with the emphasis turning to how participation in GVCs was associated with stronger economic performance (and the need for developing economies to make themselves attractive to lead firms). (For a comprehensive listing of reports from multilateral economic institutions that made use of the GVC concept, see Neilson [2014, Appendix A]).

Taking Stock of the GVC Literature

Over the past quarter of a century, the increasingly sophisticated study of GVCs has identified key questions that have to be addressed to understand the political economy of development in the contemporary global economy. Moreover, it has led to a complete rethink on how trade should be measured – building on case studies that showed that "Made in Thailand" or "Made in China" had little meaning when the vast majority of the value of an export was previously imported from elsewhere (Bernard & Ravenhill, 1995; Dedrick et al., 2010; Xing & Detert, 2011). Data on trade in value-added provide a much more accurate picture of what is produced where than do conventional customs data on exports and imports (Organisation for Economic Co-operation and Development [OECD], 2018).

Despite this progress in analysing the role of GVCs, the literature has a number of weaknesses and several major gaps. First, although hundreds of studies into various GVCs have been conducted, and more sophisticated typologies developed, there is a marked lack of cumulation in the literature and a great paucity of theory (Pipkin and Fuentes [2017], however, provide a rare stock-taking exercise). Many analysts seemed content to provide empirically rich case studies that mapped participation in GVCs but eschewed (and, indeed, explicitly resisted) synthesis and generalization (Bair, 2008).

A second set of criticisms arose from questions whether "global value chain" was an appropriate metaphor for capturing the complexities of contemporary production – in particular, whether GVCs were either *chains* or *global*. The concept of a chain implied a linearity – in which goods moved in a sequential manner from upstream to downstream stage – that was absent in all but the most simple of interfirm linkages. For this reason, many analysts, especially geographers, preferred the concept of network, which allowed for the possibility that there were multiple hubs and spokes within any production process; the economists, Baldwin and Venables (2010), offered the alternative metaphor of a "spider" in which multiple limbs came together, but this provides no obvious advantages over the network metaphor. This was not merely an arcane dispute over terminology. Much of the literature on GVCs that focused on governance issues appeared to assume that a single dominant mode prevailed throughout the value chain. Such an assumption was appropriate where the focus was on the relationship between a lead firm and first-tier suppliers (Sturgeon, 2008). But across the whole of the production process (network) one might find a mix of different forms of interfirm relationships that were obscured by analogies based on a linear relationship.

The concept of *global* value chains owed much to the preoccupation with the new geography of production brought about by contemporary globalization. The reality, however, is that relatively few production processes are genuinely global. Rather, they are concentrated within a particular region – Baldwin's (2012) concepts of "Factory Asia," "Factory Europe," and "Factory North America" are particularly apt descriptions of how production is currently organized in the majority of industries (for a comprehensive demonstration of the regional character of value chains, see Baldwin and Lopez-Gonzalez [2015]). Again, the distinction had importance beyond terminological disputes. Much of the content of contemporary trade agreements reflects the efforts of companies to organize their production networks on a regional basis, while recent negotiations of "mega-regionals" such as the Trans-Pacific Partnership often pit the interests of competing regional value chains against one another (Ravenhill, 2017).

A third set of concerns with much of the literature relates to questions of what benefits are generated by participation in GVCs and how they are shared. The essence of the arguments from business economics was that the imperfect competition characteristic of many sectors in contemporary economies produced "rents" (above-normal returns). This literature overlaps with the recognition in mainstream economics of the impact of imperfect competition on international trade (Krugman, 1979). For students of GVCs, the question was whether these rents would be captured solely by lead firms in value chains or if others over time would be able to increase their share of the value added within any chain.

The spread of value chains did dramatically transform the composition of exports of less developed economies: in the quarter century after 1985, the share of manufactures in their total merchandise exports increased from 30 per cent to over 70 per cent (Hoekman, 2014, p. 16). But the distribution of gains from participating in GVCs remained markedly unequal with relatively few companies from lower-income economies succeeding in capturing a larger share of value added (and few national economies moving from low- to middle-income let alone high-income status [Wade, 2017]).

Some of the more enthusiastic proponents of GVCs appeared to lose sight of the origins of contemporary globalization in the outsourcing to lower wage economies of the less profitable stages in production. The issues involved had been concisely illustrated as early as 1992 by Stan Shih, the founder of the Acer computer company. Shih had suggested that distribution of value creation in contemporary industry took the form of a U-shaped "smiling curve" in which high value-added activities were concentrated at the two ends of the curve (OECD, 2013). At one end were upstream activities such as research and development, product design, and the manufacture of key components. At the other end were downstream activities such as branding, marketing, and customer service. In the middle were the core manufacturing activities that lead firms outsourced because they generated relatively low returns on capital investment.

Evidence suggested that with globalization the smiling curve had become steeper, that is, fabrication and assembly (the middle part of the curve) were accounting for a lower share of total value added. The incorporation of more countries (and, of course, especially China from the 1990s onwards) in GVCs produced increasing competition among less developed economies seeking to capture a share of the production process. One consequence was that the price of manufactured exports from developing economies fell relative to that of their manufactured imports (Wood, 1997, Fig. 4) – a decline in their terms of trade reminiscent of the arguments of Raul Prebisch (1950) and Hans Singer (1950) a half century earlier about the problems faced by commodity exporters in developing countries. Growth may not have been "immiserizing," but it seemed that many economies were having to run faster just to stay in place.

To be sure, there were outstanding success stories of firms that had succeeded in increasing their share of value added in GVCs and grown into significant players – most notably a number of the electronic contract manufacturers that were able to offset relatively low margins by producing in huge volumes. But few authors attempted to explore how common such success stories were in contrast to experiences where companies were unsuccessful in upgrading. Again, the work of Pipkin and Fuentes (2017) is an exception in providing a systematic review of a number of GVC case studies, discerning the reasons for some companies' success in upgrading while others, in their terms, underwent a process of "treadmilling" in which attempts to upgrade were followed by "backsliding, decay and obsolescence." In line with the arguments of authors who were sceptical of the opportunities GVCs provided for accelerating economic development, Pipkin and Fuentes found that "treadmilling" was a frequent outcome from the involvement of companies from lower income economies in GVCs.

The huge variance in experiences in upgrading indicated that the enthusiasm of the multilateral economic institutions (MEIs) for participation in GVCs at least needed to be tempered. Studies (e.g., United Nations Conference on Trade and Development, 2013) that purportedly demonstrated a correlation between participation in GVCs and positive outcomes on various developmental indicators had to be interpreted cautiously: even if a correlation existed, the direction of the causal relationship was not necessarily clear. To be sure, the MEIs had warned that gains from participation in GVCs were not automatic

(OECD, World Trade Organization, & World Bank, 2014, p. 4). Their more recent publications have even gone so far as to acknowledge not only that the benefits from participation in GVCs are very unevenly distributed, depending upon where on the "smiling curve" the activities are located, but also that GVCs do not always produce win-win outcomes (World Bank, 2017, p. 6). But other than advising developing economy governments to make their countries attractive locations for GVCs, the MEIs have offered few policy prescriptions for economies seeking to upgrade their participation in value chains. Despite the supposed official abandonment of the "Washington Consensus" (a set of free market economic policies advocated by international organizations and many Western governments in the 1980s), the programs proposed were essentially unchanged from those proffered a quarter of a century previously – where the principal prescription was to get the state out of the economy.

Little attention has been paid to the crucial role that the state could play in upgrading by supplying complementary institutions (Ravenhill, 2014). A key finding of the Pipkin and Fuentes (2017) synthesis of studies of upgrading in GVCs was that the state's capacity to craft supportive policies was critical to successful upgrading. Mayer and Phillips (2017) and Horner (2017) have built on this argument to discuss the various roles that the state can play in shaping GVCs.

Another set of issues that has received relatively little attention in the GVC literature also relates to distributional questions – but this time within countries. One dimension was at the firm level. A focus on firm upgrading often overlooked the possibility that some companies might increase their share of value added at the expense of other *local* firms. Of far greater import, however, has been the failure of much of the literature to examine whether participation in and upgrading within GVCs had a positive impact on working conditions in developing economies. The early outsourcing of athletic footwear manufacturing to Korean and Taiwanese companies demonstrated that local firms were no better employers than subsidiaries of multinational corporations – and, with the growth of the corporate social responsibility (CSR) movement in the twenty-first century in Western countries, workers in many developing economies were likely to receive better treatment at the hands of foreign rather than local corporations. The collapse of the Rana Plaza building in Dhaka, Bangladesh, in 2013 was a reminder that CSR had not percolated through the multiple tiers of the supply chains of many firms in the garment and footwear industries.

Upgrading in firms thus was no guarantee that "social" upgrading would follow. Increasingly, the focus has been on how lead firms in GVCs can be encouraged to ensure that their activities lead to "decent work" (International Labour Office, 2016), an objective incorporated in Goal 8 of the United Nation's 2030 Agenda for Sustainable Development.

Reference to labour brings us to two additional topics frequently missing from discussion of GVCs: services and people.

Bringing Services and People into the Study of GVCs

The vast majority of the literature on GVCs focuses on the production of commodities and manufactured goods. To coin a pun, services have been largely invisible (for readers without a background in economics, the services trade is often referred to as the invisible trade or trading in "invisibles"). This neglect occurred despite a growing recognition of the increasing role of services in international trade.

Even when references to services are made in the GVC literature, the discussion is frequently limited to the role they play in supporting trade in manufactures; the flagship joint report by World Bank, OECD, WTO, and IDE-JETRO entitled *Global Value Chain Development Report 2017* (World Bank, 2017), for instance, views services primarily in terms of how they facilitate linkages between different stages of the manufacturing process and as direct inputs (such as research and development) into manufacturing.

The development of databases on trade in value-added, however, has highlighted the growing importance of services themselves in international trade (Low, 2013). Measured in conventional customs data, the share of services in world exports has remained fairly constant at around 20 per cent since 1980. When measured in terms of value-added, however, the share of services in world trade increased from below 30 per cent to above 40 per cent in this period (World Bank, 2017, p. 146). These data almost certainly underestimate the importance of the international provision of services because they do not take into account all of the means through which trade in services occurs. The General Agreement on Trade in Services recognizes four modes of supply: cross-border trade (e.g., international provision of accountancy services); consumption by nationals abroad (for instance, tourism or students enrolled at foreign institutions); through commercial presence in a foreign country (when corporations supplying services establish foreign subsidiaries, a dimension not captured in trade statistics); and through the presence of natural persons in a foreign country (such as nurses working outside their home countries).

Some of the increased prominence of services in global trade has come about in part because of the "servicification" of manufacturing, an ugly but nonetheless accurate term coined by Vandermerwe and Rada (1988). An illustration is provided by Lodefalk (2013) in his study of a Swedish multinational engineering company that used over forty types of services – ranging from accountancy to audio-visual services – within its supply chain. But servicification was not confined to the production of this company's goods. It offered a further fifteen types of services to customers such as design, research and development, and logistics. The "Smiling Curve" suggests that it is in services that many of the higher value-added activities in GVCs are located. Some of these service functions increasingly are outsourced not just nationally but also internationally. Moreover, as production in value chains has become more fragmented geographically with revolutions in transportation and information and communications technology, so demand has grown for services such as the management of logistics.

Viewing services only within the context of manufacturing provides a partial picture of their contemporary importance in the global economy. There are, of course, many GVCs that consist entirely of different forms and stages of services production, which at most have only an indirect link to manufacturing. Tourism is a traditional example; transnational provision of medical care is another. Increasingly, however, the growth in international trade in services has been driven by knowledge-intensive business services such as telecommunications, accountancy, IT services, and other professional services. Like their counterparts in manufacturing, lead firms in services are increasingly seeking to break up their value chains and to outsource the less profitable components (Drake-Brockman & Stephenson, 2012).

For some authors, the growth of trade in services offers significant new opportunities for less developed economies to diversify their international trade and to accelerate their economic development. The emergence of GVCs has made it possible to enter supply chains by contributing only a small component of the final product. The investment and technical requirements involved are much smaller than if an economy had to master all stages of the value chain. Moreover, economies of scale are typically less important in the provision of services than in manufacturing. Stephenson and Drake-Brockman (2014) note that, consequently, even small economies can create "niche" opportunities for themselves. And, in services, the principal costs come from improving the educational system rather than investing in physical plant upgrades or technology, where the start-up costs (and consequently barriers to entry) are often substantially higher. Studies show that more small and medium-sized enterprises (SMEs) are involved in international value chains in services than in manufacturing. And work in services is often more open to women.

While these are all plausible arguments, some caveats are in order. First and foremost is not to forget that, as in manufacturing value chains, it is those tasks that generate the least value-added that are the first to be outsourced. There is a risk that the (low value-added) service activities that are captured internationally will be "ghettoized" with no opportunities for upgrading – the establishment of international call centres providing non-technical information being a prime example. Stephenson and Drake-Brockman (2014, p. 16) suggest that participation in services value chains may offer less developed economies the possibility of "leap-frogging" in their development, skipping the manufacturing stage and moving directly to the post-industrial world of services. But one has to compare the potential of the two forms of activity for providing the foundations for broader development: manufacturing may offer opportunities to create externalities, to generate spin-offs and benefit from learning-by-doing. The service sector may not provide the same possibilities. Moreover, because barriers to entry into services are relatively low, other economies may enter the market, generating a "race to the bottom" in terms of earnings and working conditions. And, because little investment in physical facilities/infrastructure is involved, it is relatively easy for lead firms to "re-shore" service positions in response to political or economic pressures at home. A final concern with entry into services value chains is that it may produce distortions in the allocation of educational resources – with money directed into creating skills that primarily address the needs of foreign economies rather than those that will contribute most to local development. This argument applies equally to the training of individuals who will pursue careers abroad. It is an empirical question whether this allocation of resources is more problematic for a developing country than, say, the oft-criticized use of scarce arable land to grow fruit, flowers, and vegetables for the markets of industrialized countries.

People in Value Chains

Although studies of GVCs in service industries and studies of social upgrading bring people into the discussion, typically it is in terms of aggregates: the labour force in social upgrading, and firms in service industries. Individuals seldom feature in the literature: the emphasis is on things rather than on people.

The major exception to this generalization is studies of value chains in professional sports, most notably of the "production" of baseball players in the Caribbean for export to Major League Baseball in North America (Klein, 2011a, 2011b), and of soccer players in Africa for export to the European professional leagues (Darby, 2012, 2013). Many readers will be uncomfortable with the idea that the "commodity" being produced in these instances is human labour. And, to be sure, the value chain metaphor is far from perfect. For a start, it is debatable whether it is individuals or the "club" or the "game" that is the commodity being produced in these sporting value chains. Moreover, unlike manufacturing value chains, the product is not homogenized. Indeed, the emphasis in sports value chains is usually exactly the opposite: to produce individuals with unique skills. The literature on baseball players, however, does note that a function of the local academies is to socialize individuals into particular *patterns* of behaviour expected of professional players. Klein (2011b) suggests that the most significant factor hampering Dominican baseball players from realizing their fullest potential is the cultural divide that separates them from their North American teammates and from a sense of belonging (p. 99).

Another significant problem in applying the value chain metaphor is that it overlooks the potential for agency on the part of the players being "produced." Darby's (2013) comments summarize the issues succinctly:

> Considering athletes through the sort of mechanical language applied to inert commodities such as agricultural produce, clothing or electrical components requires

> caution. The risk is that the migrant sports person is portrayed in a distant and reductionist manner rather than as a sentient, social actor whose ambitions in, relationship with and mobility through sport are shaped by the exercise of agency as well as the broader structural context that they find themselves embedded in. (p. 44)

Despite these reservations about the appropriateness of applying the value chain metaphor to transactions involving people, the literature on GVCs provides relevant questions to pose about these social networks – both in terms of mapping relevant participants and the distribution of gains within the networks. Interestingly, the two leading writers in this field, Paul Darby and Alan Klein, in their early work viewed their subject matter through the lens of dependency theory and world systems analysis, perceiving the development of local players as a vertically integrated supply chain closely controlled by the "lead firm," that is, the foreign club. In the words of the former president of the world governing body for soccer (the Fédération Internationale de Football Association [FIFA]), Sepp Blatter, these clubs acted as "neo colonialists ... who engage in social and economic rape by robbing the developing world of its best players" (as cited in Darby, 2012, p. 266). In later publications, however, both Darby and Klein argued that local economies were starting to capture an increasing share of gains from the transactions.

Darby (2012, 2013) pointed to the development by FIFA of a new set of international transfer regulations in 2001 that helped limit the more exploitative aspects of the trade. Moreover, he saw the development of football academies as promoting inward investment and bringing new resources to the local economy. Remittances from migrant footballers contributed further to the resource inflow. Darby's (2013) case study of the Ghanaian Right to Dream football academy showed how it had gone beyond the honing of skills for potential professional footballers to place an emphasis on the provision of a quality education to all its recruits. Moreover, it developed a vocational stream for individuals interested in pursuing a football-related professional career in coaching or physiotherapy. In a similar manner, Klein (2011a) argued that in the first decade of this century, local talent-spotters played an increasingly important part in the development of young players before they entered formal baseball academies, providing two to four years of training. A new, largely unregulated, stage was added at the beginning of the production chain in which local talent-spotters became gatekeepers to entry into the whole system – and were able to command a share of the signing bonus when fledgling players entered the formal baseball academies.

The work of both Darby and Klein demonstrates how local economies can capture a share of the income generated within sports value chains. But neither author comments on how significant overall the documented changes are for the local economy. Most of the gains are captured by a few individuals. Even allowing for the possibility that the players who succeed in professional leagues overseas send remittances back to their home countries, the impact on local economies in aggregate is likely to be small. On the other hand, so too is the detrimental impact on economic development: local clubs may be deprived of star players but the overall impact on development prospects will be minimal.

The same conclusion cannot be drawn regarding the impact of the migration of health services professionals, a central topic in controversies over globalization over the last two decades. This chapter is not the place to explore the extensive literature on contending arguments pitting critical shortages of medical professionals in some developing economies – and the loss of (typically) state investment in the production of these professionals – against the opportunities for individuals to upgrade their skills, to enjoy the right to free international movement, and to the potential for their home countries to benefit both from remittances and from their enhanced skills should the workers return home (for an overview of the arguments and some of the supportive evidence see, for instance, Bach, 2006; Cometto et al., 2013; Grignon et al., 2012).

Rather, the question posed here is how relevant the GVC metaphor is to the study of the migration of health services professionals. As with professional sportspeople, one has to address the appropriateness of a commodity-based metaphor when dealing with individuals who exercise agency. On the other hand, in contrast to the development of footballers or baseball players, professional development leading to the certification of a particular level of proficiency in the health professions is more akin to the production of a homogenized commodity. Questions that are central to the GVC literature can be posed: Where is value-added captured? How can more value-added be captured locally? What role can the state play in local upgrading (Yeates, 2009)? How can international agreements change the balance of bargaining power among the parties?

If, in the study of many value chains there is often a conflation between upgrading at the firm level, on the one hand, and economic development on the other, then in the study of the migration of professionals a central challenge is differentiating private and public gain. A key question is, What policy instruments can effectively ensure that states recoup the investment of scarce resources in training professionals? For instance, should higher fees be charged to those receiving training (should the training process be privatized)? Should there be obligations imposed on beneficiaries to work domestically for a specific period before they are permitted to emigrate? Can one build in an obligation on migrants to return at least temporarily to share their skills? Can international agreements (whether at the global level, such as the WHO Global Code of Practice on the International Recruitment of Health Personnel or through bilateral trade agreements) ensure that the benefits from the migration of medical professionals be more balanced between supplying and recipient countries)? Lorenzo et al. (2007) provide a useful discussion of some of these questions through a study of Philippine migrant nurses.

Gender

Discussion of individuals in value chains leads us to a final significant omission in much of the literature on GVCs: gender. The original work on the topic within a global commodity chain context notably focused on all inputs into the chain, including unpaid household labour (Dunaway, 2014a). With the emphasis shifting to firms in the GVC literature, households disappeared from the picture. Recent trends in economic analysis, especially the development of data on value-added trade, reinforced the tendency to exclude unpaid labour from the study of GVCs. The measurement of trade in value added rests on input-output tables, that is, data collected on sales and purchases by producers and consumers in an economy. By definition, unpaid transactions are excluded.

Issues pertaining to gender did figure in the analysis of some value chains, such as those in horticulture involving African producers (Barrientos et al., 2003; Dolan et al., 2004; Tallontire et al., 2007; for an early general survey of the literature, see Mayoux & Mackie, 2007). As Dunaway (2014b) notes, however, despite the numerous investigations of women's work in global production, very few of these analyses have used a GVC perspective. Conversely, very few studies in the literature on GVCs have examined gender issues: Dunaway estimated that at the end of 2012, these constituted less than one per cent of the accumulated research on the subject (p. 3).

The Literature on Global Care Chains Is a Notable Exception to These Arguments

Introduced by Hochschild (2001), the concept of global care chains was popularized – and placed explicitly within a GVC perspective – through the work of Yeates (2005, 2012). Yeates makes a persuasive argument that the global care chain has parallels with GVCs: it involves an outsourcing of labour, it frequently is internationalized (reflecting trends towards globalization), and – very much in the tradition of the original global commodity chain analysis – it involves the extraction and transfer of resources from the Global South to the industrialized world. Rather than the focus being on networks of firms, in global care chains (GCC) it is on networks of families. Like the world systems approach, however, the GCC focuses on unpaid as well as paid labour. Gender is at the heart of the GCC analysis, the participants in networks being predominantly female. And, whereas in much GVC analysis, as noted above, the state is peripheral, for Yeates, the state plays a key role in the GCC. It is prominent in the development of chains (for instance, in providing relevant education and facilitating contacts between producers and consumers). The state also regulates these chains both through national rules on professional registration and through international negotiations such as the inclusion of provisions in preferential trade agreements on the migration of care workers.

As with the application of GVCs to professional sports migration, the metaphor when used to study global care chains is imperfect (for criticisms, see Murphy, 2014). Most fundamentally, as Yeates (2012) notes, because many of the care services provided are often intangible, it is impossible to calculate their value. Acknowledging that care involves both physical ("caring for") dimensions – that can reasonably be measured – and emotional ("caring about") – that cannot, introduces a complexity absent from most trade in goods and services (pp. 138–9). These reservations notwithstanding, the GCC concept represents an important application and extension of some of the core ideas of the value chain literature.

Conclusion

Perhaps the most disappointing aspects of the substantial literature that now exists on GVCs are the lack of cumulation and an associated failure of authors to build on work in related fields and disciplines. That economists should fail to read or reference the work of non-economists is not unusual but not, for this reason, less inexcusable. More surprising is the failure of writers with other disciplinary backgrounds to draw on work from related fields. Very few authors who study GVCs make reference, for instance, to the literature on the value chains in professional sports or to the innovative application of the concept to GCCs.

The fundamental contribution of the GVC literature is that it helps to answer, in the contemporary era of globalization, the questions of who gets what, when, and how. Importantly, it suggests that the answers to these questions are not static but reflect the evolving relationships among the multiple players in GVCs and the institutional contexts in which they interact.

REFERENCES

Bach, S. (2006). International mobility of health professionals: Brain drain or brain exchange? Research Paper 2006/82. UNU-WIDER. https://www.wider.unu.edu/sites/default/files/rp2006-82.pdf.

Bair, J. (2005). Global capitalism and commodity chains: Looking back, going forward. *Competition and Change*, *9*(2), 153–80. https://doi.org/10.1179/102452905X45382.

Bair, J. (2008). Analysing economic organisation: Embedded networks and global chains compared. *Economy and Society*, *37*(3), 339–64. https://doi.org/10.1080/03085140802172664.

Baldwin, R. (2012). Global supply chains: Why they emerged, why they matter, and where they are going. CEPR Discussion Papers 9103. https://ideas.repec.org/p/cpr/ceprdp/9103.html.

Baldwin, R., & Lopez-Gonzalez, J. (2015). Supply-chain trade: A portrait of global patterns and several testable hypotheses. *The World Economy, 38*(11), 1682–721. https://doi.org/10.1111/twec.12189.

Baldwin, R., & Venables, A. (2010). Spiders and snakes: Offshoring and agglomeration in the global economy. *NBER Working Papers 16611*. National Bureau of Economic Research. https://ideas.repec.org/p/nbr/nberwo/16611.html.

Barrientos, S., Dolan, C., & Tallontire, A. (2003). A gendered value chain approach to codes of conduct in African horticulture. *World Development, 31*(9), 1511–26. https://doi.org/10.1016/S0305-750X(03)00110-4.

Bernard, M., & Ravenhill, J. (1995). Beyond product cycles and flying geese: Regionalization, hierarchy, and the industrialization of East Asia. *World Politics, 45*(2), 179–210. https://doi.org/10.1017/S0043887100016075.

Borrus, M., & Zysman, J. (1997). Globalisation with borders: The rise of Wintelism as the future of global competition. *Industry and Innovation, 4*(2), 141–66. https://doi.org/10.1080/13662719700000008.

Coase, R.H. (1937). The nature of the firm. *Economica, 4*(16), 386–405. https://doi.org/10.1111/j.1468-0335.1937.tb00002.x.

Cometto, G., Tulenko, K., Muula, A.S., & Krech, R. (2013). Health workforce brain drain: From denouncing the challenge to solving the problem. *PLoS Med*. https://doi.org/10.1371/journal.pmed.1001514.

Darby, P. (2012). Gains versus drains: Football academies and the export of highly skilled football labor. *Brown Journal of World Affairs, 18*(2), 265–77. https://www.jstor.org/stable/24590876.

Darby, P. (2013). Moving players, traversing perspectives: Global value chains, production networks and Ghanaian football labour migration. *Geoforum, 50*(1), 43–53. https://doi.org/10.1016/j.geoforum.2013.06.009.

Dedrick, J., Kraemer, K.L., & Linden, G. (2010). Who profits from innovation in global value chains? A study of the iPod and notebook PCs. *Industrial and Corporate Change, 19*(1), 81–116. https://doi.org/10.1093/icc/dtp032.

Dolan, C., Opondo, M., & Smith, S. (2004). Gender, rights and participation in the Kenya cut flower industry. *NRI Report No. 2768*. Natural Resources Institute. https://assets.publishing.service.gov.uk/media/57a08d49ed915d622c0018d5/R8077a.pdf.

Drake-Brockman, J., & Stephenson, S. (2012). Implications for 21st century trade and development of the emergence of services value chains. *Working Paper*. International Centre for Trade and Sustainable Development. https://g20-tirn.org/wp-content/uploads/2020/04/implications-for-21st-century-trade-and-development-of-the-emergence-of-services-value-chains.pdf.

Dunaway, W.A. (2014a). Bringing commodity chain analysis back to its world-systems roots: Rediscovering women's work and households. *Journal of World-Systems Research, 20*(1), 64–81. https://doi.org/10.5195/jwsr.2014.576.

Dunaway, W.A. (2014b). Introduction. In W.A. Dunaway (Ed.), *Gendered commodity chains: Seeing women's work and households in global production* (pp. 1–24). Stanford University Press.

Gereffi, G. (1994). The organisation of buyer-driven global commodity chains: How US retailers shape overseas production networks. In G. Gereffi & M. Korzeniewicz (Eds.), *Commodity chains and global capitalism* (pp. 95–122). Praeger.

Gereffi, G. (2001). Beyond the producer-driven/buyer-driven dichotomy: The evolution of global value chains in the internet era. *IDS Bulletin*, *32*(3), 30–40. https://doi.org/10.1111/j.1759-5436.2001.mp32003004.x.

Gereffi, G., Humphrey, J., Kaplinsky, R., & Sturgeon, T.J. (2001). Introduction: Globalisation, value chains and development. *IDS Bulletin*, *32*(3), 1–8. https://doi.org/10.1111/j.1759-5436.2001.mp32003001.x.

Granovetter, M. (1985). Economic action and social structure: The problem of embeddedness. *American Journal of Sociology*, *91*(3), 481–510. https://doi.org/10.1086/228311.

Grignon, M., Owusu, Y., & Sweetman, A. (2012). The international migration of health professionals. *Discussion Paper No. 6517*. Institute for the Studies of Labor (IZA). http://ftp.iza.org/dp6517.pdf.

Hochschild, A.R. (2001). Global care chains and emotional surplus value. In W. Hutton & A. Giddens (Eds.). *On the edge: Living with global capitalism* (pp. 130–46). Vintage.

Hoekman, B. (2014). *Supply chains, mega-regionals and multilateralism: A road map for the WTO*. CEPR Press. https://voxeu.org/sites/default/files/file/WTO_Roadmap.pdf.

Hopkins, T., & Wallerstein, I. (1977). Patterns of development of the modern world-system. *Review: A Journal of the Fernand Braudel Center for the Study of Economies, Historical Systems and Civilizations*, *1*(2), 111–45. https://www.jstor.org/stable/40240765.

Horner, R. (2017). Beyond facilitator? State roles in global value chains and global production networks. *Geography Compass*, *11*(2), 1–13. https://doi.org/10.1111/gec3.12307.

Humphrey, J., & Schmitz, H. (2002). Developing country firms in the world economy: Governance and upgrading in global value chains. *INEF Report 61*. Institut für Entwicklung und Frieden der Gerhard-Mercator-Universität Duisburg. http://citeseerx.ist.psu.edu/viewdoc/download?doi=10.1.1.557.1063&rep=rep1&type=pdf.

International Labour Office. (2016). *Report IV: Decent work in supply chains*. International Labour Conference, 105th Session. http://www.ilo.org/wcmsp5/groups/public/---ed_norm/---relconf/documents/meetingdocument/wcms_468097.pdf.

Klein, A. (2011a). Chain reaction: Neoliberal exceptions to global commodity chains in Dominican baseball. *International Review for the Sociology of Sport*, *47*(1), 27–42. https://doi.org/10.1177/1012690210390426.

Klein, A. (2011b). Sports labour migration as a global value chain: The Dominican case. In J. Maguire & M. Falcous (Eds.), *Sport and migration: Borders, boundaries and crossings* (pp. 88–101). Routledge.

Krugman, P.R. (1979). Increasing returns, monopolistic competition and international trade. *Journal of International Economics*, *9*(4), 469–79. https://doi.org/10.1016/0022-1996(79)90017-5.

Lasswell, H.D. (1936). *Politics: Who gets what, when, how*. Whittlesey House, McGraw-Hill Book Company.

Lodefalk, M. (2013). Servicification of manufacturing--evidence from Sweden. *International Journal of Economics and Business Research*, *6*(1), 87–113. https://doi.org/10.1504/IJEBR.2013.054855.

Lorenzo, F.M.E., Galvez-Tan, J., Icamina, K., & Javier, L. (2007). Nurse migration from a source country perspective: Philippine country case study. *Health Services Research*, *43*(2), 1406–18. https://doi.org/10.1111/j.1475-6773.2007.00716.x.

Low, P. (2013). The role of services in global value chains. In D. K. Elms & P. Low (Eds.), *Global value chains in a changing world* (pp. 61–81). World Trade Organization.

Mayer, F.W., & Phillips, N. (2017). Outsourcing governance: States and the politics of a "global value chain world." *New Political Economy*, *22*(2), 134–52. https://doi.org/10.1080/13563467.2016.1273341.

Mayoux, L., & Mackie, G. (2007). Making the strongest links: A practical guide to mainstreaming gender analysis in value chain development. International Labour Organization. https://www.ilo.org/wcmsp5/groups/public/---ed_emp/---emp_ent/documents/instructionalmaterial/wcms_106538.pdf.

Murphy, M.F. (2014). Global care chains, commodity chains, and the valuation of care: A theoretical discussion. *American International Journal of Social Science*, *3*(5), 191–9. http://www.aijssnet.com/journals/Vol_3_No_5_October_2014/19.pdf.

Neilson, J. (2014). Value chains, neoliberalism and development practice: The Indonesian experience. *Review of International Political Economy*, *21*(1), 38–69. https://doi.org/10.1080/09692290.2013.809782.

Organisation for Economic Co-operation and Development. (2013). Who's smiling now. *OECD Observer*, *3*, 61. https://doi.org/10.1787/observer-v2013-3-en.

Organisation for Economic Co-operation and Development, World Trade Organization, & World Bank. (2014, July 19). Global value chains: Challenges, opportunities, and implications for policy: Report prepared for submission to the G20 Trade Ministers Meeting Sydney, Australia. http://www.oecd.org/tad/gvc_report_g20_july_2014.pdf.

Organisation for Economic Co-operation and Development. (2018). Trade in value added. https://www.oecd.org/sti/ind/measuring-trade-in-value-added.htm.

Pipkin, S., & Fuentes, A. (2017). Spurred to upgrade: A review of triggers and consequences of industrial upgrading in the global value chain literature. *World Development*, *98*, 536–54. https://doi.org/10.1016/j.worlddev.2017.05.009.

Ponte, S. (2002). The "Latte Revolution?" Regulation, markets and consumption in the global coffee chain. *World Development*, *30*(7), 1099–22. https://doi.org/10.1016/S0305-750X(02)00032-3.

Prebisch, R. (1950). *The economic development of Latin America and its principal problems*. United Nations.

Quinn, J.B., & Hilmer, F.G. (1994). Strategic outsourcing. *MIT Sloan Management Review, 35*(4), 43–55. https://sloanreview.mit.edu/article/strategic-outsourcing/.

Ravenhill, J. (2014). Global value chains and development. *Review of International Political Economy, 21*(1), 264–74. https://doi.org/10.1080/09692290.2013.858366.

Ravenhill, J. (2017). The political economy of the Trans-Pacific partnership: A "21st century" trade agreement? *New Political Economy, 22*(5), 573–94. https://doi.org/10.1080/13563467.2017.1270925.

Singer, H.W. (1950). The distribution of gains between investing and borrowing countries. *American Economic Review, Papers and Proceedings, 40*(2), 473–85.

Stephenson, S., & Drake-Brockman, J. (2014). The services trade dimension of global value chains: Policy implications for commonwealth developing countries and small states. *Commonwealth Trade Policy Discussion Paper 2014/04*. Commonwealth Secretariat, London.

Sturgeon, T.J. (2008). From commodity chains to value chains: Interdisciplinary theory building in an age of globalisation. In J. Bair (Ed.), *Frontiers of commodity chain research* (pp. 110–35). Stanford University Press.

Tallontire, A., Dolan, C., Smith, S., & Barrientos, S. (2007). Reaching the marginalised? Gender, value chains and ethical trade in African horticulture. *Development in Practice, 15*(3–4), 559–71. https://doi.org/10.1080/09614520500075771.

United Nations Conference on Trade and Development. (2013). *World investment report 2013: Global value chains: Investment and trade for development*. United Nations. https://unctad.org/system/files/official-document/wir2013_en.pdf.

Vandermerwe, S., & Rada, J. (1988). Servitization of business: Adding value by adding services. *European Management Journal, 6*(4), 314–24. https://doi.org/10.1016/0263-2373(88)90033-3.

Wade, R.H. (2017). Global growth, inequality and poverty: The globalization argument and the "political" science of economics. In J. Ravenhill (Ed.), *Global political economy* (5th ed., pp. 319–55). Oxford University Press.

Williamson, O.E. (1975). *Markets and hierarchies, analysis and antitrust implications: A study in the economics of internal organization*. Free Press.

Wood, A. (1997). Openness and wage inequality in developing countries: The Latin American challenge to East Asian conventional wisdom. *The World Bank Economic Review, 11*(1), 33–57. https://doi.org/10.1093/wber/11.1.33.

World Bank. (2017). *Measuring and analyzing the impact of GVCs on economic development: Global value chain development report 2017*. https://www.wto.org/english/res_e/booksp_e/gvcs_report_2017.pdf.

Xing, Y., & Detert, N. (2011). How the iPhone widens the United States trade deficit with the People's Republic of China. *ADBI Working Paper No. 257*. Asian Development Bank Institute. https://www.adb.org/sites/default/files/publication/156112/adbi-wp257.pdf.

Yeates, N. (2005). Global care chains: A critical introduction. *Global Migration Perspectives, 44*. http://www.refworld.org/docid/435f85a84.html.

Yeates, N. (2009). Production for export: The role of the state in the development and operation of global care chains. *Population, Space and Place, 15*(2), 175–87. https://doi.org/10.1002/psp.546.

Yeates, N. (2012). Global care chains: A state-of-the-art review and future directions in care transnationalization research. *Global Networks, 12*(2), 135–54. https://doi.org/10.1111/j.1471-0374.2012.00344.x.

2 Circulation of Love: Care Transactions in the Global Health Care Market of Transnational Medical Travel

HEIDI KASPAR

Introduction

The Global South is subsidizing the Global North in the caring industries of childcare, housekeeping, and nursing. The global care chain (GCC) approach has been essentially contributing to providing evidence and empirically grounded narratives about this trade in care labour. The migration of (more or less) formally trained and experienced workers from countries of the Global South to take care of people in the Global North constitutes a key engine and manifestation of this transaction. As it transfers labour resources from underprivileged to privileged places, the migration of care workers from the Global South to the Global North has been named as a "perverse subsidy" (Mackintosh et al., 2006).

However, other types of care mobilities are emerging. They constitute inversed mobilities, yet yield similar kinds of "perverse subsidies." First, elderly people in the North/West with long-term intensive or extensive care needs move to (or are moved to) regions where good care is cheaper, or available in the first place (Bender et al., 2017; Horn & Schweppe, 2017). Second, patients in the search for a cure, alleviation from suffering, prolonged life, increased beauty, or wellness travel to places distant from their place of residence to receive treatment (Chee et al., 2018; Connell, 2016; Holliday et al., 2015; Kangas, 2010; Kaspar, 2019; Pian, 2015; Scheper-Hughes, 2011). One common characteristic of both forms of "patient" mobility is the tight link between the care and the sojourn. But while for elderly people requiring long-term care, the move is set up as permanent, medical treatment for the later type is temporary and so is the sojourn abroad. This type of care mobility is commonly called "medical tourism" or "medical travel." Research on medical travel thus far has focused on patients from the Global North travelling to so-called medical tourism destinations in the Global South, such as Costa Rica, Thailand, or India, suggesting that medical travel largely is a North-to-South movement (Kaspar, 2019). More recently, though, there has been increasing evidence of prevalent intra-South patient mobility (Bochaton, 2015; Connell, 2013; Crush & Chikanda, 2015; Kaspar & Reddy, 2017). Against the backdrop of long-term migration of care workers, the migration and travelling of patients constitutes an inversed mobility. It is those who seek care who become mobile and those who give care who remain moored, and the direction of mobility is reversed.

The transfer of care, though, is the same as for "classical" care mobilities: people from the Global South provide care to people from the Global North.

This chapter follows Parvati Raghuram's (2012) call to challenge and enrich the concept of GCCs through the inclusion of a more variegated sample of empirical cases. I demonstrate that transnational medical travel offers valuable insights that enrich the debate on global transfers of care. On the one side, transnational medical travel[1] has been celebrated as boosting low- and middle-income countries' national economies and modernizing their ailing health care systems (Bookman & Bookman, 2007). On the other side, medical travel has been criticized as the continuation of imperialist practices (Buzinde & Yarnal, 2012). More recently, though, research reveals a more nuanced picture of the multilayer processes and kinds of mobilities commonly subsumed under the notion of medical travel.

My own empirical study on the metropolitan region of Delhi (India), one of the leading medical travel destinations in Asia (Reddy & Qadeer, 2010), contributes to this emerging body of research. Based on a multi-sited ethnography on transnational medical travel, including destination and source sites, I argue that a range of very different transfers are simultaneously at play in medical travel to Delhi. Diversity refers both to the entity transferred as well as the direction of the transfer. This chapter shows that the transfers involved in transnational medical travel are less linear and occur along other than classical neocolonial connections, too.[2]

Global Care Chains (GCC) and the Transfer, Circulation, and Accumulation of Love

"To care is to relate: to fellow human beings, to the environment, to the self, as individuals and members of society, consciously, existentially, and over time" (Jochimsen, 2003, as cited in Madörin, 2010, p. 87). The proposition of this chapter to link the GCC approach and research on transnational medical travel is twofold. First, as outlined in the introduction, I suggest that transnational medical travel constitutes a useful case to broadening the GCC perspective. Second, I suggest that GCC provides a useful analytical lens for medical travel research, because it allows bringing medical travel into fruitful discussions with other forms of transnational care arrangements. As such, it supports the efforts to understand medical travel as part of the larger picture that constitutes transnational healthcare (Bell et al., 2015). GCC offer two more propositions that I consider particularly useful for the study of transnational patient mobility. First, it inspires the analysis of the actual care work performed in everyday care encounters. Within medical travel research, embodied care work has rarely been studied in detail. Second, drawing on feminist theories, GCC offers a nuanced concept of care work with a focus on hierarchies and asymmetries as they unfold on various levels, from the global to the intimate. A focus on power relations includes examining the different positionalities and vulnerabilities involved in transnational care as well as the transfers of care (and other entities).

Global Care Chains – The Concept in a Nutshell

The GCC concept describes "a series of personal links between people across the globe based on the paid or unpaid work of caring" (Hochschild, 2000, p. 33). Drawing on a study on migrant domestic workers, and later expanded to other forms of care work such as unpaid housework and childcare (Aggarwal & Das Gupta, 2013) or skilled and unskilled health workers (Schwiter et al., 2014; Walton-Roberts, 2012; Yeates, 2013), the GCC concept points out that the migration of care workers leaves a care deficit not only in the migrants' households but also in the migrants' domestic pool of workers. This lacuna is filled with a paid domestic worker, or family member, who sometimes moves in from distant (national or transnational) places, and therefore leaves yet another gap in their homes, hence the term "chain." This transnational perspective of the approach addresses care as a nested notion that affects society at its various scales: from the intimate sphere of families and households to global labour markets.

GCC research is very clear about the power relations involved in this "production chain": the migration of care workers engenders a transfer of care resources along the gradient of economic strength on the global as well as national level. Often, these transnational connections follow colonial trajectories; which is why the migration of care workers is flagged as neocolonial phenomenon (England, 2015).

Care Work: A Two-Component Model of Asymmetric Relations

Asymmetry is indeed a fundamental characteristic of care situations, regardless of whether or not they involve a migrant and whether or not it is paid or unpaid work when a person's limited capability to function is the primary trigger for care work and care situations (Jochimsen, 2003). Limited capacity to function assigns the involved persons dissimilar and unequal starting positions, resulting in potential and actual dependences. A caring motivation might result in a person to remain in a caring situation although valid reasons to leave exist.

For example, if a person feels responsible for the wellbeing of another person, the caregiver's own wellbeing depends on how well the person cared for is. In this case, the dependence of the person in need of care is transferred to the caregiver. The person receiving care can use this obligation as leverage, and in so doing considerably shift power relations between involved persons.

Feminist economists state that care work is composed of two components: one that is instrumental and hands-on, and one that is emotional (Folbre & Nelson, 2000). Lynch and McLaughlin (1995) have called establishing and maintaining a supportive relationship that is characterized by sympathy and affinity as love labour, defining it as a particular kind of care work. "Love labour," they state, is "the labour involved in developing solidary bonds" (p. 258). It can be argued, therefore, that the effort to create, develop, and sustain relationships constitutes a third component of care work, which I call "relational." The caring motivation mentioned above corresponds to and nurtures emotional and relational work.

Transfer, Circulation, and Accumulation of Care

Care requires giving and taking and therefore constitutes a transfer. But rather than simply transplanting something from one person to another, it involves practices and situations that are co-produced by those involved in the care situation. Within families, care often is coordinated as reciprocity, that is, some return is expected, given that a corresponding situation occurs, but unlike immediate and equivalent exchanges, the type and amount of the return as well as when and where it is given usually remains undefined (Baldassar & Merla, 2014). Baldassar and Merla (2014) argue that due to this vagueness, care keeps circulating among family members over time and space. Paid care work (usually including non-family members), however, is at least partially arranged as an immediate and equivalent exchange: labour is traded for money. Usually, not all components of care work are valued and factored in; relational and emotional work remaining unpaid or underpaid (Folbre, 2006; Madörin, 2010). Therefore, Hochschild (2000) calls the emotional and relational component of care work as surplus value. I take this argument a step further: If the emotional and relational component is not levelled out, it becomes either a gift or theft, respectively, or becomes reciprocity. In the latter case, care circulates beyond family members and includes paid caregivers.

In her article "Economies of Affect," Ahmed (2004) states that emotions circulate between bodies, they move sideways and bind them together or separate them. Love, compassion, and sympathy are clearly binding emotions productive and products of relationships of affinity. On top of this, emotions accumulate (Ahmed, 2004). Watkins (2010) highlights the "capacity of affect ... to accumulate" (p. 269), to be stored and remembered, though not necessarily in the sense of a consciously retrievable memory. Accumulation happens through repetition: "Through the iteration of similar experiences, and therefore similar affects, they accumulate in the form of what could be considered dispositions that predispose one to act and react in particular ways" (p. 278). However, "this ability of affect to accumulate is either denied or rarely made explicit" in much of the diverse literature on affect (p. 278). Given the relational work that emotions are able to perform and the prominent role they play in care work, the circulation and accumulation of emotion is of crucial interest for an accurate analysis of transnational care arrangements. A GCC analysis is particularly well-positioned to address the non-material aspects of care work (Yeates, 2005).

Transnational Medical Travel – Another Global Care Chain?

Raghuram (2012) states that in the "last decade, there were intense debates on the transfer of care from the global South to the North" (p. 155). However, according to Raghuram, this research has been limited to particular sites and has been adopting the commonly used understanding of global care chains, rather than enriching it conceptually. She therefore calls for a more variegated grounding of transnational care arrangements.

Transnational Medical Travel Offers a Different Kind of Care Arrangement Altogether

Compared to the types of mobilities GCC studies are conventionally concerned with, transnational medical travel clearly constitutes a different case. Foremost, with respect to the mobile agent, it constitutes an inversed mobility: It is those who receive care who are mobile, rather than those who give care. Apart from the swapped mobile agent and inversed mobility pattern, medical travel apparently follows the same pattern of care transfer as care worker migration. There are two remarkable parallels. First, residents of high-income countries extract care resources in low- and middle-income countries that in turn aggravate care deficits in low- and middle- income countries' health care systems. Second, the medical travel industry benefits from and increasingly capitalizes on emotional surplus value.

Medical Travel Constitutes a Reverse Subsidy of the Already Disenfranchised

Medical travel has commonly been portrayed as patients from high-income countries such as the US or the UK travelling to low- and middle-income countries such as Costa Rica, Thailand, or India to access state-of-the-art specialized medical care. The reasons for doing so are predominantly seen in the strikingly cheaper costs in destination countries (Amodeo, 2010), long waiting lists

In October 2013, Kevin and Adele took a twenty-nine-hour flight for knee replacement surgery. For the first time in their lives, the couple travelled from Columbia, South Carolina in the US to Kuala Lumpur in Malaysia. Friends and relatives were sceptical about Kevin's idea. After some research, however, for Kevin the decision seemed obvious. First of all, he made the calculation of costs. Although being insured via his employer, he would still have to pay US$5,000 as deductibles and copay if he underwent surgery in Columbia. His insurer would cancel this amount if he opted to fly to Kuala Lumpur for the intervention. Furthermore, Kevin figured out that he was actually curious about the Malaysian culture and – being employed in the health care insurance business himself – curious about the way health care was done over there, too.

Kevin's blog (see source below) leaves no doubt that, for him, the idea worked out. In an interview with me, Kevin repeats his enthusiasm about health care in Malaysia, foremost the much more patient-focused way of doing health care. He states that small things make a big difference. As examples he offers the following: doormen call you by your name; fruit baskets are delivered to your room; care coordinators help you navigate the hospital's rooms and hallways as well as administrative procedures; and staff knock on the your room's door before entering. Even Adele, who accompanied Kevin as his caretaker and who at the beginning had been doubtful of the endeavour, writes about the trip to Malaysia as a great experience; she concludes: "I heartily recommend that if given the opportunity, you strongly consider taking advantage of medical tourism. It was a great adventure and a wonderful vacation with the added benefit of a new knee for Kevin!"

Figure 2.1. A narrative of savvy patient-consumers who embark on medical tourism
Sources: *Kevin's New Knee* (blog), 10 November 2013, https://kevinsnewknee.wordpress.com, and video conference interview conducted by the author on 20 February 2014.

in source countries (Connell, 2006), or legal restrictions prohibiting access to or existence of specific treatments such as stem cell therapies and artificial reproductive technologies (Gunnarsson Payne, 2015; Petersen et al., 2017). The story of Kevin Wyatt (see figure 2.1), a US resident who needed a knee replacement and had it done in Malaysia, represents the narrative of the savvy patient-consumer the industry is invoking (Ormond & Sothern, 2012).

Kevin Wyatt's story shows that medical travellers get "first world treatment at third world prices," just as some destinations claim it (FPJ Bureau, 2017).

While some scholars highlight medical travellers' distress, arguing they are *forced* to seek options outside domestic territories because waiting times are intolerable, costs are prohibitive, or treatments are foreclosed by restrictive jurisdictions (Inhorn & Patrizio, 2009), others claim that transnationally mobile patients extract care and subsequently health from the already disenfranchised (Sengupta, 2011). Particularly with regard to organ transplantation and surrogacy, scholars have criticized medical travel as an exploitative practice that capitalizes on poor people's desperation to create opportunities for those who can afford them (Tanderup et al., 2015). Cohen (2011) develops the concept of supplementarity to capture transactions between organ donors and receivers. Some people are in a position to supplement malfunctioning body parts and hence improve their own well-being by disaggregating others, "through the mobilization or acquisition of the organic form of others" (p. 31), which often happens at the expense of these others' health (see also Inhorn, 2015). Deploying postcolonialism as an analytical lens, Buzinde and Yarnal (2012) highlight the historical trajectories that link medical travel to former modes of colonial extractions.

There is a similar debate regarding the impact of medical travel to the destinations' societies, namely access to and quality of local health care. While proponents of the medical tourism industry claim that it generates work places, income and economic growth and modernizes health care (Reisman, 2010), opponents point out the risks for aggravating inequalities and deteriorating healthcare (Reddy & Qadeer, 2010). While private hospitals in India indeed dispose of cutting-edge medical technology and highly specialized doctors return to or remain in the country, there is no evidence thus far of some sort of trickling-down effect. The main barrier can be seen in the concentration of technology and skill in private institutions in metropolitan regions that foreclose treatment to those who cannot afford the immense costs through aggressive pricing.

Transnational Medical Travel Capitalizes on Emotional Surplus Value

The emotional and relational component of care work mentioned by Kevin Wyatt (see figure 2.1) points towards the second parallel between care migration and medical travel. In the interview, Kevin clarifies that he could not have expected such warm care at home. This resonates with Ormond (2013) when he highlights that it is not merely rational factors such as cheaper prices and shorter waiting times that drive patients across national borders for medical treatment but also the desire for comfort, care, and dignity, which are not warranted in the North's health care under austerity. Kevin describes how in Malaysia he had been treated as a wholesome human being with a physical ailment rather than as the host of a dysfunctional body part. Indeed, featured patient testimonials on medical travel facilitators' and hospitals' websites praise not only medical competence and affordable cost but also the compassionate, empathic, and comprehensive way of caregiving.

Such narrations present the emotional component of care as a unique selling proposition that differentiates the service of a medical travel destination from the service you are getting at home. I argue that this embodied emotional component of care in medical travel contexts can be understood as what Hochschild (2000) has called the emotional surplus value with respect to the emotion work provided by migrant workers in domestic child and elderly care in the households of high-income countries.

Research Methods and Data Base

The research is based on a multi-sited ethnography (Marcus, 1995) that includes fieldwork at the medical travel destination of the metropolitan region of Delhi, and in a Central Asian source country. In Delhi, research was conducted during various visits during the years 2014 to 2016, adding up to a total of six months, while in the Central Asian country fieldwork took place in three different cities in a two-week-long visit in 2016.

The data comprises interviews and conversations with patients and family caregivers who were travelling to Delhi, as well as care professionals. The latter include doctors, nurses, coordinators, language interpreters, marketing and management staff, and medical travel facilitators. Some of the patients and family caregivers we met once, in India or in Central Asia, others we met several times in both countries. Observations in hospitals' international lobbies, common waiting areas, cafeterias, wards, and doctors' offices as well as documents and media reports related to health care in India and Central Asia complement the data set.

Interviews were jointly conducted by the author and research assistant Jyotishmita Sharma, and Dr. S., an Indian doctor working and living in Delhi who had been trained in Russia. For informants not proficient in English, a language interpreter hired by the hospital, the patient, or Dr. S. translated the conversation. In Delhi, interviews were largely conducted in and facilitated through three corporate hospitals, all of them leading brands in the medical tourism business. The interviews with people from Central Asia were facilitated by Dr. S. who helped Russian-speaking patients and their families with language translation and guidance through the city's daily life, apart from medical counselling.

Interviews were designed as open-ended, semi-structured interactions. Inter alia, we asked health professionals about their tasks and responsibilities towards international patients and how they experienced working with foreigners, as well as what challenges they faced while doing so. Interviews with patients and family caregivers focused on their therapeutic journeys in their countries and to Delhi and on their experiences of these travels.

Analysis of the data was carried out using initial coding as described by Charmaz (2006). Following the two-component concept of care as asymmetric relations, the analysis focused on descriptions of practical and emotional work performed for someone else's well-being, of experiencing the same, and of the effects on the self, on others, and on relations in between.

Transfers of Care in Transnational Medical Travel to Delhi

Sideways Transfers of Care Work

Medical travel is seen as a neocolonial practice (Buzinde & Yarnal, 2012) and a "perverse subsidy" of the rich (Mackintosh et al., 2006), because it is commonly understood as a practice embodied by the "classical" neocolonial positionalities, that is, powerful and demanding people from the Global North and less privileged and serving people from the Global South.

However, the empirical case of medical travel to the Delhi metropolitan region presents itself as more complex, and indeed quite different. In fact, foreign patients who travel to Delhi in order to "extract" health care services by and large belong to the Global South (Kaspar & Reddy, 2017). Our own empirical research in corporate hospitals – the key players in the Indian medical travel industry – reveals that the bulk of foreign patients reaching Delhi in 2014 came from Iraq and Afghanistan. In more general terms, the majority of foreign patients come from the Middle East, Africa, Central Asia, and neighbouring countries. The exact composition differs for each hospital, depending on the transnational networks entertained to recruit foreign patients.

Furthermore, patient flows are intertwined with larger political and economic events and processes. For example, interviews conducted in 2015 revealed that in one hospital, the number of patients from Iraq had dropped from around 80 per cent of all of their foreign patients to a few with the fall of oil prices that same year. What has immediately become clear, though, is that Western patients are rare in Delhi corporate hospitals.

Medical travel to Delhi, hence, is a foremost South-South mobility. This resonates with more recent literature highlighting South-South relations within medical travel (Bochaton, 2015; Crush & Chikanda, 2015; Inhorn, 2011; Kangas, 2002; Ormond & Kaspar, 2018; Yeoh et al., 2013). As the Global South is anything but a homogenous region, foreign patients arriving in Delhi compromise the entire socio-economic spectrum. The majority of the people we interacted with, however, are not wealthy. Many have taken huge financial efforts to facilitate the travel, such as selling their farm land or house. Furthermore, some Southern countries are wealthy, but their health care is inadequate. For these reasons, I argue that the transfer of care involved in medical travel to Delhi is not clearly an upward transfer, but rather a varied transfer that might be best summarized as sideways.

Medical travellers' positions are best described as plurivalent. With respect to their local contexts, foreign patients arriving for medical treatment in Delhi can be seen as privileged. They are embarking on a therapeutic journey and hence escaping the sometimes deadly, often painful, limitations of their locality, while others stay in place. It is important to note, though, that these medical journeys have little in common with the straightforward story of Kevin Wyatt (see figure 2.1). First, many foreigners arrive in Delhi with poorly managed conditions, including chronic, terminal, complex, and ill-diagnosed disorders. In fact, coming to Delhi, for many is the last resort after having wandered from doctor to doctor in their own country and sometimes their country's border regions. Second, regarding costs, rather than saving money or getting extra benefits, many of the foreign patients come from poor backgrounds and regions plagued by war. They sell assets, take loans, and mobilize support from their communities or via crowdfunding to collect the needed money. Third, regarding the endeavour itself, most foreign patients arriving in Delhi have no organization or company to check the reliability of hospitals, doctors, and interpreters; they are responsible for themselves. Foreign patients' obvious unfamiliarity with the local context renders them vulnerable to fraud and abuse (Kaspar, 2015). For most of our informants, medical travel is a desperate, effortful, and risky odyssey rather than a neat combination of medical intervention and vacation, coming with extra benefits.

Language Interpreters as Care Workers

I have argued elsewhere (Kaspar, 2015; Kaspar & Reddy, 2017) that for medical travel to Delhi, language interpretation is both critical and pivotal, since many of the foreign patients arriving in Delhi do not speak English or Hindi, the two common local languages. This constitutes a language barrier in a field in which communication is essential. Hospitals catering to a foreign clientele therefore engage language interpreters, mostly Arabic, but also Russian, Persian, and French. Some patients engage their own interpreters, though.

Andrews and Evans (2008) call for "a wider understanding of health care as a multi-professionally defined and reproduced endeavor" and suggest taking into account "the full range of clinical professions and their practices," including non-medical care work such as social work for inpatients (p. 772). This is particularly true for medical travel, where foreign patients benefit from a supposedly all-encompassing service package (Hartmann, 2017), and where language interpreters and medical travel intermediaries are part of the hospital's freelance and employed staff.

Instrumental Care and Emotion Work

The interpreter's job encompasses much more than mere language translation (Kaspar, 2015). In every hospital I worked in and with every interpreter I talked to, the same story emerged.

Interpreters complete or help with a bundle of different tasks. Interpreters pick up patients from the airport; change money; organize SIM cards; arrange accommodation; guide patients and family members around the hospital as well as through or to the local market; deal with bureaucratic necessities; make appointments with doctors; collect medical reports; and more. In sum, their task portfolio largely overlaps with those of medical travel facilitators (see Hartmann, 2017).

Ziyba,[3] a freelance Russian interpreter, who has been interpreting for Central Asian patients and their family members for the last two years, pointedly states: "The full responsibility is with me. Patients and attendants don't know anything, they know nothing. And they can't speak the language; they are helpless. I am their tongue, I am their body, I am their action." Foreign patients' ignorance about the local context and functioning coupled with their inability to communicate with people in the hospital, the guesthouse, the cafeteria, on the street – basically, anywhere – renders them highly dependent on an intermediary person, who speaks for them and who also carries out certain acts on their behalf. For example, collecting a medical report for a person not familiar with the place and unable to ask for directions is troublesome.

Ziyba does it to spare patients and their families from this burden. The emotional component of an act of instrumental care, particularly when it stretches the mere scope of an agreed-on service (language translation, in this case), constitutes an amenity, but it also produces comfort, warmth, and the feeling of being cared for, that is, it has an emotional surplus value.

Furthermore, given the stark knowledge imbalance, Ziyba is not merely carrying out her clients' will but also shapes decisions that have to be taken, such as whether to be admitted or not, which doctor to choose, or which diet is best. Each interpreter might handle these many minute and sometimes consequential decisions differently. Yet it shows the crucial role and far-reaching impact interpreters accept. They are, at large, foreign patients' advocates. Foreign patients who are unable to communicate, therefore, completely rely on the attitude and aptitude of their interpreters.

Many of the language interpreters embody a cultural bridge between the source and the destination country, that is, between the receivers and providers of medical care. Many of them are foreigners themselves but have been living in the city for several years. They came as students or refugees, found an income in the medical tourism industry, and stayed – or vice versa. They are "one of them," yet familiar with the place. As compatriots on alien terrain they embody affinity, they offer a relationship and familiarity. Their position as knowledgeable foreigners enables interpreters to offer both practical help and sympathy in confusing and wearing situations.[4]

Ziyba's mobile phone is always on. She is one of the few female interpreters. She tells me that patients and attendants will call her for literally anything – to buy a bottle of water, ask for a washroom, or find out prices while shopping. "I am there for them till the end," she concludes. She is available for her patients and their families at the hospital by day, and she will stay into the night or return to the hospital in an emergency. She moved closer to the hospital in order to meet her job's demands better. Although she is a freelancer, she works with one hospital only to maximize her availability for "her" patients.

Being there for the patient and her/his family members "till the end" contains a double meaning. It means that her job does not end with the discharge of the patient, but only when she/he leaves the country. However, 10 to 20 per cent of her clients die in Delhi. In her understanding, her job includes being with the family and the patient, when her/his life comes to an end. Since she is the translator, it is her, conveying the message to the family. With this, she's taking over emotion work that conventionally a doctor would do. She relates how she has to control her own feelings before facing the family so as to be in a position to offer emotional support. Performing what Hochschild (1983) has called emotion management, Ziyba stays in the intensive care unit (ICU) to cry after having heard the bad news from the doctor, and only then steps out to meet the family members and convey them the bad news with composure. Sadness expresses her personal affinity towards her clients, while emotional stability manifests her professional position as caregiver. Dying is but the most extreme example; in a hospital there are countless incidences of despair and relief, joy and grief. An interpreter sees her/his patients and relatives on a daily basis. Whatever happens, she/he is likely to be a part of it.

Love Work

For Ziyba, becoming an interpreter for foreign patients has evolved gradually and unintentionally into a full-time job. During her study time, friends came to visit Delhi and she showed them around for shopping and sightseeing. More and more people came, including people whom she didn't know; her phone number circulated through her hometown. Two years ago, the first patient called her. And from there on, her phone number has been circulating among patients and their families. A phone number circulates not nakedly, though. Attached to it come stories and experiences. Narrations transport emotions. Even if there are few stories exchanged along with the phone number, the mere act of sympathetically giving the phone number to a neighbour, friend, or relative equals a recommendation – that with this person you are in safe hands. Word-of-mouth recommendations are the predominant way that patients connect to interpreters, facilitators, and hospitals.

Hence, the care, sympathy, knowledge, and love given to one patient, travels with her/him as a memorable and recounted experience. Love moves sideways, first as a hope and promise, then as an experience. In so doing, love establishes bindings between bodies (Ahmed, 2004). Through this practice of comprehensive care and emotion work, interpreters accumulate love (Watkins, 2010). And the accumulated love generates further business, not only for the interpreter but also for the hospital that profits from the interpreter bringing in patients. Hospitals pay commissions for referring patients, but pay low salaries to their in-house interpreters. In so doing, corporate hospitals capture the emotional surplus value added by interpreters.

However, love is not the only thing circulating. Mistrust and suspicion are travelling too. For Ziyba, foreign patients' helplessness is binding. But stories of misuse and fraud circulate among interpreters and facilitators in Delhi as well as among future and past foreign patients. Narrations of fraud work are the opposite direction to circulations of love work: they caution patients to get connected and hence keep bodies apart; rather than mobilizing, they halt patients. Revealing the ugly grotesque face of medical travel, narrations of fraud render the benign connections even more precious and further enforce the accumulation of trust and business for interpreters who perform good care work.

Lynch and McLaughlin (1995) called establishing and maintaining a supportive relationship, characterized by sympathy and affinity, love work and defined it as a particular kind of care work. Love labour, they state is "the labour involved in developing solidary bonds" (p. 258). The distinction between care and love resonates with the distinction of "caring for" and "caring about." "Love labour … refers not only to a set of tasks but to a set of perspectives and orientations integrated with tasks" (p. 258f.). Love labour involves reciprocity, that is, the person who receives love is no passive object, even though the relationship is all but balanced (p. 264f.).

When I ask Ziyba if she stays in touch after the patients' departure, her face lightens up while she responds that yes, she would. She sends good wishes for seasons' greetings and other occasions, too. They stay in touch via Whatsapp and Viber. When Ziyba announces that she's coming to her hometown for a holiday, former patients invite her to their homes and give her gifts. Invitations combined with gifts are clear expressions of the reciprocity involved in love work; both involved parties perform love work to form and maintain the relationship.

Lynch and McLaughlin (1995) state that due to its fundamentally personal character, love work cannot be commodified. Based on the presented empirical findings, however, we can argue that love work sells. My data do not allow assessing whether interpreters strategically deploy love work to prompt phone numbers to circulate and hence generate business, nor would I dare to judge this. Conversations with Ziyba revealed that for her, personal attachment and provision of professional care are not two contradictory things, but mutually constitutive, intrinsically linked. It has also become clear that medical travellers tremendously benefit from the produced emotional surplus value as they receive good care and minimize the risk of being misused.

Conclusions

I am proposing that we think of transnational medical travel as a global care chain. This is not an obvious suggestion as care migration and medical travel entertain inversed movement patterns. In care migration, care workers travel *from the Global South to the North to provide care*, whereas in medical travel, patients travel *from the North to the South to receive care*. I have argued that there are two key parallels between care migration and medical travel that justify this conceptual shift against the backdrop of inversed mobilities. The two parallels are as follows:

First, transnational patient mobility yields exploitative care relations as care is transferred up the economic gradient, that is, from places and people coined by economic and medical deprivation to places and people endowed with economic strength and medical sophistication.

Due to its welding with privatization, medical travel furthermore fosters the concentration of therapeutic capacity in central places (globally and within national health care landscapes) and for-profit hospitals. This further forecloses access to health care for people situated in geographical, social, and economic peripheries and hence further aggravates gaps in local health care systems. Second, as in other global care chains, in medical travel emotional surplus value is produced to benefit those who are in a position to supplement local care deficits with transnational care transactions.

With this proposition I follow Raghuram's (2012) call to expand the sample of empirical cases to develop the GCC concept. So, what does the inclusion of medical travel yield to GCC research? The common portrayal of medical travel constitutes a narrative of "perverse subsidies" up the economic gradient, just like care migration.[5] Yet a closer inspection of actual mobility patterns and care relations reveals more complex relations on the global as well as the intimate level. On the global level, recent research highlights the relevance of South-South medical travel in which care work is transferred between places of plurivalent positions of intersecting economic and medical/health care hierarchies. Geopolitical relations emerging from the medical travel industry oscillate between South-South cooperation and global competition (Ormond & Kaspar, 2018), and they co-exist with neocolonial relations (Buzinde & Yarnal, 2012).

On the intimate level of personal interactions, however, the picture becomes even more variegated as individual actors' positionalities do not simply mirror global orders. Most striking is the presented evidence of a caregiver's powerful position vis-à-vis those cared for, resulting from the inversed mobility pattern that renders those demanding care strangers who are unfamiliar with the place and isolated from supportive networks. The presented case indicates that such an asymmetry can be a reason to exploit or prompt sympathy and solidarity.

For medical travel research, the embedding within GCC is extremely fruitful, as the focus on care work contributes to closing a relevant research gap. Up to date, the question of how care work unfolds in the highly commodified field of transnational medical travel and what this means for those who receive as well as those who give care has hardly been addressed (but see Bochaton, 2015; Kaspar & Reddy, 2017).

In sum, conceptualizing transnational medical travel as a global care chain enriches both strands of literature. This chapter contributes to both fields of study by widening our focus regarding the question of who counts as caregivers and what situations count as care encounters. Language interpreters are hardly seen as care workers (nor are medical travel intermediaries, but see Hartmann, 2017), but our research reveals that they clearly provide care. This chapter has focused on emotion work from actors that usually are not even considered care workers, that is, language interpreters. The findings highlight that the emotional surplus value of love is a key motor for medical travel; it provides potential medical traveller patients with the needed comfort and protection from fraud to dare this travel. This insight points to yet another key differentiation: The dependency between the actors involved in the care encounter is mutual (though not equivalent). To keep their jobs, language translators depend on patients' satisfaction. Circulating testimonials are language translators' potential – and jeopardy. This is equally true for the industry at large. An orientation towards patient satisfaction might moor a caregiver in an exploitative or abusive relation that she/he otherwise would terminate. As Jochimsen (2003) states, it is not untypical, though often overseen that asymmetry and dependence usually affect both the person in need of care and the caregiver. In care encounters in the context of transnational medical travel, intricate asymmetries are clearly the norm, not the exception.

The conceptual shift suggested here contributes to acknowledging the diversity and intricate nature of global care relations. This chapter has provided but one example of transnational medical travel. In the vibrant global health care market, existing destinations are rapidly developing and new ones emerging, including a diverse set of geopolitical patterns such as following colonial lines (Pian, 2015) or South-South collaborations (Crush & Chikanda, 2015) or occurring within borderlands (Bochaton, 2015), and including a diverse set of health issues and medical procedures as, for examples, experimental therapies such as in globally orchestrated clinical trials (Rajan, 2010) or stem cell therapies (Petersen et al., 2017) or traditional medicine (Cyranski, 2016). There is, hence, ample room for further medical travel case studies to enrich GCC research and push our understanding of global care relations.

ACKNOWLEDGMENTS

I am deeply indebted to all the research participants we have met and who so generously took time to meet with us and who dared to entrust the strangers we were with their experiences. I am furthermore indebted to the Indian doctor who provided unique insights and facilitated countless encounters with patients and their families in Delhi and Central Asia. I thank Lawrence Cohen and Sunita Reddy who received me as a visiting fellow in their respective research units and provoked novel lines of thought. Thanks also go to the Swiss National Science Foundation (SNSF) and Indo-Swiss Joint Research Programme (ISJRP) for financing this research.

NOTES

1 I use *transnational* – instead of *global* or *international* – because the notion invokes an "image of the world made up of flows that transverse national boundaries," which are understood as productive and performative, rather than fixed and pre-existing (Wilson, 2012, p. 49).
2 See also Henry (chapter 6) and Peralta (chapter 14) in this volume. Henry suggests the concept of a global intimate workforce to avoid the linearity prevalent in global care chain literature. Peralta draws our attention to the work of the pioneering Filipina nurse Julita Sotejo, who was a key figure in building nurse education in the Philippines, but whose merit has been largely overseen because it did not fit the common understanding of nursing as being brought to the Philippines through colonialism.
3 All informants' names are pseudonyms to protect their identity.
4 For Arabic interpreters, the situation is different but works similarly, if not as seamlessly. Arabic interpreters working in the medical travel business mostly are Muslim Indians. Hence, what they share with their clients is a common religious (rather than national or ethnic) identity. On a terrain that is generally perceived as predominantly Hindu, a Muslim identity is likely to spark proximity, solidarity, and sympathy; the warm, almost intimate, character of the relations I observed between Indian interpreters and foreign patients and their family members took me by surprise.
5 The chapters in this volume provide an excellent documentation of this mechanism; see particularly, Chikanda, chapter 8, and Thompson, chapter 9, who emphasize the detrimental effects of care migration to sending societies.

REFERENCES

Aggarwal, P., & Das Gupta, T. (2013). Grandmothering at work: Conversations with Sikh Punjabi grandmothers in Toronto. *South Asian Diaspora*, *5*(1), 77–90. https://doi.org/10.1080/19438192.2013.722382.

Ahmed, S. (2004). Affective economies. *Social Text, 22*(2), 117–39. https://doi.org/10.1215/01642472-22-2_79-117.

Amodeo, J. (2010). Medical refugees and the future of health tourism. *World Medical & Health Policy, 2*(4), 65–81. https://doi.org/10.2202/1948-4682.1103.

Andrews, G.J., & Evans, J. (2008). Understanding the reproduction of health care: Towards geographies in health care work. *Progress in Human Geography, 32*(6), 759–80. https://doi.org/10.1177/0309132508089826.

Baldassar, L., & Merla, L. (2014). Introduction: Transnational family caregiving through the lens of circulation. In L. Baldassar & L. Merla (Eds.), *Transnational families, migration and the circulation of care: Understanding mobility and absence in family life* (pp. 3–24). Routledge.

Bell, D., Holliday, R., Ormond, M., & Mainil, T. (2015). Transnational healthcare, cross-border perspectives. *Social Science & Medicine, 124*, 284–9. https://doi.org/10.1016/j.socscimed.2014.11.014.

Bender, D., Hollstein, T., & Schweppe, C. (2017). The emergence of care facilities in Thailand for older German-speaking people: Structural backgrounds and facility operators as transnational actors. *European Journal of Ageing, 14*(4), 365–74. https://doi.org/10.1007/s10433-017-0444-1.

Bochaton, A. (2015). Cross-border mobility and social networks: Laotians seeking medical treatment along the Thai border. *Social Science & Medicine, 124*, 364–73. https://doi.org/10.1016/j.socscimed.2014.10.022.

Bookman, M.Z., & Bookman, K.R. (2007). *Medical tourism in developing countries.* Palgrave Macmillan.

Buzinde, C.N., & Yarnal, C. (2012). Therapeutic landscapes and postcolonial theory: A theoretical approach to medical tourism. *Social Science & Medicine, 74*(5), 783–7. https://doi.org/10.1016/j.socscimed.2011.11.016.

Charmaz, K. (2006). *Constructing grounded theory. A practical guide through qualitative analysis*. Sage.

Chee, H.L., Whittaker, A., & Por, H.H. (2018). Sociality and transnational social space in the making of medical tourism: Local actors and Indonesian patients in Malaysia. *Mobilities, 14*(1), 87–102. https://doi.org/10.1080/17450101.2018.1521124.

Cohen, L. (2011). Migrant supplementarity: Remaking biological relatedness in Chinese military and Indian five-star hospitals. *Body & Society, 17*(2–3), 31–54. https://doi.org/10.1177/1357034X11400766.

Connell, J. (2006). Medical tourism. Sea, sun, sand and ... surgery. *Tourism Management, 27*(2006), 1093–100. https://doi.org/10.1016/j.tourman.2005.11.005.

Connell, J. (2013). Contemporary medical tourism: Conceptualisation, culture and commodification. *Tourism Management, 34*, 1–13. https://doi.org/10.1016/j.tourman.2012.05.009.

Connell, J. (2016). Reducing the scale? From global images to border crossings in medical tourism. *Global Networks, 16*(4), 531–50. https://doi.org/10.1111/glob.12136.

Crush, J., & Chikanda, A. (2015). South–South medical tourism and the quest for health in Southern Africa. *Social Science & Medicine, 124*, 313–20. https://doi.org/10.1016/j.socscimed.2014.06.025.

Cyranski, C. (2016). *Purifying purges and rejuvenating massages: Ayurvedic health tourism in South India* [Unpublished doctoral dissertation]. Heidelberg University.

England, K. (2015). Nurses across borders: Global migration of registered nurses to the US. *Gender, Place & Culture, 22*(1), 143–56. https://doi.org/10.1080/0966369X.2013.832658.

Folbre, N. (2006). Measuring care: Gender, empowerment, and the care economy. *Journal of Human Development, 7*(2), 183–99. https://doi.org/10.1080/14649880600768512.

Folbre, N., & Nelson, J.A. (2000). For love or money – or both? *The Journal of Economic Perspectives, 14*(4), 123–40.

FPJ Bureau. (2017, July 19). Medical value travel – First World treatment at Third World prices. *The Free Press Journal*. https://www.freepressjournal.in/cmcm/medical-value-travel-first-world-treatment-at-third-world-prices. https://doi.org/10.1257/jep.14.4.123.

Gunnarsson Payne, J. (2015). Reproduction in transition: Cross-border egg donation, biodesirability and new reproductive subjectivities on the European fertility market. *Gender, Place & Culture, 22*(1), 107–22. https://doi.org/10.1080/0966369X.2013.832656.

Hartmann, S. (2017). *The work of medical travel facilitators: Caring for and caring about international patients in Delhi* (Vol. 6). CrossAsia-eBooks.

Hochschild, A.R. (1983). *The managed heart: Commercialization of human feeling*. University of California Press.

Hochschild, A.R. (2000). Global care chains and emotional surplus value. In W. Hutton & A. Giddens (Eds.), *On the edge: Living with global capitalism* (pp. 130–46). Jonathan Cape.

Holliday, R., Bell, D., Jones, M., Hardy, K., Hunter, E., Probyn, E., & Taylor, J.S. (2015). Beautiful face, beautiful place: Relational geographies and gender in cosmetic surgery tourism websites. *Gender, Place & Culture, 22*(1), 90–106. https://doi.org/10.1080/0966369X.2013.832655.

Horn, V., & Schweppe, C. (2017). Transnational aging: Toward a transnational perspective in old age research. *European Journal of Ageing, 14*(4), 335–9. https://doi.org/10.1007/s10433-017-0446-z.

Inhorn, M.C. (2011). Globalization and gametes: Reproductive "tourism," Islamic bioethics, and Middle Eastern modernity. *Anthropology & Medicine, 18*(1), 87–103. https://doi.org/10.1080/13648470.2010.525876.

Inhorn, M.C. (2015). *Cosmopolitan conceptions. IVF sojourns in global Dubai*. Duke University Press.

Inhorn, M.C., & Patrizio, P. (2009). Rethinking reproductive "tourism" as reproductive "exile." *Fertility and Sterility, 92*(3), 904–6. https://doi.org/10.1016/j.fertnstert.2009.01.055.

Jochimsen, M.A. (2003). *Careful economics. Integrating caring activities and economic science*. Kluwer Academic Publishers.

Kangas, B. (2002). Therapeutic Itineraries in a global world: Yemenies and their search for biomedical treatment abroad. *Medical Anthropology: Cross-Cultural Studies in Health and Illness*, *21*(35), 35–78. https://doi.org/10.1080/01459740210620.

Kangas, B. (2010). Traveling for medical care in a global world. *Medical Anthropology*, *29*(4), 344–62. https://doi.org/10.1080/01459740.2010.501315.

Kaspar, H. (2015). Private hospitals catering to foreigners underestimate interpreters' role. *Hindustan Times*. hospitals-catering-to-foreigners-underestimate-interpreters-role/article1-1333370.aspx.

Kaspar, H. (2019). Searching for therapy, seeking for hope: Transnational cancer care in Asia. *Mobilities*, *14*(1), 120–36. https://doi.org/10.1080/17450101.2018.1533688.

Kaspar, H., & Reddy, S. (2017). Spaces of connectivity: The formation of medical travel destinations in Delhi National Capital Region (India). *Asia Pacific Viewpoint*, *58*(2), 228–41. https://doi.org/10.1111/apv.12159.

Lynch, K., & McLaughlin, E. (1995). Caring labour and love labour. In P. Clancy, S. Drudy, K. Lynch, & L. O'Dowd (Eds.), *Irish society. Sociological perspectives* (pp. 250–92). Institute of Public Administration.

Mackintosh, M., Mensah, K., Henry, L., & Rowson, M. (2006). Aid, restitution and international fiscal redistribution in health care: Implications of health professionals' migration. *Journal of International Development*, *18*(6), 757–70. https://doi.org/10.1002/jid.1312.

Madörin, M. (2010). Care Ökonomie – eine Herausforderung für die Wirtschaftswissenschaften [Care economy – A challenge for economic sciences]. In C. Bauhardt & G. Çaglar (Eds.), *Gender and economics. Feministische Kritik der politischen Ökonomie* [*Feminist critique of political economy*] (pp. 81–104). VS Verlag.

Marcus, G.E. (1995). Ethnography in/of the world system: The emergence of multi-sited ethnography. *Annual Review of Anthropology*, *24*(1), 95–117. https://doi.org/10.1146/annurev.an.24.100195.000523.

Ormond, M. (2013). *Neoliberal governance and international medical travel in Malaysia*. Routledge.

Ormond, M., & Kaspar, H. (2018). South-south medical tourism. In E. Fiddian-Qasmiyeh & P. Daley (Eds.), *Routldege handbook of south-south relations* (pp. 397–405). Routledge.

Ormond, M., & Sothern, M. (2012). You, too, can be an international medical traveler. Reading medical travel guidebooks. *Health & Place*, *18*(5), 935–41. https://doi.org/10.1016/j.healthplace.2012.06.018.

Petersen, A., Munsie, M., Tanner, C., MacGregor, C., & Brophy, J. (2017). *Stem cell tourism and the political economy of hope*. Palgrave Macmillan.

Pian, A. (2015). Care and migration experiences among foreign female cancer patients in France: Neither medical tourism nor therapeutic immigration. *Journal of Intercultural Studies*, *36*(6), 641–57. https://doi.org/10.1080/07256868.2015.1095712.

Raghuram, P. (2012). Global care, local configurations – challenges to conceptualizations of care. *Global Networks*, *12*(2), 155–74. https://doi.org/10.1111/j.1471-0374.2012.00345.x.
Rajan, K.S. (2010). The experimental machinery of global clinical trials. Case studies from India. In A. Ong & N.N. Chen (Eds.), *Asian biotech: Ethics and communities of fate* (pp. 55–80). Duke University Press.
Reddy, S., & Qadeer, I. (2010). Medical tourism in India: Progress or predicament. *Economic and Political Weekly*, *45*(20), 69–75. https://www.jstor.org/stable/27807028.
Reisman, D. (2010). *Health tourism: Social welfare through international trade*. Edward Elgar Publishing Limited.
Scheper-Hughes, N. (2011). Mr Tati's holiday and João's safari: Seeing the world through transplant tourism. *Body & Society*, *17*(2–3), 55–92. https://doi.org/10.1177/1357034X11402858.
Schwiter, K., Berndt, C., & Schilling, L. (2014). Ein sorgender Markt. Wie transnationale Vermittlungsagenturen für Seniorenbetreuung Im/mobilität, Ethnizität und Geschlecht in Wert setzen [A caring market. How transnational agencies for senior care capitalize on im/mobility, ethnicity and gender]. *Geographische Zeitschrift*, *102*(4), 212–31. https://www.zora.uzh.ch/id/eprint/105336/.
Sengupta, A. (2011). Medical tourism: Reverse subsidy for the elite. *Signs*, *36*(2), 312–19. https://doi.org/10.1086/655910.
Tanderup, M., Reddy, S., Patel, T., & Brunn Nielsen, B. (2015). Reproductive ethics in commercial surrogacy. Decision-making in IVF-clinics in New Delhi, India. *Journal of Bioethical Inquiry*, *12*(3), 491–501. https://doi.org/10.1007/s11673-015-9642-8.
Walton-Roberts, M. (2012). Contextualizing the global nursing care chain: International migration and the status of nursing in Kerala, India. *Global Networks*, *12*(2), 175–94. https://doi.org/10.1111/j.1471-0374.2012.00346.x.
Watkins, M. (2010). Desiring recognition, accumulating affect. In M. Gregg & G.J. Seigworth (Eds.), *The affective theory reader* (pp. 269–85). Duke University Press.
Wilson, J. (Ed.). (2012). *The Routledge handbook of tourism geographies*. Routledge.
Yeates, N. (2005). *Global care chain: A critical introduction* (Global Commission on International Migration (GCIM) No. 44). http://de.slideshare.net/conormccabe/yeates.
Yeates, N. (2013). Global care chains: Bringing in transnational reproductive laborer households. In W. Dunaway (Ed.), *Gendered commodity chains. Seeing women's work and households in global production* (pp. 175–89). Stanford University Press.
Yeoh, E., Othman, K., & Ahmad, H. (2013). Understanding medical tourists: Word-of-mouth and viral marketing as potent marketing tools. *Tourism Management*, *34*, 196–201. https://doi.org/10.1016/j.tourman.2012.04.010.

SECTION 2

Conceptualizing Workplace Integration and Stratification: Immigration Policy, International Credentials, and Intersectional Disadvantage

3 The Migration of Health Professionals to Canada: Reducing Brain Waste and Improving Labour Market Integration

ARTHUR SWEETMAN

Globally, there exist appreciable long-standing imbalances in the health workforce. Many low- and middle-income countries face shortages of key health professionals with the World Health Organization (WHO) and the Organisation for Economic Co-operation and Development (OECD) calling it a crisis. In terms of attention to this issue from the developed world, international policy discussion probably peaked near the end of the first decade of the twenty-first century. For example, in 2007, the OECD's *International Migration Outlook* pointed to several developed countries "formulating policy recommendations to overcome the global health workforce crisis" (p. 162), with a fuller discussion in the 2008 OECD report, *The Looming Crisis in the Health Workforce*. In 2010, the WHO adopted a global code of practice on the international recruitment of health personnel with a focus on ethics and protecting less-developed immigrant-sending countries (WHO, 2010). Aligned with this initiative, several developed countries devised protocols taking the ethics of international health professional migration into account (for Canada, see Canadian Federal/Provincial/Territorial Advisory Committee on Health Delivery and Human Resources [ACHDHR], 2009; Dumont et al., 2008; Health Canada, 2004), although the extent to which such protocols have had any effect remains unclear. This chapter proposes a policy innovation to reduce Canada's contribution to the ongoing growth of this global crisis.

In contrast to the shortages in developing countries, many developed countries – especially Canada – faced, and continue to face, a surplus of individuals reporting health credentials who are not licensed to practise. Despite numerous popular discussions of physician shortages in Canada, the stock of internationally educated health professionals not working in their trained professions accumulates. While provincial governments actively welcome some foreign-trained health professionals (e.g., http://www.healthforceontario.ca), the popular press and other outlets regularly report on the challenges faced by others. For example, an article by CTV News.ca Staff (2014) indicated that only half of international medical graduates residing in Canada work in Canada as physicians, and see Adekola (chapter 15 in this collection) on nursing.

Focusing on the eight largest regulated occupations, Owusu and Sweetman (2015) nuance the latter observation, pointing out dramatically different propensities for immigrants and the Canadian born to be trained in, and to work in, a range of health occupations. More broadly, Lofters et al. (2014) point out both the brain drain from developing countries and the brain waste in Canada, and Chikanda (chapter 8 in this collection) discusses the crucial role of intermediaries in active recruitment. As an indication of the magnitude of the problem, in early 2020, HealthForceOntario reported 13,000 physicians and 6,000 nurses in Ontario alone, comprising both immigrants and Canadians by birth, who had foreign training and were not working in their trained occupation (see Olaizola & Sweetman, 2019, for a discussion of Canadian accreditation exam pass rates for the foreign trained.)

Background regarding some of the origins of these contradictions can be found in the overviews of regulatory and economic issues related to the international migration of health professionals by Bourgeault and Grignon (2013), Grignon et al. (2013), and van Riemsdijk (chapter 10 in this collection). Building on these rationales, the argument put forward here is that at the heart of the Canadian problem is a misalignment of government responsibilities coupled with what can be interpreted as substantial transactions costs for federal and provincial governments attempting to coordinate policies within the federation. This coordination problem matters since most immigrant selection occurs at the federal level whereas those in health occupations practice almost entirely within the provincial domain.

Two policy directions seem relevant. The first, which is already underway, involves improved provincial oversight of regulatory colleges to ensure that the education and skills of internationally educated health professionals are appropriately evaluated for licensure and that, where modest gaps relative to an appropriate standard are observed, bridging programs are made available where warranted by occupational demand (see Bourgeault et al., chapter 5 in this collection, on bridging programs). Improved information, especially premigration, also falls into this category and is being pursued at present.

Second, and more novel, new immigrants in relevant occupations should be shifted out of federal immigrant selection programs to the Provincial Nominee Program, or some comparable new provincial government led immigration stream within the Economic Class. This latter proposal would concentrate decision-making responsibility within one level of government, thereby reducing complexity and alleviating the need for costly policy coordination. It also more clearly identifies accountability.

The next section of this chapter provides a brief and non-technical description of the economic structure of Canadian health labour markets, which is essential background for this policy discussion. Attention next focuses on immigrants in health occupations and the relevant Canadian immigration and labour market policy that they need to navigate. Finally, the details of policy proposals to address these problems are put forward together with a discussion of implementational challenges.

Background: The Economic Structure of Canadian Health Labour Markets

An understanding of the economic structure of Canadian health labour markets is required to appreciate the relevant policy context for immigration. These labour markets have numerous complex features (Baumann et al., chapter 4 in this collection) by contrast to the relatively straightforward classical competitive model discussed in most introductory undergraduate economics courses that seems to motivate much policy analysis. The latter provides insight and is useful as a touchstone, but is highly stylized and unrealistic for most labour markets, and particularly for those with numerous distortions such as the health labour markets at issue in this study. In the stereotypical competitive market, there are many workers independently offering their labour services, and many firms independently looking to purchase labour. Labour market equilibrium is defined by the wage and number of employed individuals (i.e., the number of potential jobs filled) where the minimum acceptable wage of the marginal worker is equal to the maximum wage offered by the marginal employer. Because workers and firms each seek out the best match they can find for themselves, and are presumed to have full information and to be able to change employers/workers and/or renegotiate without transactions costs, in the long run all workers within a particular skill level end up being paid the same wage to do comparable jobs. Since the economic structure of the labour market is influenced by the economic structure of the product/output market, it is similarly useful to distinguish between classical/naïve views of those markets compared to the realistic characteristics of Canadian health care markets.

Although each health occupation's labour market has unique elements, several broadly common features that deviate from the stylized model are relevant for particular subsets of Canada's health professions.[1] Therefore, to facilitate this brief discussion, attention will be drawn to selected issues that are common to multiple occupations.

Focusing first on the employer/payer side, one grouping of health occupations – including physicians, nurses, laboratory technologists, and medical radiation technologists – effectively has a single (or at least a dominant) employer or purchaser of services in each jurisdiction: the provincial government.[2] This is sometimes termed a monopsony (a monopoly purchaser), but – unsurprisingly given the complexity of Canadian healthcare – it's not your standard monopsony.

Beyond (effectively) either directly or indirectly controlling funding for worker remuneration, provinces control or strongly influence the number of spots in training programs (especially medical schools). Making the situation still more complex, the payers (provinces) do not have a profit motive but rather some hard-to-specify and frequently changing objective function, with the latter sometimes interpreted as a type of myopic behaviour. Changes in provincial governments can lead to changes in policies that, were the payer a single individual, might be interpreted as irrational by observers looking for consistency across time. Further, for extended periods of time (decades), the payer need not balance the books, so private sector fiscal discipline – negative profits or even bankruptcy – has only a very weak influence.[3] Finally, although this discussion of institutional complexity is far from exhaustive, there are substantial transfers from the federal to the provincial governments through, in particular, Canada's Equalization program and the Canada Health Transfer program that induce gaps between accountabilities on the revenue and expenditure sides (Rosen et al., 2008). While there are good reasons for the existence of this division of responsibilities, it nevertheless introduces substantial complexity into health care labour markets.

A second cluster of health professions – including dentists, optometrists, and pharmacists – primarily work in what is nominally referred to as the private sector, although there are certainly some who work in the public sector. However, the funding of the private sector in health care has far more public sector elements than are present in most other parts of the private economy. A very high fraction of utilization is government-subsidized through employer-provided health insurance or (especially in Quebec, e.g., Wang et al., 2015) other government mandated programs, and much out-of-pocket expenditure is subsidized through tax programs such as the Medical Expense Tax Credit, the Medical Expense Supplement, and the Disability Tax Credit (Department of Finance Canada, 2016; Smart & Stabile, 2005, 2006). There are also direct refund programs in many provinces for individuals on social assistance, over the age of sixty-five, or with catastrophic health care costs. Clearly, there are good reasons for these programs, but they nevertheless have important implications for the structure of relevant labour markets. While their impacts on various dimensions of expenditures go beyond the scope of this brief discussion, a key outcome is to increase demand (and I suspect also make demand less sensitive to various economic drivers such as the business cycle). Of course, the whole purpose of these programs is to increase demand since it is believed that left to their own devices markets would undersupply essential health care goods and services to some with low ability to pay. But, there is a large literature suggesting that overshooting occurs (e.g., Law et al., 2011; Zhang & Sweetman, 2018).

These features are mostly about the output/product market, but they have implications for the structure of the labour market for relevant occupations (and other input markets).

Turning to the worker side of the labour market, many health care professions are unionized, especially physicians and nurses.[4] There is a large literature in economics discussing differences between (public sector) unionized and nonunionized labour markets. A coherent introduction to these issues in the Canadian context is by Benjamin et al. (2012) and Card et al. (2020). Empirical analyses of centralized bargaining by a monopoly union has long been interpreted to mean – for example, Kuhn (1998) – that monopoly "rents" (i.e., excess wages above the competitive norm) exist. While some argue that the cost of this premium is a reduction in employment and service provision, on the whole the evidence is unclear.

Context is, of course, important for interpretation since there are particular design features to the institutional setting of health labour markets in Canada, whereas most of the literature studying the economics of unionization looks at the economy as a whole and does not focus on the health sector.

Moreover, much of the research literature does not distinguish between industrial and trade unions, whereas in the Canadian health context most unions are trade unions (i.e., occupation-specific unions historically referred to as guilds) that are the sole provider of relevant services within the jurisdiction. This contrasts with industrial unions where, frequently, a specific union does not represent all the firms in the relevant product market, which imposes some limitations on collective bargaining since the firm can lose business to competitors.[5] Also, these are public-sector unions and, as discussed above regarding employers, there are a wide array of factors that differentiate the public from the private sector: public sector collective bargaining does not face bankruptcy as a constraint, governments wield legislative power, and the demands of patient health limit the ability of workers to withdraw services and employers to lock out workers.

Formally, economists refer to the situation where there is both a monopoly purchaser and a monopoly seller as a bilateral monopoly, and outcomes are difficult to model in this context. The outcome of collective bargaining depends on the relative bargaining power of each side, which is in large part dictated by relative opportunity costs, impatience (i.e., rates of time preference), risk aversion and similar characteristics. Glimpses of these ideas in practice can be seen when, for example, in relation to opportunity costs some physicians threaten to leave a province during bargaining.

A third institutional structure adds even more complexity to most health profession labour markets: they are government regulated. In the Canadian context this almost universally takes the form of self-regulation by legislatively established regulatory colleges (emergency medical technicians being an exception). Relatively little Canadian research studies the relationship between self-regulation and labour and product market outcomes, and no research of which I am aware studies causal impacts. There is somewhat more research activity in the US with overviews by Kleiner (2000, 2006, 2013), and a recent analysis by Kleiner and Vorotnikov (2017). Regulation, though less studied, is thought to have a similar impact on wages as unionization. However, while there is appreciable overlap, the two labour market institutions tend to affect quite different types of workers. Regulated occupations tend to involve more highly educated, and frequently own-account, workers who have occupation-specific skills.

Physicians, of Course, Fall into Both Categories

Controlling for the type of characteristics commonly observed in survey data, research by Kleiner and Krueger (2013) suggests that occupational licensing in the US is associated with wages that are about 18% higher than similar individuals without such occupational regulation. Kleiner and Vorotnikov (2017) obtain a larger dataset with a more extensive set of control variables, including (what appear to be self-reported) measures of math and reading skills, and find that this reduces the premium to around 11 per cent. It is worth noting that they find the premium for occupational regulation in the US with no statistical controls whatsoever to be just over 29 per cent. Clearly, occupational regulation is reasonably highly correlated with characteristics that are associated with high earnings since adding statistical controls reduces the premium (i.e., the ordinary least squares coefficient). Interestingly for the context of this chapter, Kleiner and Vorotnikov (2017) perform conditional quantile regressions and observe the coefficient on the licensure variable to increase systematically across the quantiles of the conditional earnings distribution. It is 4 or 5 per cent at the 30th quantile and below, but increases to over 25 per cent at the 90th percentile. Broadly speaking, this range of estimates is comparable in magnitude to the impact associated with unionization, but unionization has a stronger impact on earnings at the low end of the conditional earnings distribution. Several other studies find similar relationships, as surveyed by Sweetman et al. (2015).

Background: Selected Empirical Findings

For Canada, Zhang (2017, 2019) observes a cross-sectional pay premium of about 14 per cent for occupational regulation and about 7 per cent for unionization controlling for variables observed in Statistics Canada data and averaging across the entire economy. Interestingly, these two premia appear to operate independently in the sense that the coefficient on the interaction of the two is not usually statistically significantly different from zero. (However, in some specifications the coefficient on the interaction term is modestly negative and marginally statistically significant.) Hence, individuals in a regulated occupation that is also unionized would experience (almost) the sum of the two. When

Zhang focuses on those who change occupations (both voluntarily and involuntarily) and looks for differences across those moving into/out of a regulated occupation or a unionized position, she finds that the estimates are reduced to 2.6 per cent for licensing and 4.4 per cent for unionization. Of course, she has no source of exogenous variation to identify a causal impact and the selection of individuals into and out of these categories is not random.

Evidence on the total causal effects of the various labour market institutions and related economic structures around health occupations is rare since there are few contexts where causal impacts can be credibly estimated (see Sweetman et al., 2015). I am unaware of work on this topic for Canada; however, a credible estimate of the premium for physicians[6] in the Netherlands is provided by Ketel et al. (2016) who exploit the fact that medical schools in the country employed lotteries to allocate positions among similar applicants. By comparing outcomes of those accepted into the program versus those participating in their next best occupation, and following individuals for twenty-two years, they observe that every single year after graduation, physicians (i.e., those who win the lottery) earn, on average, at least 20 per cent more than those denied entry to medical school, and after twenty-two years, the gap is just under 50 per cent. They examine the degree to which this gross earnings difference can be attributed to working hours and/or human capital investments, and find that these explain only a very small portion of the difference. Also, the gap does not vary with sex or measured ability among the selected set of those eligible for the lottery. If immigrants are excluded from these occupations, they clearly do not benefit from the monopoly rents (i.e., excess profits) available in these bilateral monopolies.

Focusing directly on immigrants, Kugler and Sauer (2005) also examine plausibly causal, though more narrowly defined, impacts by studying the flow of trained physicians from the former Soviet Union to Israel following the dissolution of the Soviet Union. They exploit substantial differences in the Israeli regulatory system in the probability of being licensed as a function of years of pre-migration work experience to estimate the impact of medical licensure on earnings. They estimate a premium of at least 90 per cent (i.e., an almost doubling of earnings) as a result of licensure, but unlike the above-mentioned Dutch study which focuses on new domestic graduates, they are looking at the average premium for older physicians who have recently migrated, so their next best opportunity is quite different from that of those entering medical school.

So what does this section have to do with immigration? My view is that it is not credible (indeed, it may promote serious misperception) to analyse immigrant outcomes, and especially to formulate policy, without understanding the relevant context. That is, this description is an essential precursor – especially given the degree of government intervention in selected markets such as health care.

Immigrants in Health Professions and Immigration Policy in Canada

The fundamental issue for health occupations can be seen table 3.1. Drawing on Owusu and Sweetman (2015), Panel A presents the percentage of individuals not working in their trained occupation in each of the four place of birth and place of education categories whose highest credential is relevant to one of the eight regulated occupations studied. A first observation is that in no occupation, and in none of the four categories, is everyone working in the occupation for which they hold a credential. It's also obvious that for all four groups there is substantial heterogeneity across occupations in the probability of working in the trained occupation. Focusing on the Canadian-born, Canadian-educated category, at the low end, just over 10 per cent of those reporting medical degrees are not working as physicians, whereas somewhat over 60 per cent of psychologists are not working in their trained field. This baseline provides some perspective in considering the other three categories.

A second observation is that place of education is more important than place of birth for predicting the probability of working in one's trained occupation. With the exception of pharmacists, the Canadian born, Canadian educated look very similar to the foreign born, Canadian educated group, although the foreign born, Canadian educated have slightly higher probabilities of working for almost all the occupations.

A third observation represents a large portion of the motivation for the policy proposals contained in this chapter. A remarkably high percentage of the foreign born, foreign educated in our sample are not working in their trained occupation. The numbers range from a high of 76 per cent of medical laboratory technologists to a low of 52 per cent of physiotherapists. About 60 per cent of foreign born, foreign educated individuals who self-report a medical degree are not working in their trained field despite ongoing discussions of physician shortages in Canada. Of course, the categorization of foreign-born individuals as having Canadian or foreign education is endogenous, and this selection effect needs to be recognized and is potentially quite important. Since one of the mechanisms to obtain licensure and practice in one's trained profession in Canada is to obtain some amount of Canadian education, the foreign born, foreign educated able to obtain such education end up being categorized as foreign born, Canadian educated. Nevertheless, a very substantial share of the foreign born, foreign educated are not practising in the occupation for which they trained.

Panel B of table 3.1 puts some of this into perspective. Each row in this table sums to 100 per cent, and it illustrates the percentage of the workforce in each occupation that is in each of the four categories. Keep in mind that the sample in Panel B is restricted to those actually working in their trained occupation. The bottom row of Panel B presents the share of each of the four categories in the Canadian labour force and serves as a point of comparison. Focusing on the Canadian born, Canadian educated column, this table can be interpreted to mean that this group

Table 3.1. Health professionals by place of birth and education

	Canadian Born	Canadian Born	Foreign Born	Foreign Born
	Canadian Education	Foreign Education	Canadian Education	Foreign Education
Panel A: Per Cent Reporting Qualification but Not Practicing by Location of Highest Training				
Dentist	14.5	–	14.5	68.7
Med. Lab. Tech.	51.3	–	58.3	76.1
Med. Radiation Tech.	36.7	42.9	38.4	62.1
Pharmacist	18.2	45.0	30.2	59.6
Physician	10.6	38.3	12.9	59.3
Physiotherapist	19.4	18.6	23.4	52.4
Psychologist	61.3	67.7	66.3	85.4
Registered Nurse	36.3	47.3	35.6	55.4
Panel B: Workforce Distribution by Place of Birth and Education (each row sums to 100%)				
Dentist	67.1	–	24.1	8.7
Med. Lab. Tech.	78.2	–	11.7	10.1
Med. Radiation Tech.	84.4	0.4	8.6	6.5
Pharmacist	71.1	0.2	16.5	12.2
Physician	66.7	0.7	19.9	12.7
Physiotherapist	81.0	1.2	10.7	7.2
Psychologist	82.0	1.3	14.0	2.7
Registered Nurse	80.9	0.3	11.3	7.6
Canadian Labour Force	73.1	0.5	13.5	12.9
Panel C: Share Reporting Degree Relative to the Share for the Canadian Born, Canadian Educated				
Dentist	NA	–	1.9	2.0
Med. Lab. Tech.	NA	–	0.9	1.5
Med. Radiation Tech.	NA	0.8	0.6	0.7
Pharmacist	NA	0.7	1.5	2.0
Physician	NA	2.4	1.7	2.4
Physiotherapist	NA	2.2	0.7	0.9
Psychologist	NA	2.9	1.1	0.5
Registered Nurse	NA	0.6	0.7	0.8

Notes: NA = not applicable; "–" = cell suppressed since sample too small.
Source: Owusu & Sweetman (2015).

represents 67.1 per cent of dentists, but 73.1 per cent of the Canadian labour force, so the Canadian born, Canadian educated are underrepresented in dentistry relative to their share of the entire workforce. This category is similarly underrepresented in pharmacy and medicine. It is overrepresented in the remainder of the occupations. Sometimes the gaps are quite small, and sometimes they are substantially larger. Overall, aggregating the final two columns together, the foreign-born are overrepresented in several of the professions, for example medicine, dentistry, and pharmacy, and underrepresented in others, for example psychology and medical radiation technology.[7] Looking at the foreign born, foreign educated, they are approximately proportionally represented in medicine and pharmacy, although substantially underrepresented in psychology and medical radiation technology.

Panel C in table 3.1 reconciles the above two panels by showing the relative probability of reporting credentials in each of the eight occupations relative to the Canadian born, Canadian educated for the entire immigrant population regardless of employment in the occupation associated with the reported credential. Hence the "2.0" in the foreign born, foreign educated column for dentists indicates that this latter group is twice as likely to report holding a degree in dentistry than the Canadian born, Canadian educated. This difference is a result of both the access that provincial governments' provide to the Canadian born to enter relevant training programs (i.e., the number of spots funded), and the operation of the immigration system. The ratios that are statistically significantly larger than one are in bold. What is clear from table 3.1 is that the distribution of each of the three groups across these health professions differs dramatically from that of the Canadian born, Canadian educated. Olaizola and Sweetman (2019) update and extend this analysis for physicians and nurses. Between 2006 and 2016, despite a very high growth rate in immigrants in these two professions, the flow of immigrants holding relevant credentials for such jobs increased more quickly making the brain waste problem even more dire than that presented in table 3.1.

Some combination of provincial governments restricting access to health professional education (e.g., medical schools) for Canadians, immigration policy, and immigrants self-selecting into migration is causing the foreign born, foreign educated to be dramatically more likely to hold credentials in dentistry, pharmacy, and medicine. The end result is that immigrants in these fields are simultaneously overrepresented among those practising and also among those not practising. This combination is, to a large extent, the source of the wasted human capital noted so often in the popular press and academic discussion and forms a large part of motivation – in terms of the actions of the developed world – for the WHO's code of conduct for the international recruitment of health professionals.

Differences in the quality of educational/training outcomes, of course, exist both across and within countries. A central purpose of regulatory bodies is to protect the public interest by assessing competence and establishing a minimum threshold to practise in health care. There are a large number of reports by various professional organizations documenting pass rates for various licensure exams, as well as prerequisite language exams and the like (e.g., Canadian Residency Matching Service, 2018; CAPER/RCEP, 2017; see Sweetman et al., 2015, for a summary). Table 3.2 provides some insight into one of the hurdles

Table 3.2. Licensure exam pass rates for first-time takers (%)

	Canadian Trained	Foreign Trained
Physicians (2011)		
MCCQE Part 1	99	68
MCCQE Part 2	92	38
Registered Nurses (2011)		
CRNE	86.8	48.4
Pharmacists (2002–6)		
Part I: MCE	94	44
Part II: OSPE	96	36
Physiotherapists (2010)		
MCQ Written	92	48.5
Clinical Practical Exam (OSCE)	93.3	57.2
Medical Radiation Technologists (2012)		
CAMRT Certification	87.5	24.4
Optometrists (2011–12)		
CACO	97.2	75.8

Source: Owusu & Sweetman (2015).

faced by the foreign trained. On average, they have much lower probabilities of successfully passing licensure exams. Further, in a small, and unrepresentative, study of international medical graduates' integration into Ontario, Sharieff and Zakus (2006) find that language testing is more of a hurdle than the professional testing for their study group. That is, the prerequisites to sitting these exams are more challenging than the exams themselves for some. Note that the fact that foreign-born, foreign-trained individuals are sitting these exams implies that they are making substantial efforts to enter the occupation for which they were trained.

Augustine (2015a, 2015b), the former Fairness Commissioner for Ontario, and Jafri (chapter 11 in this collection), argue in favour of the need for regulators themselves to be regulated/monitored regarding the "fair" treatment of immigrants with non-Canadian credentials, and that office has put in a substantial effort on this front. Further, these gaps, and related issues – in particular Canadian training spots – have been the subject of numerous studies, and the Ontario government commissioned an independent review by Thomson and Cohl (2011a, 2011b), with one of the two reviewers a retired justice. This was in an effort to understand the situation and promote a fair residency matching process on the medical side – with other occupations learning from this review. Overall, processes have been put into place in an effort to improve the fairness of accreditation and entry to practise for new immigrants trained outside of Canada. While these processes are not necessarily complete, many regulatory bodies have clearly changed their pathways to licensure for the foreign born.

Turning to broader issues and related to immigrant integration in regulated professions (not only regulated health professions) in the Canadian context, Gomez et al. (2015) observe that occupational licensing increases wages more for immigrants, from a lower base, than for the Canadian born. However, immigrants are less likely to be employed in a regulated occupation. Owusu and Sweetman (2015) note that the earnings of the Canadian born are sometimes lower and other times higher than the earnings of immigrants, with substantial diversity across the health professions they study. Moreover, visible minority status has no association with the earnings of the Canadian trained, and it has mixed correlations with the probabilities of workers trained in health care professions working in their trained profession (sometimes higher and sometimes lower). However, it appears that place of training, not place of birth, is the variable that matters. Indeed, the Canadian born, foreign trained look more like the foreign born, foreign trained than they do the Canadian born, Canadian trained. Related to this, but in the US context, Blair and Chung (2018) show that occupational regulation can act as a signal that reduces the importance of characteristics such as race and gender in determining wages.[8] This brings us back to Augustine's (2015a, 2015b) point about fairness in foreign credential recognition.

Getting the regulatory accreditation process "right" for regulated occupations is clearly of central importance for this policy issue. Various initiatives have been undertaken to pursue this issue at the federal and provincial levels for several years. Federally, the Ministry of Immigration, Refugees and Citizenship Canada set up the Integration/Foreign Credential Referral Office, which addresses both regulated and non-regulated occupations but has something of a focus on regulated occupations. A few provincial governments have established "Fairness Commissioners" or undertaken other initiatives to the same end, focusing on regulated occupations, which facilitate the integration of the foreign educated, particularly foreign educated immigrants, into regulated occupations in Canada.

There are, however, limits to what can be achieved by addressing excessive/inappropriate regulatory hurdles and/or credential recognition. For health care, on the demand side, such actions do not increase provincial government expenditures for the purchase of health care services, nor change what the health care receiving public is willing to pay out-of-pocket (or through insurance companies). On the supply side, in professions with a private sector component, where prices are – at least to some degree – flexible, such policy initiatives likely increase labour supply and drive down the (inflation adjusted) price of labour – especially among new entrants – to the benefit of patients (as, some might argue, has occurred in pharmacy). Where there are shortages, improved credential recognition for regulated occupations may usefully speed up the process of alleviating shortages, again potentially with a beneficial effect to patients regarding the price of labour. In highly unionized labour markets, especially those where the provincial government is close to the only purchaser of occupational services, such policies have indirect impacts on collective bargaining, but predicting short-run effects is difficult given the aforementioned bargaining processes.

Improving regulatory accreditation processes improves fairness with respect to who becomes (and who does not become) licensed, and likely increases the number of licensed practitioners somewhat. However, it does not solve the overarching problem when there is an oversupply of individuals who report a relevant credential but are not working in their trained profession, especially in the monopsonistic and highly regulated quasi-monopsonistic contexts described above. Further, since only about 20 per cent of the workforce is regulated, there are limits to this approach for the economy as a whole. In the US, it appears that the percentage of the workforce that is regulated has increased appreciably over the past few decades, but Zhang (2017, 2019), using what little data exist for Canada, finds no evidence of such a trend. My view is that while this approach certainly is necessary and has value to society, even for the 20 per cent working in a regulated profession, in particular a regulated health profession, it is insufficient. Brain waste in Canada makes the brain drain from sending countries all that much more ethically and economically problematic. An additional option is required.

Selected Relevant Aspects of Canada's Immigration System

Canadian public policy regarding the immigration of health professionals, and some other professionals with patient-specific skills, has suffered from an insufficient understanding of relevant labour markets. Moreover, classic Canadian federal-provincial and interministry jurisdictional problems have aggravated the situation. In particular, health labour markets were not well served by the introduction of the federal government's Immigration and Refugee Protection Act (IRPA). While the IRPA legislation has many positive design features, the one-size-fits-all points system of the Skilled Worker Program, which until recently was the mainstay of the Economic Class, is (I argue) ill-suited to many health professions (and other occupations with occupation-specific specialized skills – including teachers, social workers, and the like – where provincial governments and organizations, including municipalities under their jurisdiction are the dominant payer/employer).[9] IRPA treats education as if the skills generated are generic whereas those in health (and some other fields) tend to be quite specific and not of much value outside the relevant occupation. Nurses, pharmacists, dentists, physicians, and many others receive field-specific training and cannot switch to other occupations without substantial cost. An example of this would be fully qualified immigrant nurses who wish to work in nursing (it is important to recognize that many nurses voluntarily exit the profession) employed as personal support workers (PSWs), or in a different field altogether.

Also, IRPA seems to implicitly assume that labour markets adjust to increasing supply, which is true of those markets studied in first-year undergraduate courses. But, health labour markets are, as discussed above, much more complex and adjustment is to a large extent a function of government action rather than in response to "traditional" economic incentives since labour demand in health occupations is influenced and even controlled by governments.

Compounding the problem is that, as discussed above, most occupations in health are regulated, and the regulation is almost entirely undertaken by regulatory authorities that are creatures of provincial governments.[10] IRPA does not understand, or at least it does not take into account, provincially regulated labour markets and the afore-described economic structure of many provincial-government dominated labour markets.[11] This contributes not only to a waste of human capital but also to the drain of health professional expertise from developing countries against which the WHO and OECD rail.

Some legislative and regulatory amendments updating the original IRPA legislation have had impacts on health professions. One example was the introduction of quotas limiting the number of individuals in particular occupations who could enter the country through relevant immigration classes that were in place prior to the introduction of Express Entry in 2015. As far as I'm aware, no study has been undertaken to determine how successful these innovations were, but casual observation suggests that many were not binding and had very little impact. While such quotas are relatively easy to implement, assuming that occupations can be identified for a sufficient number of applicants, a central problem remains: the geographic distribution of shortage and surplus. The federal government must impose one rule across the country, but provinces such as Saskatchewan and Newfoundland and Labrador might, for example, have shortages of physicians whereas Ontario might have a surplus and there is no federal mechanism by which immigrants can easily be directed to one province or another, or more ideally to a subprovincial geographic region. Or, at least, the geographic maldistribution of health care workers problem is well known and pressing, and no policy solution (federal mechanism) has been put forward to date; and while the various provincial efforts have undoubtedly – hopefully – improved the situation, it remains far from "solved."

Express Entry, to a limited degree, has also addressed the problem for most of the Economic Class inasmuch as job offers from employers play an important role in determining entry. This new immigrant application management framework encompasses three major Economic Class streams: the Federal Skilled Worker Program, the Skilled Trades Program, and the Canadian Experience Class. Applicants must satisfy the requirements of the stream under which they apply, and then the Express Entry framework is used to determine if and when candidates receive an offer to apply to immigrate. But, recent regulatory changes that are decreasing the role played by employers (i.e., reducing the number of points for having a job offer in the consolidated points system, which is distinct from the Skilled Worker Program's points system) has reduced its effectiveness for the purposes under discussion. This is because an increasing share of workers are entering without a job offer based exclusively on generic skills and other requirements of the underlying programs and the Express Entry consolidated points system. Note that, like many Canadian immigration policies under discussion, the policy changes concerning Express Entry were not enacted with health labour markets as a concern. They had other motivations. Nevertheless, they have important implications for health labour markets.

A New Way Forward

One way to address the ongoing problem of the brain waste associated with health professionals not being able to work in their trained profession is to shift responsibility for these occupations away from the federal-government-managed immigration streams and into the Provincial Nominee Program, or perhaps a new program for this purpose. This reduces the transactions costs of negotiations between the federal and provincial governments, and aligns responsibility/accountability for immigrant selection with the level of government responsible for those labour markets. Assuming that other occupations in the Economic Class remain at their current levels, though this need not be the case, this means a reduction of the influence of the federal government in selecting economic stream immigrants. This need not be a bad thing, but a federal role should not be eliminated completely even for these occupations since post-immigration interprovincial mobility means that the actions of one province may affect others (e.g., McDonald & Worswick, 2012). Alternatively, and perhaps preferably, most accommodations at the immigration levels that are needed to undertake this policy change could result from a redistribution of the Provincial Nominee Program's slots across the provinces, with only a modest shift of slots from the federal to the provincial governments.

Admittedly, one of the weaknesses of this proposal is that the federal and provincial governments will need to negotiate (coordinate) and come up with some appropriate distribution of responsibility for the annual new immigrant admission targets. Hopefully, this issue will not need to be revisited very frequently. Also, this would require only a "higher level" target and not need to get into occupation-specific details.

One challenge in implementing this proposal would be figuring out how to exclude individuals in these occupations from other immigration streams. Clearly, an absolute ban is not feasible; in the extreme, refugees should not be screened out based on their occupations. But, a substantial reduction would suffice for the proposal to be successful. First, the simple announcement that individuals intending to work in these occupations should apply through a new stream or the Provincial Nominee Program stream would have some impact. However, some individuals might still apply through other streams – especially those that are part of the Express Entry stream (including the Canadian Experience Class). A key issue is how to prevent this.

One approach to this would be for the federal government to collect information on the post-secondary fields of study and assign zero educational points to fields associated with relevant health occupations. Similarly, work experience in relevant fields should also be assigned a value of zero. This would need to apply to both the Skilled Worker Program's points system and Express Entry's consolidated points system. Other related elements of the points systems could also be amended; for example, adaptability points in the Skilled Worker Program's points system for arranged employment would not apply for the federal stream for jobs in relevant occupations – such workers would need to enter through the provincial stream where such qualifications and experience would be valued. This would affect occupations in which immigrants are currently underrepresented relative to their share of the workforce (such as nursing), as well as occupations in which immigrants are overrepresented relative to their share of the workforce (such as dentistry). Provincial governments would be able to rectify these imbalances in their approach to selection. There is also a gender lens that could be used to evaluate this policy proposal. Unfortunately, as far as I'm aware, research that provides appropriate results for this analysis does not exist, although the data for such analysis do exist. This is an important area for future research.

The combined impact of these changes would effectively make it extremely difficult, if not impossible (depending on the allocation of points across other categories), for individuals in relevant occupations to enter through federal economic class immigration streams. This is not to say that individuals with health credentials cannot enter the country as immigrants. They certainly could.

However, they should enter through the provincial immigration stream as opposed to the federal one if they intend to practise in a health profession. If a potential immigrant had another credential in addition to a health one and did not intend to work in a health occupation, then a federal stream might be appropriate with the health credential and health-occupation work experience receiving zero weight in the points calculation(s) and/or for meeting the Canada Experience Class criteria, and so on. The goal of these proposals is to ensure that individuals enter through the appropriate immigration stream so as to improve outcomes for both Canada and the new immigrants.

The changes do, of course, increase the administrative burden on applicants and the federal government, and categorizing occupations is sometimes challenging, but that is less the case for these well-defined health occupations. Current provisions regarding the "completeness and honesty" of information provided during the application process would apply. Moreover, there may be some beneficial side effects from understanding the distribution of field of study among new immigrants. Quebec's skilled worker points system already allows limited "bonus" points for particular fields of study.

Since the two points systems primarily apply to the principal applicants, a relevant question relates to the occupation of an accompanying spouse if there is one. In an environment in which both members of a family are likely to work in the paid labour market, having a points system that focuses on one member of the family is an issue with which policy analysts have been struggling for decades – especially with respect to fairness across immigrating individuals compared to families. At present, a small number of points are provided in the Skilled Worker Program for the spouse's education, and a field of study exclusions similar to that for the principal applicant could be imposed. More strongly, if either spouse intends to work, or has a relevant credential in one of the relevant provincially managed health sectors, then the entire family could be required to apply through the provincial stream, as opposed to the federal one.

Similarly, for the Canadian Experience Class, Canadian work experience in relevant health occupations should not count. Rather, individuals with such experience should apply through the new provincially administered immigration streams. Other similar changes can be made throughout the federal Economic Class immigration streams.

Beyond the federal government, the provincial governments could mandate their regulatory colleges to evaluate potential applicants' credentials prior to an invitation to immigrate being extended. This would solve many labour market issues and increase the speed with which new immigrants can enter the labour market post-arrival. This approach is just starting, with the Medical Council of Canada in particular serving this function for the physicians.

The proposed policies only affect the Economic Class, and not the Family or Refugee Classes. It seems appropriate for these issues not to affect the Refugee Class since that goes against its very purpose, and it represents a relatively small share of the total flow so the few health professionals that immigrate through the refugee stream are unlikely to undermine the current policy proposal. Moreover, the brain drain for the sending country of refugees fleeing is unavoidable. In terms of the Family Class, information could be provided by the federal ministry of Immigration, Refugee and Citizenship Canada (IRCC) to potential immigrants indicating the licensure/accreditation difficulties faced by workers in the health occupations not entering through, and being adjudicated by, the processes associated with the relevant provincially administered programs. Migrants in the family class declaring an intention to work in a relevant health occupation could be adjudicated prior to admission using the same provincially mandated credentialing process as applied to those coming through the Economic Class's provincially administered programs for those in relevant health occupations. This would provide such immigrants with important information regarding their status in Canada, and it would be provided before migration when it is most useful to the relevant individuals. Those not making such a declaration would, of course, still need some type of credentialing process, but they would be taking an appreciable risk that may not prove successful (as is currently the case). Again, this is modestly burdensome for both the federal and provincial governments and for new immigrants (since they would need to, respectively, collect and provide additional information in the immigrant application process), but it would likely lead to improved outcomes and would be worth the investment on all sides since labour market outcomes in Canada would be improved.

An additional, but important, benefit of decentralizing the immigration of those in health occupations (and perhaps other occupations beyond health where provincial governments are the dominant employer) is that it can benefit the geographic distribution of the health workforce. At present, there are simultaneous shortages in some provinces while there are surpluses in others. While a provincially controlled system cannot prevent secondary migration, it can send strong signals about employment opportunities in different geographic regions. This can be extremely important information for new immigrants.

Conclusion

In the context of a global imbalance in the distribution of health professionals, Canada has a responsibility to ensure that its immigration system not only accords with the WHO's ethical standards, to which Canada is a signatory, but also goes beyond these to ensure that relevant training does not go to "waste." Examples of processes targeting these goals are starting to become apparent; in particular, there are some bilateral agreements between Canadian provinces and sending countries such as the Philippines. However, more broadly, the current system does not appear to meet that standard. Further, it seems unlikely that a federal government managed system could do so.

For those who are already here, improving licensure processes for the foreign educated is important, but it does not address the full problem. Shifting responsibility for health professional migration to provincial governments, which have constitutional responsibility for health care, is a sensible complement that can solve several problems simultaneously. The cost of doing this is not zero, but neither is it large, and the benefits appear to outweigh the costs. A key concern would be devising a mechanism by which immigrants in relevant occupations are deterred from immigrating through the federal immigration classes while being simultaneously encouraged to migrate through the provincial ones. Again, this does not seem onerous. Relatively straightforward exclusion restrictions could be imposed on the federal side, while – beneficially for new immigrants – mechanisms could be put in place by the provincial governments in their immigration streams to facilitate a path to licensure and practice. Undoubtedly, this approach will not completely eliminate the problem of wasted human capital in health professions, but it is likely to make it a much smaller problem.

One weakness of this discussion is that there is a minimal gender lens applied to the policy proposals. Unfortunately, this remains an area for future empirical research since, as far as I am aware, no gendered analysis of these issues has been undertaken. Clearly, that is an important lacuna in the literature.

ACKNOWLEDGMENTS

Arthur Sweetman holds the Ontario Research Chair in Health Human Resources, which is endowed by the Government of Ontario's Ministry of Health; all views expressed are his own and not necessarily those of the Province of Ontario. As cited, some of the empirical work on which this chapter is based comes from joint research with Yaw Owusu.

NOTES

1 This is, of course, a brief discussion. Relevant extended discussions of the structure of Canadian health care are by Marchildon (2013), Lavis (2016), and Flood et al. (2004).

2 Formally, nurses are mostly employees of hospitals, long-term care facilities, physicians and the like; physicians are most commonly independent private sector contractors; and, medical laboratory technologists are usually employees of private medical laboratories and hospitals. These arrangements do not negate there being a single payer. Moreover, even for contexts such as the long-term care sector where residents with sufficient wealth/income pay some portion of the costs, the markets are sufficiently regulated and subsidized by provincial governments that the essence of the argument goes through.

3 Some might think that taxpayer/voter interest in future generations' welfare might play an important role in terms of government fiscal discipline reducing deficits and debts. However, the most recent research of which I am aware, by the Canadian Institute of Actuaries for Ontario, continues to find that with respect to personal investments/financial planning "bequests were generally viewed as fairly unimportant" (as reported by Carrick, 2018).

4 In many provinces physicians are, from a legal perspective, not unionized in the sense that, for example, the Ontario Medical Association (OMA) is not registered as a union. However, from an economic perspective the OMA bargains collectively on behalf of physicians and thereby serves the central economic function of a

union. Since this chapter focuses on the economic, and not the legal, structure of these markets, physicians are "as if" unionized.

5 The trade union movement clearly recognizes the issue of product market competition across firms curbing collective-bargaining and therefore often seeks to consolidate union activity within industries. Historical examples of this consolidation include the United Steelworkers Union, and the Canadian Auto Workers Union (renamed Unifor in 2013 after a merger with the Communications, Energy and Paperworkers Union).

6 A disproportionate share of the literature focuses on physicians.

7 Dentistry is a particular case, since in recent years essentially all foreign-educated potential dentists must undertake Canadian training prior to being accredited. This implies that they end up in the foreign born, Canadian educated category.

8 Although, it must be recalled that these licensure exams are minimum competency thresholds, and are not competitive processes identifying / promoting excellence.

9 Discussions and overviews of the Canadian immigration system are provided by Picot and Sweetman (2012); Sweetman (2017); Beach et al. (2011); and Ferrer et al. (2012); see also the Ministry of Immigration, Refugees and Citizenship Canada's website. Prior to IRPA, the Skilled Worker Program's points system was designed so that it was extremely difficult to obtain sufficient points unless there was (not perfectly measured but still an informative measure of) occupational demand and/or a job offer. This prevented the build-up of new immigrants trained in specific health occupations without employment – see, for example, McDonald et al. (2012).

10 Occupational regulation, with an emphasis on health occupations, is discussed in a 2016 special issue of *Canadian Public Policy* with a broad overview by Sweetman et al. (2015).

11 More broadly, the pre-Express Entry IRPA immigration selection mechanism was fundamentally unstable (indeed, this problem remains for immigration classes outside of those that are part of Express Entry). There was (and there still is) excess demand for Canadian immigration slots – that is, more people applied and met the minimum threshold criteria for admission each year than were permitted to enter the country under the "levels plan." This meant that backlogs (sometimes called inventories) accumulated with corresponding implications for processing times. See Burstein (2017) for an analysis of these issues.

REFERENCES

Augustine, J. (2015a). Employment match rates in the regulated professions: Trends and policy implications. *Canadian Public Policy/Analyse de Politiques*, Supp (July), S1–20. http://doi.org/10.3138/cpp.2014-085.

Augustine, J. (2015b). Immigrant professionals and alternative routes to licensing: Policy implications for regulators and government. *Canadian Public Policy/Analyse de Politiques*, Supp (July), S1–14. http://doi.org/10.3138/cpp.2014-022.

Beach, C.M., Green, A.G., & Worswick, C. (2011). *Toward improving Canada's skilled immigration policy: An evaluation approach*. C.D. Howe Institute.
Benjamin, D., Gunderson, M., Lemieux, T., & Riddell, W.C. (2012). *Labour market economics* (7th ed.). McGraw-Hill Ryerson.
Blair, P.Q., & Chung, B.W. (2018). Job market signaling through occupational licensing. National Bureau of Economic Research Working Paper 24791. https://www.nber.org/system/files/working_papers/w24791/w24791.pdf.
Bourgeault, I.L., & Grignon, M. (2013). A comparison of the regulation of health professional boundaries across OECD countries. *European Journal of Comparative Economics, 10*(2), 199–223. https://www.researchgate.net/publication/281580272_A_Comparison_of_the_Regulation_of_Health_Professional_Boundaries_across_OECD_Countries.
Burstein, M. (2017). Managing immigration: A short history of turbulence and public policy. In V. Esses & D.E. Abelson (Eds.), *Twenty-first-century immigration to North America: Newcomers in turbulent times* (pp. 15–53). McGill-Queen's University Press.
Canadian Federal/Provincial/Territorial Advisory Committee on Health Delivery and Human Resources. (2009). *How many are enough? Redefining self-sufficiency for the health workforce*. Government of Canada. https://www.canada.ca/en/health-canada/services/health-care-system/reports-publications/health-human-resources/redefining-self-sufficiency-health-workforce-discussion-paper.html.
Canadian Residency Matching Service. (2018). *CaRMS forum*. https://www.carms.ca/news/carms-forum-presentation-is-now-live-2/.
CAPER/RCEP. (2017). *The national IMG database report: 2017*. https://caper.ca/sites/default/files/pdf/img/2017_CAPER_National_IMG_Database_Report_en.pdf.
Card, D., Thomas L., & Riddell, W.C. (2020). Unions and wage inequality: The roles of gender, skill and public sector employment. *Canadian Journal of Economics, 53*(1), 140–73. https://doi.org/10.1111/caje.12432.
Carrick, R. (2018, June 29). New report makes case for delaying CPP till 75. *The Globe and Mail*, B6.
CTV News.ca Staff. (2014, June 26). Only half of international medical grads in Canada work as doctors: Study. *CTV News*. https://www.ctvnews.ca/health/only-half-of-international-medical-grads-in-canada-work-as-doctors-study-1.1887996.
Department of Finance Canada. (2016). *Report on federal tax expenditures-Concepts, estimates and evaluations 2016*. Government of Canada. http://www.fin.gc.ca/taxexp-depfisc/2016/taxexp16-eng.asp.
Dumont, J.-C., Zurn, P., Church, J., & Thi, C. Le. (2008). International mobility of health professionals and health workforce management in Canada: Myths and realities. OECD Health Working Papers, No. 40. OECD.
Ferrer, A.M., Picot, G., & Riddell, W.C. (2012). New directions in immigration policy: Canada's evolving approach to immigration selection (Working Paper No. 107). Canadian Labour Market and Skills Researcher Network (CLSRN).

Flood, C.M., Tuohy, C., & Stabile, M. (2004). What is in and out of medicare? Who decides? Defining the medicare basket. Working Paper No. 5. Institute for Research on Public Policy.

Gomez, R., Gunderson, M., Huang, X., & Zhang, T. (2015). Do immigrants gain or lose by occupational licensing? *Canadian Public Policy/Analyse de Politiques*, Supp (July), S1-18. http://doi.org/10.3138/cpp.2014-028.

Grignon, M., Owusu, Y., & Sweetman, A. (2013). The international migration of health professionals. In K. F. Zimmermann & A. F. Constant (Eds.), *International handbook on the economics of migration* (pp. 75–97). Edward Elgar Publishing. http://doi.org/10.4337/9781782546078.00011.

Health Canada. (2004). Health human resources. *Health Policy Research*, (8), 1–48.

Ketel, N., Leuven, E., Oosterbeek, H., & van der Klaauw, B. (2016). The returns to medical school: Evidence from admission lotteries. *American Economic Journal: Applied Economics*, *8*(2), 225–54. https://doi.org/10.1257/app.20140506.

Kleiner, M.M. (2000). Occupational licensing. *Journal of Economic Perspectives*, *14*(4), 189–202. https://doi.org/10.1257/jep.14.4.189.

Kleiner, M.M. (2006). *Licensing occupations: Ensuring quality or restricting competition.* W.E. Upjohn Institute for Employment Research.

Kleiner, M.M. (2013). *Stages of occupational regulation: Analysis of case studies*. W.E. Upjohn Institute for Employment Research.

Kleiner, M.M., & Krueger, A.B. (2013). Analyzing the extent and influence of occupational licensing on the labor market. *Journal of Labor Economics*, *31*(2), S173–202. https://doi.org/10.1086/669060.

Kleiner, M.M., & Vorotnikov, E. (2017). Analyzing occupational licensing among the states. *Journal of Regulatory Economics*, *52*(2), 132–58. https://doi.org/10.1007/s11149-017-9333-y.

Kugler, A.D., & Sauer, R.M. (2005). Doctors without borders? Relicensing requirements and negative selection in the market for physicians. *Journal of Labor Economics*, *23*(3), 437–65. https://doi.org/10.1086/430283.

Kuhn, P. (1998). Unions and the economy: What we know; what we should know. *Canadian Journal of Economics*, *31*(5), 1033–56. https://doi.org/10.2307/136458.

Lavis, J. (Ed.). (2016). *Ontario's health system: Key insights for engaged citizens, professionals and policymakers*. McMaster Health Forum. https://www.mcmasterforum.org/find-evidence/ontarios-health-system.

Law, M.R., Dijkstra, A., Douillard, J.A., & Morgan, S.G. (2011). Geographic accessibility of community pharmacies in Ontario. *Healthcare Policy/Politiques de Sante*, *6*(3), 36–46. http://www.ncbi.nlm.nih.gov/pubmed/22294990 https://doi.org/10.12927/hcpol.2011.22097.

Lofters, A., Slater, M., Fumakia, N., & Thulien, N. (2014). "Brain drain" and "brain waste": Experiences of international medical graduates in Ontario. *Risk Management and Healthcare Policy*, 7, 81–9. https://doi.org/10.2147/RMHP.S60708.

Marchildon, G.P. (2013). *Health systems in transition: Canada* (2nd ed.). University of Toronto Press.

McDonald, J.T., Warman, C., & Worswick, C. (2012). Immigrant selection systems and occupational outcomes of international medical graduates in Canada and the United States. *Canadian Public Policy*, 41 (Supplement 1), 48. https://doi.org/10.3138/cpp.2013-054.

McDonald, J.T., & Worswick, C. (2012). The migration decisions of physicians in Canada: The roles of immigrant status and spousal characteristics. *Social Science and Medicine*, *75*(9), 1581–8. https://doi.org/10.1016/j.socscimed.2012.07.009.

Olaizola, A., & Sweetman, A. (2019). Brain gain and waste in Canada: Physicians and nurses by place of birth and training. In *Recent Trends in International Migration of Doctors, Nurses and Medical Students* (pp. 123–39). OECD Publishing. https://doi.org/10.1787/5571ef48-en.

Organisation for Economic Co-operation and Development. (2007). Immigrant health workers in OECD countries in the broader context of highly skilled migration. In *International Migration Outlook Annual Report* (pp. 161–228). OECD Publishing. http://www.oecd.org/migration/mig/41515701.pdf.

– 2008. *The looming crisis in the health workforce: How can OECD countries respond?* OECD Health Policy Studies. OECD Publishing. https://doi.org/10.1787/9789264050440-en.

Owusu, Y., & Sweetman, A. (2015). Regulated health professions: Outcomes by place of birth and training. *Canadian Public Policy*, *41* (Supplement 1), S98–S115. https://doi.org/10.3138/cpp.2015-008.

Picot, G., & Sweetman, A. (2012). Making it in Canada: Immigration outcomes and policies. IRPP Study No. 29. Institute for Research on Public Policy. https://irpp.org/research-studies/making-it-in-canada/.

Rosen, H.S., Wen, J.F., Snoddon, T., Dahlby, B., & Smith, R.S. (2008). *Public finance in Canada* (3rd ed.). McGraw-Hill Ryerson.

Sharieff, W., & Zakus, D. (2006). Resource utilization and costs borne by international medical graduates in their pursuit for practice license in Ontario, Canada. *Pakistan Journal of Medical Sciences*, *22*(2), 110–15. https://pjms.com.pk/issues/aprjun06/article/article3.html.

Smart, M., & Stabile, M. (2005). Tax credits, insurance, and the use of medical care. *Canadian Journal of Economics*, *38*(2), 345–65. https://doi.org/10.1111/j.0008-4085.2005.00283.x.

Smart, M., & Stabile, M. (2006). Tax support for the disabled in Canada: Economic principles and options for reform. *Canadian Tax Journal*, *54*(2), 407–25. https://www.ctf.ca/ctfweb/Documents/PDF/2006ctj/06ctj2-smart.pdf.

Sweetman, A. (2017). Canada's immigration system: Lessons for Europe? *Intereconomics*, 52(5). https://doi.org/10.1007/s10272-017-0690-7.

Sweetman, A., McDonald, J.T., & Hawthorne, L. (2015). Occupational regulation and foreign qualification recognition: An overview. *Canadian Public Policy*, *41*, S1–S13. https://doi.org/10.3138/cpp.41.s1.s1.

Thomson, G., & Cohl, K. (2011a). *IMG selection: Independent review of access to postgraduate programs by international medical graduates in Ontario: Vol. 1: Findings and recommendations*. Government of Ontario. https://www.health.gov.on.ca/en/common/ministry/publications/reports/thomson/v1_thomson.pdf.

Thomson, G., & Cohl, K. (2011b). *IMG selection: Independent review of access to postgraduate programs by international medical graduates in Ontario: Vol. 2: Analysis and background*. Government of Ontario. https://www.health.gov.on.ca/en/common/ministry/publications/reports/thomson/v2_thomson.pdf.

Wang, C., Li, Q., Sweetman, A., & Hurley, J. (2015). Mandatory universal drug plan, access to health care and health: Evidence from Canada. *Journal of Health Economics, 44*, 80–96. https://doi.org/10.1016/j.jhealeco.2015.08.004.

World Health Organization. (2010). *WHO global code of practice on the international recruitment of health personnel*. https://www.who.int/hrh/migration/code/WHO_global_code_of_practice_EN.pdf.

Zhang, T. (2017). *The regulation of occupations and labour market outcomes in Canada: Three essays on the relationship between occupational licensing, earnings and internal labour mobility*. University of Toronto Press.

Zhang, T. (2019). Effects of occupational licensing and unions on labour market earnings in Canada. *British Journal of Industrial Relations, 57*, 791–817. https://doi.org/10.1111/bjir.12442.

Zhang, X., & Sweetman, A. (2018). Blended capitation and incentives: Fee codes inside and outside the capitated basket. *Journal of Health Economics, 60*, 16–29. https://doi.org/10.1016/j.jhealeco.2018.03.002.

4 Global Migration and Key Issues in Workforce Integration of Skilled Health Workers

ANDREA BAUMANN, MARY CREA-ARSENIO,
AND VALENTINA ANTONIPILLAI

Introduction

Economic insecurity and sociopolitical instability have contributed to heightened migration (World Health Organization [WHO], 2018). Skilled health workers constitute a substantial proportion of migrants and can play an important role in the delivery of care for changing patient demographics. Over the past decade, there has been a 60 per cent rise in the number of migrant physicians and nurses working in Organisation for Economic Co-operation and Development (OECD) countries (WHO, 2018). These health workers are increasingly relied upon to fill critical shortages, a phenomenon reported by over fifty-seven OECD countries (WHO, 2006). However, the provision of health services is heavily regulated with many prerequisites for employment, which can complicate timely and efficient hiring and integration of immigrants who are suited for positions in the health care sector. Evidence indicates these professionals face obstacles in accessing employment commensurate with their skills and qualifications (Office of the Fairness Commissioner [OFC], 2011).

In Canada, 22 per cent of the population is foreign born, and this number is expected to rise over the next ten years (Statistics Canada, 2017). Health workers are "all people engaged in actions whose primary intent is to enhance health" (WHO, 2006, p. 2), including physicians, nurses, midwives, laboratory technicians, public health professionals, community health workers, pharmacists, and other workers whose primary role involves the delivery of preventive, promotive, or curative health services.

To be both proactive and responsive, the health workforce nationwide must reflect the linguistic and cultural heterogeneity of the population. Research shows a number of adverse events and negative health care experiences among immigrants and patients with limited English-language proficiency (Agency for Healthcare Research and Quality, 2012; Divi et al., 2007; Suurmond et al., 2011). Studies also demonstrate that quality of care and delivery of services are better when patients and health care providers share the same language and culture (Ayoola, 2013; Wood, 2018). Skilled migrant health workers can mediate barriers to care, but obtaining suitable employment is a challenge and many remain in "survival jobs" rather than resuming their professional roles. One study revealed that only 17.5 per cent of IENs were working at or above their skill level as compared to 64.5 per cent of domestically educated nurses (OFC, 2011).

Various multisectoral and cross-ministerial interventions have been created in Canada to facilitate the productive employment and integration of internationally educated health professionals (IEHPs), all of whom face similar integration challenges regardless of profession. This chapter focuses on the nursing workforce and issues relevant to other skilled workforces. These include policy barriers and changes, regulatory requirements, timing and delay of the registration process, institutional responses, and societal and markets demands. Federal, provincial, and regulatory policy approaches to employment and integration and strategies implemented by health care employers are examined. In addition, best practices in the development and support of equitable employment opportunities and the integration of migrant skilled workers into the health sector are identified using the nursing workforce as an example.

Background

Definitions of cultural competence frequently draw upon the seminal work of Cross et al., (1989) who conceptualized it "as a set of congruent behaviors, attitudes, and policies that come together in a system, agency, or amongst professionals and enables that system, agency, or those professionals to work effectively in cross-cultural situations" (p. iv). On average, more than 200,000 immigrants and an estimated 25,000 refugees resettle in Canada each year, providing the nation with the highest percentage of foreign-born residents among the G8 countries (Statistics Canada, 2013). The health care sector must be able meet the needs of changing demographics.

Increasing diversity in the health workforce by hiring IEHPs can improve access to services for immigrant patients and enhance their options, choices, and satisfaction (Association of American Medical Colleges, 2018; Health Professionals for Diversity Coalition, 2012; National Conference of State Legislatures, 2014). Yet the efforts of IEHPs to secure employment that matches their qualifications and skills are often hindered (OFC, 2011), as is their successful workforce integration. Baumann et al. (2011) define workforce integration as the "the process by which ... [workers] enter the workforce efficiently, effectively and with productive employment" (p. 49). It includes full-time employment opportunities, as well as preparation for practice (skills and knowledge) and adapting to an organization's culture and system.

Effective integration maximizes all workforce supply sources and is essential for workforce sustainability (Baumann et al., 2021). Ramji and Etowa (2014) note there is an "abundance of research on the challenges faced in the earlier phases of pre-registration" and cite the need to "focus on IENs' progress and experiences over the longer term" (p. 231).

In 2017, the Canadian nursing workforce comprised 398,845 nurses and included 32,293 IENs, approximately 8 per cent of the total workforce (Canadian Institute for Health Information [CIHI], 2018). Internationally educated nurses complete their nursing education outside the country in which they reside. Compared to Canadian-born new graduate nurses, IENs who enter the workforce are usually more experienced and older, with the majority being more than thirty years of age (CIHI, 2018). Many IENs enter Canada as secondary family class migrants, immigrating on the employment of their spouses (Blythe et al., 2009). Some IENs enter Canada on visas or work permits as live-in caregivers with the intention of becoming registered to practice, which may require upgrading their education.

During the past few years, researchers have documented barriers to effective and efficient integration of IEHPs (Covell et al., 2017; Neiterman & Bourgeault, 2015; Paul et al., 2017). Much of this research concerns pre-licensure practices and predictors of registration success. The employment phase of integration is discussed in this chapter, with particular emphasis on IENs attaining their first experience as nurses in Canada. The findings are informed by a review of existing policies and regulations that enable employment for IENs and a case analysis of employer-led best practices for IEN integration. The following questions are addressed:

- What are the key federal and provincial policy initiatives that aimed to reduce barriers to skilled worker integration?
- What are the nursing regulatory changes that had an impact on the integration of IENs?
- What are the employer-led best practices that resulted in successful integration of IENs into Canada's health care system?

Current Federal Immigration Policy Overview

In 2015, Citizenship and Immigration Canada (CIC), now known as Immigration, Refugees and Citizenship Canada (IRCC), introduced the electronic Express Entry system to "manage applications for permanent residence for these programs: Federal Skilled Worker Program, Federal Skilled Trades Program and Canadian Experience Class" (Government of Canada, 2017, para. 1). The goals of the program were to provide "1) flexibility in selection and application management, 2) responsiveness to labour market and regional needs and 3) speed in application processing" (Immigration, Refugees and Citizenship Canada, 2016, p. 4).

Applicants complete online profiles that provide information on age, education, spousal factors, and skills transferability. They are evaluated using the point-based Comprehensive Ranking System. Applicants are eligible for additional points for having a Canadian degree, a job offer supported by a positive Labour Market Impact Assessment (LMIA), or a nomination by a province or territory. An LMIA is a form of labour market verification that confirms both the need for a temporary worker and the lack of Canadians available to fill a position (Government of Canada, 2018a). Points are also awarded for French-language proficiency and having a sibling in Canada who is a permanent resident or citizen (Government of Canada, 2017).

The federal government launched the Canadian Experience Class (CEC) program in 2008, expediting immigration and acceptance of permanent resident applications for highly skilled migrant professionals and international students already residing in Canada (Ali, 2014). Amendments to the Federal Skilled Worker Program in 2012 involved the addition of a "Canadian experience" clause that became a central component of immigrant selection, eliciting a process based on the values of human capital and Canadian employability (Ali, 2014; Bhuyan et al., 2015).

In 2017, the federal government implemented the Atlantic Immigration Pilot Program (AIPP) and the Global Skills Strategy (GSS). These targeted initiatives do not necessarily encourage applications from migrants seeking employment within the health care workforce. The AIPP is a three-year program intended to attract 2,000 new immigrants and their families to Atlantic Canada to fill labour market shortages (Government of Canada, 2018c). Under the AIPP, employers offer jobs to foreign workers with the provision that their businesses will assist migrant workers with their individualized resettlement plans. In return, the federal government has committed to a six- to twelve-month processing time for permanent residency applications in the intermediate-skilled, high-skilled, and international graduate programs (House of Commons, Canada, 2017). Unlike the majority of programs, the AIPP offers a one-year LMIA-exempt employer-specific work permit to those who qualify (House of Commons, Canada, 2017). The GSS values high-skilled migrant workers. It establishes a "Global Talent Stream" to recruit the best and brightest in specialized and technical areas of work and includes exceptions to immigration processing timelines to expedite workforce integration (Canadim, 2017).

These federal policy frameworks provide a guide, but provinces and territories vary in their approaches to IEHPs. Furthermore, there are additional obstacles for regulated health professionals following their entry to Canada when obtaining licensure to work. For example, educational credentials obtained from a foreign institution must be validated through an educational credential assessment (ECA) completed by an IRCC designated organization. However, the ECA does not guarantee applicants will obtain licensure in a regulated profession or find a job in their field. Nurses must complete the licensing process in the province or territory in which they plan to settle. In Ontario, IENs have three years to register with the College of Nurses of Ontario (CNO). According to the CNO (2018), an IEN "must provide proof of [their] citizenship, residency status, or authorization to practise nursing in Ontario before [they] can register as a nurse in Ontario" (para. 1). The requirement is considered met if the applicant has a work permit or study permit authorizing her/him to practise nursing in Canada.

In an effort to reduce barriers to registration, the Government of Canada's Foreign Credential Recognition Program invested in the establishment of the National Nursing Assessment Service (NNAS), a partnership between Canadian nursing regulatory bodies (NNAS, 2013). It provides IENs with a secure form to submit documents for nursing registration and was created to harmonize the initial education and work experience assessment of nurses immigrating to Canada (NNAS, n.d.). The initial assessment is subcontracted to an external service and recommendations are made to the respective regulatory bodies. Quebec, the Northwest Territories, Nunavut, and the Yukon do not participate in this assessment program.

Pre-arrival Federal Initiatives

A positive trend is the increase in pre-arrival programs that provide support for potential health care workers. Projects relevant to the nursing sector that have received funding from the IRCC Foreign Credentials Referral Office include Planning for Canada, formerly known as the Canadian Immigrant Integration Program (CIIP), the Pre-Arrival Supports and Services (PASS) Program at the CARE Centre for Internationally Educated Nurses in Ontario, and the Canadian Culture and Communication for Nurses (CCCN) program offered through the Manitoba Nurses Union.

Planning for Canada has offices in China, India, and the Philippines that provide free services to final stage economic class candidates and their families. Sector-specific services include preliminary online equivalency of credentials, online employer and job search workshops, live online mentoring and job matching services. The PASS Program focuses on language and information barriers in the early stages of immigration (PASS, n.d.). It aims to inform IENs about the process of obtaining licensure and provides online training and support services prior to their arrival in Canada. These include (a) nursing-specific language and communication training; (b) webinars on employment-related skill development; (c) educational workshops on Canadian immigration, health care, and workplace systems to facilitate entry into and navigation of the health care field; (d) mentorship opportunities linking pre-arrival IENs to expert nurses in receiving provinces or with similar areas of specialization; and (e) employment opportunities that connect national employers to IENs through online sessions. The program provides services to IENs destined for anywhere in Canada, while CCCN provides

English-language support, cultural workshops, individualized needs assessment, and mentorship services for IENs interested in migrating to Manitoba.

Although pre-arrival services are offered in almost every Canadian province and territory, IENs frequently struggle to find services that meet their needs. Many of the services are designed and implemented independently of other organizations, resulting in a fragmented system in which redundancy and delays exist because of duplication. Additionally, very few organizations have affiliations and partnerships in the various countries of origin for IENs, and many organizations rely on Canadian embassies to deliver a letter outlining the voluntary pre-arrival services available. The Philippines is one of the only countries that provides a government-mandated seminar for internationally destined skilled workers, but these seminars are organized unilaterally by the Filipino government with no partnerships with receiving countries.

Siloed initiatives and the lack of collaboration among organizations within Canada and abroad have a considerable impact on the efficacy and ability of pre-arrival programs to meet the needs and expectations of IENs. The current landscape of patchwork services can be enhanced via pre-arrival programs that are client centred, targeted towards specific professions, and work in cooperation with post-arrival services. Pre-arrival programs educate and prepare migrating workers before leaving their country of origin (Covell et al., 2016).

Understanding the range of health care practices in Canada, enhancing language skills, and learning workforce-specific terminology has contributed to the successful licensure of IENs in Canada (Newton et al., 2012). If these services were provided to IENs within their countries of origin, delays in obtaining licensure upon arrival in Canada might be reduced.

Provincial Policy and Regulatory Approaches

Provinces and territories in Canada, except Quebec and Nunavut, can nominate immigrants through the Provincial Nominee Program (PNP), but there are variations in policies, criteria and "streams." The latter pertain to "individual immigration programs targeting precise demographics of skilled workers, students, and business professionals" (Canada Express Entry Application Service, 2018, para. 1). Immigrants must be able to meet the economic needs of the nominating province or territory, contribute to the national economy and be willing to relocate to the nominating province or territory (Government of Canada, 2018b). There are twenty-nine regulated health professions in Ontario. Skilled health workers applying to these professions must navigate the associated regulatory processes, which are an additional layer of complexity.

Beginning in 2013, the CNO has made significant changes to the registration process for nurses. These changes include introducing the NNAS for education credentialing, reducing the evidence of practice requirement from five to three years, and replacing the Canadian Registered Nurse Examination (CRNE) with the National Council Licensure Examination for Registered Nurses (NCLEX-RN®). Registered practical nurses (RPNs) still complete the Canadian Practical Nurse Registration Examination (CPNRE). These changes have adversely affected the ability of IENs to become registered in Ontario (Keung, 2012; OFC, 2011; Walton-Roberts et al., 2014).

The OFC cites a lack of transparency about the licensing changes, poor communication with applicants, delays in processing applications, and insufficient grandparenting of applications that were already in progress when the new rules came into effect. However, the CNO has implemented strategies to expedite the process, such as increasing the number of times a nurse can attempt to write and pass the NCLEX-RN® from a maximum of three to unlimited. Additionally, government investments have been made in bridge training programs, non-governmental organizations (NGOs), and targeted research projects at large health care organizations.

Educational Initiatives

Bridge training programs offered by post-secondary educational institutions are funded by provincial governments and intended to bridge gaps between the credentials of IENs and regulatory requirements to practise nursing. The programs include supervised clinical experiences of varying lengths and academic study. Blythe and Baumann (2009) found the experience obtained through clinical placements and job shadowing helped IEHPs pass registration examinations, secure employment, and achieve success in the workplace. The financial responsibility for and support of these programs relies on the intermittent commitments of different ministries, including those responsible for education, health, and immigration. An outstanding issue with education is a lack of consistent offerings over an extended period of time. There may be available programs within regions that are largely populated but they will have limited slots. For example, at one university in an urban centre, enrollment is capped at fifty per annum, but the program receives over 400 applicants each year. In this case, the demand far exceeds the availability of educational programs provincially.

EMERGING ISSUES

The Government of Canada, like that of other countries, is struggling with increased immigration and effective ways to integrate newcomers into society. According to a study conducted by the OECD, 60 per cent of highly skilled immigrants in Canada were working in high-skilled jobs compared to an average of 71 percent among other OECD countries (McMahon, 2013). Policies and initiatives to address settlement, education, and employment are paramount yet lacking.

Regulatory bodies within the health sector play a predominant role in employment, and programs pertaining to the workforce integration of IEHPs are few. Existing programs are heavily influenced by regulatory obligations and delay employment.

It is interesting to note that workforce integration processes are even more complex in certain female-dominated occupations. Additional issues arise because of gender-related biases associated with the objectives of immigrant selection policies. Strongly supported immigration initiatives that offer employment opportunities tend to disproportionately favour men. For example, the GSS focuses on high-skilled and technical or "hard" talents primarily associated with men. This initiative could be perceived as emphasizing the gender and class inequities inherent in Canada's immigrant employment system.

According to Kofman (2000), notions of skill are gendered. Work linked to women in the caring professions is devalued because it is not considered highly skilled. The overwhelming majority of IENs are women, many of whom emigrated from countries such as India and the Philippines (PASS, n.d.). These IENs tend to register as dependents despite having their own skill set. Their exclusion from labour market opportunities contributes to the redomestication of gender roles, wherein the entry of women as dependents results in their deskilling and "being confined by work in the home and isolated from the community" (Dobrowolsky, 2017, p. 212).

It is vital to recognize that women across the globe do not have access to human capital or have the same capital as men and are more likely to be responsible for dependents within the family (Dobrowolsky, 2017). A system that prioritizes capital while gendering high skills and independence provides employment opportunities for immigrant men, perpetuating gender inequity. This underlying gender disparity complicates employment pathways for IENs, impeding their integration as culturally competent members of the nursing workforce. The cultural and linguistic diversity of IEHPs is a valuable asset to Canada's health care system given the increasing population movement worldwide and the changing patient demographics within the country.

EMPLOYMENT POST-LICENSURE

The pathway to practise as a nurse in Canada is lengthy and costly for IENs. Once registration requirements are met, IENs are challenged to find employment that matches their qualifications and skills. The Ontario Ministry of Health and Long-Term Care (MOHLTC) developed the innovative Nursing Career OrIENtation (NCO) initiative to stimulate full-time employment for IENs (Baumann et al., 2012). Although some employers used the NCO to hire IENs, the uptake was reported to be slow.

On 1 April 2017, the NCO initiative was aligned with the Nursing Graduate Guarantee (NGG) initiative. The NGG was initially intended to stimulate full-time employment for new graduate nurses educated in Ontario, but it was expanded to include nurses educated in other provinces and territories and IENs. The NGG funds health care employers to offer full-time employment to eligible nurses in Ontario and supports extended orientation to facilitate transition and integration to practice.

Nonetheless, integration can be a challenge as practice requirements vary across countries. Nurses in Canada may have more responsibility for their own practice and greater liability than do nurses in some other countries. Internationally educated nurses may also be unfamiliar with the procedures, technologies, expectations, and behaviours in the Canadian workplace. Employers must recognize that unfamiliarity does not equal incompetence and acknowledge the value of cultural competence and the need for diversity in their workforces. In one study, IENs felt that managers provided them with little or no opportunities to undertake or demonstrate leadership, develop confidence, and be included in activities organized by colleagues and superiors (Newton et al., 2012).

CASE EXEMPLARS: CANADIAN INTEGRATION BEST PRACTICES

Integration is a two-way process. Nurses prepare for practice via the acquisition, use, and expansion of skills and knowledge and become team members by adapting to an organization's culture and system, and the organization provides nurses with opportunities for full-time employment and professional development. Internationally educated nurses have unique skills, knowledge, and experience from which others can learn and organizations can benefit. However, their employment and integration cannot be achieved without the support of senior leadership.

By implementing strategies for hiring and integrating IENs, employers are acknowledging their value and working towards establishing culturally competent workforces and fostering community connections. Professional development opportunities must also be provided to ensure that workplace practices are equitable and inclusive.

Nine exemplary health care organizations across Canada were selected and interviewed about their strategic practices in hiring and integrating IENs (Baumann et al., 2017). They ranged from organizations serving small- to medium-sized communities to acute-care teaching hospitals in large urban settings. All had a formalized program in place to integrate IENs and were willing to share it with other organizations. Program coordinators from each organization were interviewed via teleconference and asked to provide details regarding their practices and programs. A summary of their responses is provided in table 4.1.

The core elements across the organizations were (1) a vision that includes care of diverse patient populations and having a workforce that reflects cultural and linguistic diversity, (2) innovative approaches to creating a more representative staff complement, and (3) awareness training for managers and nursing

Table 4.1. Best practices for the integration of IENs – Core elements

Best Practice	Description
Mandate and vision that includes diversity	Vision of serving a global community and a strategic plan committed to equity and inclusion
Targeted recruitment	Recruitment that considers language requirements, staffing needs, and the diversity of clients
Extended orientation and mentorship	Three- to six-month orientation program with mentorship
Senior leadership committed to a diverse workforce	Goal of creating a culture of engagement, respect, and inclusiveness that attracts top talent
Focused education to address a changing patient population and workforce	Training around language and culture issues targeted towards managers, IENs, and other staff
Enhanced community support	Partnerships with local bridge training programs, NGOs, and other groups
Dedicated committees	Focus on equity, culture, and inclusiveness

staff. All organizations had a strong culture of inclusivity and diversity. The employment and integration of IENs was supported by corporate strategies and senior leadership commitment. Many organizations had committees dedicated to equity, culture, and inclusion and collaborated with stakeholder groups to encourage the integration of IENs.

Discussion and Conclusions

The complexity and consequences of population movement are significant. To keep pace with the changes stemming from increased migration, governments, governance strategies, approaches to health care provision and decision-making practices must change. Cross et al. (1989) observe, "A culturally competent system of care acknowledges and incorporates – at all levels – the importance of culture, the assessment of cross-cultural relations, vigilance towards the dynamics that result from cultural differences, the expansion of cultural knowledge, and the adaptation of services to meet culturally-unique needs" (p. v).

To reduce disparities and inequities, improve efficacy and encourage sustainability, the health workforce in Canada must reflect the people it serves. However, as demonstrated in this chapter, the employment and workforce integration of skilled health workers is adversely affected by various factors, including communication issues and difficulties obtaining required documentation.

In Canada, there is both a historical and continued commitment to immigration with large investments being made annually by the provinces, territories, and the country. Nevertheless, the transition to employment for many skilled workers is not seamless. The current approaches involve time-limited funding for programs to support migrant integration and often evolve out of immediate need rather than long-term planning. Considering the valuable contributions that skilled migrants in general and nurses in particular make to the demographic, economic, and cultural realms of Canadian society, it may be more effective for governments to embed programs and practices into existing institutions.

Overall, there is a divide between the goals of policy initiatives supported by the federal, provincial, and territorial governments and the need for workforce integration programs targeting skilled health workers. The immigrant employment system is economized and gendered, enabling the integration of a few specialized, high-skilled workers and leaving the majority of skilled health professionals behind. Subsequently, these workers encounter the harsh realities of a fragmented system compounded by complicated regulatory procedures.

Implementing best practices to facilitate workforce integration for skilled health professionals through continued and stable commitments will advance the provision of care for Canada's diverse population now and in the future.

REFERENCES

Agency for Healthcare Research and Quality. (2012). *Improving patient safety systems for patients with limited English proficiency: A guide for hospitals*. https://www.ahrq.gov/sites/default/files/publications/files/lepguide.pdf.

Ali, L.A.J. (2014). Welcome to Canada? A critical review and assessment of Canada's fast-changing immigration policies. In H. Bauder (Ed.), *RCIS Working Papers*. Ryerson Centre for Immigration and Settlement.

Association of American Medical Colleges. (2018, February 15). *Promoting a diverse and culturally competent health care workforce*. https://news.aamc.org/for-the-media/article/diverse-healthcare-workforce/.

Ayoola, A. (2013, May 7). *Why diversity in the nursing workforce matters*. Robert Wood Johnson Foundation. https://www.rwjf.org/en/blog/2013/05/why_diversity_inthe.html.

Baumann, A., Crea-Arsenio, M., Ross, D., & Blythe, J. (2021). Diversifying the health workforce: a mixed methods analysis of an employment integration strategy. *Human Resources for Health, 19*(62). https://doi.org/10.1186/s12960-021-00606-y.

Baumann, A., Hunsberger, M., & Crea-Arsenio, M. (2011). Workforce integration of new graduate nurses: Evaluation of a health human resource employment policy. *Healthcare Policy, 7*(2), 47–59. https://doi.org/10.12927/hcpol.2011.22662.

Baumann, A., Hunsberger, M., & Crea-Arsenio, M. (2012). Impact of public policy on nursing employment. *Canadian Public Policy, 38*(2), 167–79. https://doi.org/10.3138/cpp.38.2.167.

Baumann, A., Ross, D., Idriss-Wheeler, D., & Crea-Arsenio, M. (2017). Strategic practices for hiring, integrating and retaining internationally educated nurses: Employment manual. Nurses Health Services Research Unit, McMaster University.

Bhuyan, R., Jeyapal, D. Ku, J., Sakamoto, I., & Chou, E. (2015). Branding "Canadian experience" in immigration policy: Nation building in a neoliberal era. *Journal of International Migration and Integration, 18*, 47–62. https://doi.org/10.1007/s12134-015-0467-4.

Blythe, J., & Baumann, A. (2009). Internationally educated nurses: Profiling workforce diversity. *International Nursing Review, 56*, 191–7. https://doi.org/10.1111/j.1466-7657.2008.00699.x.

Blythe, J., Baumann, A., Rhéaume, A., & McIntosh, K. (2009). Nurse migration to Canada pathways and pitfalls of workforce integration. *Journal of Transcultural Nursing, 20*(2), 202–10. https://doi.org/10.1177/1043659608330349.

Canada Express Entry Application Service. (2018). *Canada Provincial Nominee Program guide*. http://www.canadaexpressentry.org/provincial-nominee-program/.

Canadian Institute for Health Information. (2018). *Regulated nurses, 2017*. https://www.cihi.ca/sites/default/files/document/regulated-nurses-2017-pt-highlights-en-web.pdf.

Canadim. (2017). Canada launches Global Skills Strategy. Canada Immigration News, 12 June. http://www.canadim.com/global-skills-strategy-canada-immigration-news/#1497284422499-ecede2a8-7350.

College of Nurses of Ontario. (2018). Citizenship, permanent residency or authorization to practise nursing. http://www.cno.org/en/become-a-nurse/registration-requirements/citizenship/.

Covell, C., Neiterman, E., & Bourgeault, L. (2016). Scoping review about the professional integration of internationally educated health professionals. *Human Resources for Health 14*(38), 1–12. https://doi.org/10.1186/s12960-016-0135-6.

Covell, C., Primeau, M. D., Kilpatrick, K., & St-Pierre, I. (2017). Internationally educated nurses in Canada: Predictors of workforce integration. *Human Resources for Health, 15*(26), 1–16. https://doi.org/10.1186/s12960-017-0201-8.

Cross, T., Bazron, B., Dennis, K., & Isaacs, M. (1989). *Towards a culturally competent system of care (Vol. 1)*. CASSP Technical Assistance Center, Center for Child Health and Mental Health Policy, Georgetown University Child Development Center.

Divi, C., Koss, R.G., Schmaltz, S.P., & Loeb, J.M. (2007). Language proficiency and adverse events in US hospitals: A pilot study. *International Journal for Quality in Health Care, 19*(2), 60–7. https://doi.org/10.1093/intqhc/mzl069.

Dobrowolsky, A. (2017). Bad versus big Canada: State imaginaries of immigration and citizenship. *Studies in Political Economy, 98*(2), 197–222. https://doi.org/10.1080/07078552.2017.1343001.

Government of Canada. (2017). *How Express Entry works*. https://www.canada.ca/en/immigration-refugees-citizenship/services/immigrate-canada/express-entry/works.html.

Government of Canada. (2018a). *Find out if you need a Labour Market Impact Assessment*. https://www.canada.ca/en/immigration-refugees-citizenship/services/work-canada/hire-foreign-worker/temporary/find-need-labour-market-impact-assessment.html.

Government of Canada. (2018b). *Immigrate as a provincial nominee*. https://www.canada.ca/en/immigration-refugees-citizenship/services/immigrate-canada/provincial-nominees.html.

Government of Canada. (2018c). *International Mobility Program: Atlantic Integration Pilot Program*. https://www.canada.ca/en/immigration-refugees-citizenship/corporate/publications-manuals/operational-bulletins-manuals/temporary-residents/foreign-workers/special-initiatives-pilot-project/exemption-code-c18.html.

Health Professionals for Diversity Coalition. (2012). *Fact sheet: The need for diversity in the health care workforce*. http://www.aapcho.org/wp/wp-content/uploads/2012/11/NeedForDiversityHealthCareWorkforce.pdf.

House of Commons, Canada. (2017, November). Immigration to Atlantic Canada: Moving to the future. http://publications.gc.ca/collections/collection_2017/parl/xc64-1/XC64-1-1-421-14-eng.pdf.

Immigration, Refugees and Citizenship Canada. (2016). *Express Entry: Year-end report 2016*. https://www.canada.ca/content/dam/ircc/migration/ircc/english/pdf/pub/ee-2016-eng.pdf.

Keung, N. (2012, December 18). Foreign-trained nurses face abrupt game change in Ontario licensing. *Toronto Star*. https://www.thestar.com/news/gta/2012/12/18/foreigntrained_nurses_face_abrupt_game_change_in_ontario_licensing.html.

Kofman, E. (2000). The invisibility of skilled female migrants and gender relations in studies of skilled migration in Europe. *International Journal of Population Geography*, *6*(1), 45–59. https://doi.org/10.1002/(SICI)1099-1220(200001/02)6:1<45::AID-IJPG169>3.0.CO;2-B.

McMahon, T. (2013, April 24). Why the world's best and brightest struggle to find jobs in Canada: Why do skilled immigrants often fare worse here than in the U.S. and U.K.? *Maclean's*. https://www.macleans.ca/economy/business/land-of-misfortune/#.

National Conference of State Legislatures. (2014, August). Racial and ethnic health disparities: Workforce diversity. http://www.ncsl.org/documents/health/Workforcediversity814.pdf.

National Nursing Assessment Service. (n.d.). About NNAS. https://www.nnas.ca/about-nnas/.

National Nursing Assessment Service. (2013, November 14). Funding streamlines international nursing recruitment. https://www.crpnbc.ca/wp-content/uploads/2013/09/2013-11-14-NNAS-news-release-Final.pdf.

Neiterman, E., & Bourgeault, I.L. (2015). Professional integration as a process of professional resocialization: Internationally educated health professionals in Canada. *Social Science & Medicine*, *131*, 74–81. https://doi.org/10.1016/j.socscimed.2015.02.043.

Newton, S., Pillay, J., & Higginbottom, G. (2012). The migration and transitioning experiences of internationally educated nurses: A global perspective. *Journal of Nursing Management, 20*, 534–50. https://doi.org/10.1111/j.1365-2834.2011.01222.x.

Office of the Fairness Commissioner. (2011). *Licensing outpaces employment for internationally educated professionals*. Government of Ontario. https://www.fairnesscommissioner.ca/en/Publications/PDF/Infographics/Licensing_outpaces_employment_for_internationally_educated_professionals.pdf.

Paul, R., Martimianakis, M.A., Johnstone, J., McNaughton, N., & Austin, Z. (2017). Internationally educated health professionals in Canada: Navigating three policy subsystems along the pathway to practice. *Academic Medicine, 92*(5), 635–40. https://doi.org/10.1097/ACM.0000000000001331.

Pre-Arrival Supports and Services (PASS) Program. (n.d.). What is PASS? https://pass4nurses.org/.

Ramji, Z., & Etowa, J. (2014). Current perspectives on integration of internationally educated nurses into the healthcare workforce. *Humanities and Social Sciences Review, 3*(3), 225–33. https://www.hhr-rhs.ca/index.php?option=com_mtree&task=viewlink&link_id=11130&lang=en.

Statistics Canada. (2013). *Immigration and ethnocultural diversity in Canada, National Household Survey, 2011*. Government of Canada. http://publications.gc.ca/collections/collection_2013/statcan/CS99-010-2011-1-eng.pdf.

Statistics Canada. (2017). *Number and proportion of foreign-born population in Canada, 1871 to 2036*. Government of Canada. https://www.statcan.gc.ca/eng/dai/btd/othervisuals/other006.

Suurmond, J., Uiters, E., de Bruijne, M.C., Stronks, K., & Essink-Bot, M-L. (2011). Negative health care experiences of immigrant patients: A qualitative study. *BMC Health Services Research, 11*(10), 1–8. https://doi.org/10.1186/1472-6963-11-10.

Walton-Roberts, M., Guo, J., Williams, K., & Hennebry, J. (2014). Immigration policy changes and entry to practice routes for internationally educated nurses (IENs). Integration Migration Research Center. https://scholars.wlu.ca/cgi/viewcontent.cgi?article=1020&context=imrc.

Wood, D. (2018, April 26). *Diversity in nursing: Will it ever match patient demographics?* diversitynursing.com. https://diversitynursing.com/diversity-in-nursing-will-it-ever-match-patient-demographics/.

World Health Organization. (2006). *The World Health Report 2006: Working together for health*. http://www.who.int/whr/2006/whr06_en.pdf.

World Health Organization. (2018). *Health workforce migration*. https://www.who.int/hrh/migration/14075_MigrationofHealth_Workers.pdf.

5 Gendering Integration Pathways: Migrating Health Professionals to Canada

IVY LYNN BOURGEAULT, JELENA ATANACKOVIC,
AND ELENA NEITERMAN

Introduction

The barriers to labour market integration of internationally educated professionals have been well-documented in the literature (Baumann et al., 2010; Spitzer & Torres, 2008; WHO, 2017). According to the World Health Organization (WHO; 2017), many of the approximately 17,500 internationally educated nurses (IENs) migrating to Canada annually (from countries such as the Philippines, India, and China) experience difficulty securing licensure. This prevents them from practising under provincial and territorial professional regulations, impeding their integration process. Incongruence between immigration policies and health care licensure and practice requirements in countries such as Canada, is one of the key reasons why many internationally educated professionals cannot find employment in their field (some of these barriers to labour market integration of internationally educated health professionals have been discussed and some solutions proposed by Sweetman, in chapter 3, as well as Baumann and colleagues in chapter 4 of this volume). Consequently, many of these highly skilled individuals need to rely on alternative, two-step immigrant routes to enter their profession. This includes, for instance, arriving in Canada as an international student with an intention to complete education and stay in the country, or deciding to enter lower-skilled jobs in the health care sector, such as care aide/personal support worker and home care aide (WHO, 2017).

There is growing recognition of the important role that gender plays in the integration of migrating health professionals. This is unsurprising given the predominance of women's participation in the health workforce, which reaches 70 per cent globally and over 80 per cent in Canada (Bourgeault, 2018). Indeed, the literature speaks to a growing feminization of migration (Camlin et al., 2014; Ryan, 2002) where women outnumber men in transnational mobility, even though some countries continue to restrict women's migration due to patriarchal ideologies (Kingma, 2006). Acknowledgment of the importance of gender in the migration process for health workers was not always the case. Some of the migration literature has been based on the assumption that migration decisions and migration experiences were homogeneous, with men's experiences assumed to be universal. Recognizing that "seemingly gender-neutral process of movement is, in fact, highly gender-specific and may result in differential outcomes for men and women" (Boyd & Grieco, 2003, p. 1), many scholars advocate for the inclusion of gender in international migration theory and migration studies (Mahler & Pessar, 2006).

In this chapter, we tease apart some of the gender influences on the integration process of migrating health professionals, based on the review of the relevant literature and our research, with a focus on the different and unique pathways to integration. Informing our analysis is a conceptual framework of the dynamic structures and processes of migration and integration that recognizes the interconnectedness of professional identities and intersecting influences of gender, racialization, and class (Bourgeault et al., 2016). Briefly, this framework distinguishes between migration – the decision to leave a country for another – and integration into a particular local professional labour market. We focus here primarily on integration, which is conceptualized more broadly than the attainment of licensure to also include social and cultural integration. As we argue, integration processes are tied to local contexts in a way that can be described as *glocalization,* where global forces and local adaptations intersect (Drori et al., 2014). Our focus in this chapter is on the integration experience of migrating health workers in the glocalized Canadian context.

After a brief review of the literature, our first case study will focus on the role played by bridging programs in helping to address the typical deskilling that occurs with migration. This draws upon research conducted in Canada on the migration and integration experiences of physicians, nurses, and midwives (e.g., Covell et al., 2016; Neiterman & Bourgeault, 2013, 2015) and the role of bridging programs in these and other professions, including medical laboratory technology, medical radiation technology, diagnostic medical sonography, respiratory therapy, and physical therapy (Neiterman, Bourgeault, et al., 2018). We also reflect on the gendered experiences of the international students migrating to Canada to join highly competitive health professional training programs. The third pathway is the stepping-stone approach through the older adult care sector, a sector dominated by women and internationally educated women, in particular. Across these cases, we find that gender can influence the integration processes and experiences for both international students and those who are planning to stay and live in Canada. Our analysis also reveals that migrating health professionals are susceptible to social exclusion, in part as a result of their remittance obligations, of which women are more consistent and generous to the point of poverty.

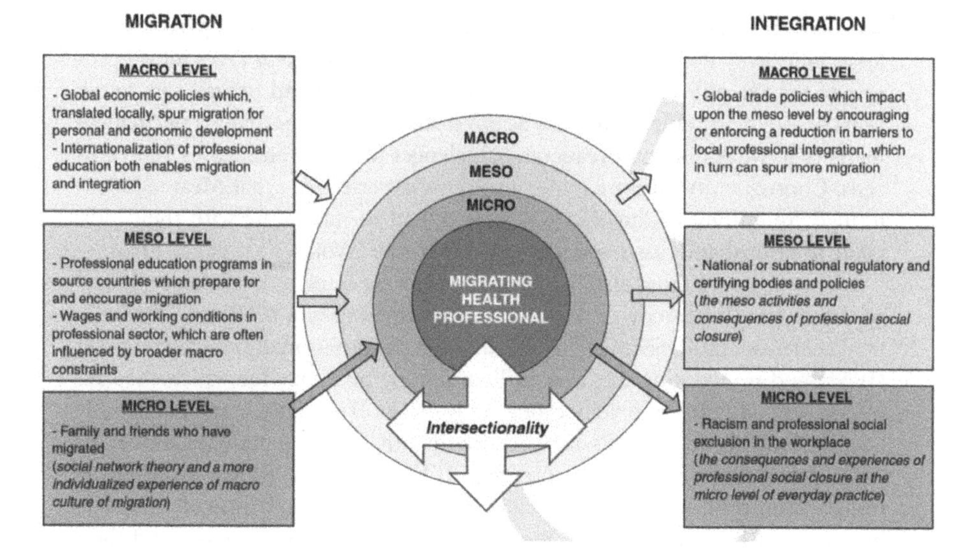

Figure 5.1. Pluralist conceptual framework of professional migration and integration
Source: Bourgeault et al. (2016).

Gender and the Integration Context for Migrating Health Professionals

Gender can affect labour market integration of migrating health workers in their host country. While both men and women experience problems in terms of recognition of their professional credentials in destination countries, the process can be uniquely challenging for women, given the family responsibilities ascribed through traditional gender roles (Neiterman & Bourgeault, 2015; Salami et al., 2018). Some researchers have found, for example, that women physicians seem to adjust to a new system better than men (Remennick & Ottenstein-Eisen, 1998), but others have noted psychological distress among female health immigrants (Factourovich et al., 1996). Given family responsibilities, many female migrating health professionals may delay the process of professional accreditation, retraining, and occupational integration, making it more difficult for them to re-establish new positions in their destination countries (Aure, 2013; Baumann et al., 2010; Bernstein & Shuval, 1999; Fossland, 2013; Iredale, 2005; Neiterman & Bourgeault, 2015). The gendered nature of the deskilling among migrant health care professionals has received increasing attention in the literature (Atanackovic & Bourgeault, 2014; Bourgeault et al., 2010). In the case of physicians, research shows that many female international medical graduates (IMGs) experience deskilling or permanent inability to work in the destination country. For instance, in their research with female physicians migrating from Sub-Saharan Africa to the UK, Belgium, and Austria,

Wojczewski and colleagues (2015) found that many physicians experienced "at least temporary deskilling or inability to work due to a long recognition process of their qualifications" (p. 8). While some women had to wait two to ten years before being able to practise as physicians, some experienced permanent inability to work due to bureaucratic challenges for non-European Union citizens. Consequently, many female physicians experienced what Meares (2010) termed "re-domestication" (p. 473) as they had to stay home with their kids, while their husbands assumed a breadwinner role (Wojczewski et al., 2015).

For nurses, deskilling also occurs when they are integrated at the lower end of the nursing hierarchy or as unregulated care workers (O'Brien, 2007; Salami & Nelson, 2014; Salami et al., 2018). Many internationally educated nurses (IENs) end up working in lower-skilled nursing positions for which they are often overqualified. For instance, one study on the experience of internationally educated nurses in Canada with a predominantly female sample (Salami et al., 2018) found that baccalaureate prepared IENs experience downward occupational mobility finding work as licensed practical nurses. This is often driven by the many barriers to workforce integration as registered nurses, and their redirected path to the licensed practical nurse registration process, which is often encouraged by nursing recruitment agencies (Salami et al., 2018). Indeed, many female IENs end up working in the older adult care sector as personal support workers or domestic/live-in caregivers (Atanackovic & Bourgeault, 2014; Bourgeault et al., 2010). An extreme case of downward occupational mobility occurs when IENs end up working as domestic workers or unskilled caregivers in destination countries.

Nurse migrants are often aware of the difficulties in transferring their qualifications and work experience to other countries, and some seem prepared to accept this demotion for the sake of bringing their children to safer and more prosperous environments (Bourgeault et al., 2010; Piper, 2005). In these cases, the gendered nature of professional migration necessitates that women/mothers "pay the price" of facing challenges in the professional realms for the benefit of their families (Runnels et al., 2016). Because the well-being of the family is tightly linked to the gendered experiences of self-sacrifice, the foregoing of their professional autonomy and career might seem appropriate and, indeed, acceptable to female immigrant health care workers (Runnels et al., 2016).

Gender has also been shown to shape working conditions of migrant health workers once they are integrated into the labour market (Bach, 2003; Batnitzky & McDowell, 2011; Hussein et al., 2011; Neiterman & Bourgeault, 2015; Oikelome & Healy, 2013). Women experience lower pay (Bernstein & Shuval, 1998; Ribeiro, 2008), lack of promotion to specialties (Bernstein & Shuval, 1998), employment in areas with less professional prestige (Ribeiro, 2008), and deskilling (Atanackovic & Bourgeault, 2014; Bourgeault et al., 2010).

While the literature unequivocally concludes that female health professionals face deskilling upon migration to a destination country, not all women experience it in the same way. Indeed, one small Canadian study that explored identity and resilience strategies employed by immigrant women facing deskilling reveals that such women's experiences differ according to particular cultural prisms (Cardu, 2013). Thus, it is important to note that experiences of professional integration are not only gendered but are also shaped by the dynamics of the health care sector and the professional hierarchy, which is itself built upon racialized and gendered norms and expectations (Dicicco-Bloom, 2004; A. George, 2007).

Comparing the experiences of professional discrimination among immigrant physicians and nurses in the Canadian health care context revealed that gender, racialization, and professional status intersect in complicated ways (Neiterman & Bourgeault, 2015). Female nurses from visible minority groups are most vulnerable to the instances of discrimination and are exposed to negative interactions with patients, physicians, and fellow nurses (Neiterman & Bourgeault, 2015). On the other hand, both female and male physicians seemed to be more immune to the experiences of discrimination from patients and nurses, but did report being discriminated against by their fellow physicians. The professional status of a physician, established by the traditionally gendered male hierarchical organization of health care, seems to serve as a shield protecting immigrant physicians from experiencing workplace discrimination to the same extent it is experienced by nurses. That is, the higher status of the traditionally "masculine" profession of medicine within the health care sector and the lower status of nursing, a profession inherently "feminine," suggests that the role of gender, racialization, and professional status intersect in the experiences of discrimination.

In addition to barriers to economic integration, certain female health care professionals also face difficulties with social integration in a new country which can leave them feeling isolated and excluded (Bourgeault, 2015). In many cases, such difficulties can be traced to certain state immigration policies and programs that limit the possibilities for social integration for professional women. This is particularly the case where health workers immigrate through the unique pathways of live-in caregiver, where female health workers predominate. Indeed, the length of the live-in requirement for participants in this program (an immigration policy) made it near impossible to meet requirements of regulatory bodies of active practice (a regulatory policy). Similarly, Taylor and Foster (2015) examined IENs' recruitment through the Temporary Foreign Worker Program (TFWP) in the Canadian province of Alberta, which they argue "encourages low trust and sense of belonging among migrant workers and resistance from domestic workers because it promotes inequality and exclusion" (p. 153). They explain that "the inability of most migrant workers to access settlement services, to bring families, to change employers, or to enroll in further education and training overtly discourages their integration into the local community" (p. 153).

Thus, gender intersects with a number of other factors in unanticipated ways for the migrating health professional seeking integration into the Canadian health care context. Gender, as we shall see, also affects the pathway to integration through bridging programs.

Gender and the Bridging Programs Pathway to Integration

Bridging programs have been developed in Canada to assist migrating health professionals to "bridge" their formal training in another country with the educational, professional, or language requirements necessary to become licensed to practise in Canada (Sattler et al., 2015). Bridging programs vary in format, style, and duration but have an overarching goal to prepare internationally trained individuals for the licensure process and professional integration (Neiterman, Bourgeault et al., 2018). Content and delivery formats were identified as some of the key factors that can contribute to the success of bridging programs (Neiterman, Bourgeault, et al., 2018; Sattler et al., 2015). Despite showing promising success rates, the funding for many of these programs remains particularly precarious in the Canadian context.

While the research on bridging programs is growing, the focus on gender is often not explicit in the analysis (Covell et al., 2016). Given that bridging programs are developed to prepare internationally educated health care professionals for practice, a context inherently gendered, analysing the utility of the bridging programs from a gender lens could provide us with an excellent opportunity to critically examine content and delivery of bridging programs as well as their social function. Applying a gender lens to an analysis of the content of bridging programs, researchers can inquire how to optimize the programs considering cultural variation in gendered norms and responsibilities of internationally educated health care professionals.

For example, nursing stakeholders in Canada sometimes expressed their concerns about the readiness of internationally educated nurses to practise and advocate on behalf of the patient (Neiterman & Bourgeault, 2013). They noted that nurses in many other health care systems have more subservient positions within the gendered health care hierarchy and thus upon immigrating to Canada, internationally trained nurses may not feel comfortable challenging physicians as they are expected to do in Canada. By way of contrast, bridging programs for internationally trained physicians have integrated cultural competency training to attenuate the more hierarchical gendered nature of their home health care system. Both these professions exist within the uniquely Canadian configuration of a gendered and hierarchical health care landscape, which shapes the structure and the content of interactions with patients and other health care professionals (Neiterman & Bourgeault, 2015). Integrating the content of gendered cultural competence into bridging programs' curricula can begin to address these dynamics with the intent of improving the cultural integration of internationally educated health professionals. Considering that women's care responsibilities are also inherently gendered can also be instrumental in developing and delivering bridging programs that are accessible by both female and male internationally educated health care providers.

Another consideration for gender analysis is the financial sustainability of bridging programs. Many bridging programs are offered with start-up funding and then are oriented towards achieving sustainability through cost-recovery models that rely on student tuition fees (Neiterman, Bourgeault, et al., 2018). For female health care professionals, cost associated with bridging education might pose significant barriers as women's professional integration is often considered to be secondary to that of men.

In brief, the research into the integration pathway through bridging programs has only begun to scratch the surface of its gendered nature, but emerging findings are of growing interest.

Gender and the International Student Migration Pathway

Another potential pathway for migrating health professionals is as an international student. In 2014, Canada was the seventh top destination country for international students (Canadian Bureau for International Education, 2015). International students represented approximately 8 per cent of the total student population in Canada, reaching 353,570 students in 2015, which represents an increase of 92 per cent in the period of just seven years (Canadian Bureau for International Education, 2016). International students in health professional education programs have been less prominent, though this is growing for some professions and in some schools.

In one of our studies (Covell et al., 2015; Neiterman, Atanackovic, et al., 2018), we explored the study-migration pathway for international students studying to become regulated health professionals in Canada. In interviews we conducted with key stakeholders, there was overwhelming agreement that international students bring critical skills to the Canadian workforce which could enable improved access to health care services for newcomers with whom they share language and culture. Moreover, there was a sense that international students could gain cultural competence through their local training in Canada, necessary for more successful integration. As we have discussed above, this could be linked to gendered and racialized hierarchies in the health workplace.

Nevertheless, international students in the health professions were not universally seen as being a potential source for the health labour markets. This was an issue owing in part to ethical concerns about increasing the potential for the health care brain drain[1] from students' countries of origin (see Kapur & McHale, 2005). It was also related to the acknowledgment that health workforce shortages in Canada tend to be in rural and remote locations where international students' cultural skills are seen as less of a match to local populations.

Similar to what is found for integrating health professionals, the key barriers for the integration of international students is the policy disconnect that exists at the juncture of the services offered by educational institutions, professional regulatory bodies, and immigration policy. Indeed, each of these policy communities has specific goals and sees international students from its own particular perspective. While all consider international students a potential benefit for the health workforce (and the Canadian society as a whole), the lack of dialogue within and between these three policy communities creates challenges for those international students who would like to remain in Canada. For instance, our analysis suggests that most post-secondary institutions make decisions on their own about their degree of involvement with the immigration community (Covell et al., 2015). Those institutions that consider it appropriate can partner with immigration bodies to offer information to the students interested in applying for permanent resident status (Covell et al., 2015).

Other research that offers insight into the gendered migration pathway of students explores visa trainees in Canadian post-graduate medical residency training programs. Visa trainees are IMGs who are foreign nationals who are funded, often by a foreign government, to complete a residency program in Canada. Visa trainees are expected to return home upon completion of their training, but Mathews et al. (2017) found that 16 per cent remain in Canada up to eleven years after they conclude their programs. The majority of these trainees are male (72 per cent), which contrasts with the gender representation in residency training reaching up to 60 per cent of female physicians. Mathews et al. (2017) found no gender differences in the retention of visa trainees following their exit from residency and fellowship programs. That is, gender was not a significant predictor for remaining in Canada, and predictors did not vary in gender-segregated analyses.

Perhaps the most directly relevant research is Walton-Roberts's (2015) study on Indian-trained nurses enrolled in a post-graduate college-based training program in geriatric care. She describes how international migration has increased the autonomy of the female migrants, but it can still be constrained by overall family migration strategies. In a recent paper, Walton-Roberts (2019) traces migration trajectory of nurses from India to Canada. She concludes that compared to their female counterparts, male nurses benefit more from migration in terms of their occupational success. Clearly, however, the state of research into the gendered nature of the student integration pathway for health professionals is still an untapped and potentially important area for further exploration.

Gender and the "Stepping Stone" Pathway through Older Adult Care

Many internationally educated health professionals see work in the older adult care sector as a stepping stone to better jobs inside or outside the sector. Many, especially women, end up in these lower paid, seemingly lesser skilled, and female-predominated positions often avoided by local workers (Atanackovic & Bourgeault, 2014; Bourgeault et al., 2010). Despite the low pay in this sector, the working conditions are quite challenging (Atanackovic et al., 2009; Cuban, 2013). The lower wages often result in these workers being more likely to have more than one job, heavy workloads, overtime work, and night shifts (Atanackovic et al., 2009; Atanackovic & Bourgeault, 2014; Bourgeault et al., 2010). Other research has also shown they experience mistreatment (Hussein et al., 2011), abuse (Pratt, 2001; Spitzer & Torres, 2008), and exposure to health risks after the migration journey (Hennebry et al., 2016; Spitzer, 2016). In some parts of the world migrant workers caring for older adults have to adapt to new challenges such as the introduction of robot labour in Japan's aged care sector that raises some issues in terms of human-robot interaction and migrant workers' rights (for more details on this issue, please refer to Perrote and Walton-Roberts, chapter 13 in this volume).

Most immigrant care workers in this sector come to Canada through immigrant admission policies that do not target older adult care workers (Atanackovic & Bourgeault, 2013). One of the exceptions to the lack of admission categories that target workers in the older adult sector in Canada has been the Canada Caregiver Program (CCP; its predecessor being the Live-in Caregiver Program) (Salami et al., 2016). This program targets immigrants living in private homes as caregivers of the elderly or of children. In one of our studies (Atanackovic & Bourgeault, 2014), we examined economic and social integration of immigrant live-in caregivers in Ontario, Canada, both during and after participation in the Live-in Caregiver program. Based on our interviews with five stakeholders and focus groups and interviews with fifty-eight live-in caregivers (95 per cent of whom were women and more than half of whom had obtained credentials in nursing, midwifery, and social work in their country of origin), we identified the barriers that hamper economic and social integration of immigrant live-in caregivers.

While caregivers of both children and older adults experienced inequitable working conditions and even exploitation due to live-in requirements and temporary migration status, older adult caregivers reported unique challenges related to the age of their clients and their medical conditions such as disability, dementia, depression, aphasia, or Alzheimer's disease. Indeed, some older adult caregivers reported that employers expected them to use their nursing skills (often not outlined nor required in job contracts), without additional compensation. In fact, the interviews with caregivers revealed that employers often used gender, racial status, and citizenship as a way to mark the difference between them and live-in caregivers, which helped them to justify discrimination against and abuse of these workers (Atanackovic, 2014). Such discrimination was evident in the greater expectations placed on the worker in terms of workload. For instance, the employers, overrepresented by white and Canadian-born individuals, placed greater expectations on female immigrant caregivers as they expected them not only to take care of their clients but also to clean and do other household chores. Such expectations were not echoed by male caregivers (Atanackovic, 2014).

Gender and other social categories (race, ethnicity, and citizenship) continued to influence the live-in caregiver's labour market integration even after completing the program (Atanackovic, 2014). Indeed, our analysis revealed that once they completed the program, many live-in caregivers (most of which are Filipinas) lost the recognition of their skills (Atanackovic & Bourgeault, 2014). Still, most of them did not plan to upgrade their nursing or other health professional skills, mainly due to it being a long and expensive process, made all the more difficult by their need to send remittances to their families back home (Atanackovic, 2014; Atanackovic & Bourgeault, 2014). The gendered cultural norms in the Philippines prescribe that women should migrate abroad and assume the role of family breadwinner in light of the precarious economic situation in their country. Indeed, some female live-in caregivers end up taking the first available job once they are done the program so as to begin the long, costly, and arduous process of family reunification (Atanackovic, 2014). Some thought that the requirement under the program to obtain a study permit for credit courses longer than six months represents a huge barrier to upgrading, as it prohibits access to credit courses in the domain of post-secondary and professional skills. Consequently, as a result of the interaction of the different factors (i.e., strict immigration policies that impose educational restrictions while under the program, gendered expectations, and financial barriers), many of these workers remain stuck in lower-paid jobs in the older adult care sector for which they are overqualified (Atanackovic, 2014; Atanackovic & Bourgeault, 2014).

We also found that caregivers felt socially isolated both during and after the program. In particular, while in the program, caregivers of older adults encountered significant restrictions on their personal movement due to their client's care needs and as such felt socially isolated, given that they had to care for their clients all the time (Atanackovic & Bourgeault, 2014). While the old age of their clients played a critical role in restricting the personal movement of caregivers of older adults, the live-in requirement of the program was certainly another important factor contributing to their isolation. The biggest barriers to their successful social integration after the program seemed to be their attachment to ethno-specific networks and their difficulties with the procedures for obtaining permanent residence and family reunification (Atanackovic & Bourgeault, 2014).

We described this as a double isolation by their immigration status and by their work in a sector that is all but invisible and undervalued (Bourgeault, 2015). A number of intersecting factors lead to this double isolation. First, the undervalued nature and poor working conditions in the older adult care sector pushes employers to rely upon migrant health professionals who are not integrated at their previous skill level. Second, the negative effects of the working conditions in this sector are exacerbated for migrant health professionals in two ways: (1) they are isolated within the workplace because of their migrant status, and (2) they are isolated in their communities because of the constraints of their work. This double isolation can be exacerbated when migrant health professionals are racialized. As Piper (2005) describes, markers of class, race, and ethnic differences can intersect to produce compounded disadvantage, such that many migrant women are "triply disadvantaged" (as migrants, women, and ethnically different), and therefore "more likely to be overrepresented in marginal, unregulated, and poorly paid jobs" (p. 2). The burden of sending remittances back home can serve to exacerbate the exclusion effects on the lives of these women in terms of heavy workloads and social isolation (Eckenwiler, 2014).

Conclusion

In conclusion, there is a growing body of research that examines the ways in which gender influences the integration process of migrating health professionals, whether it is directly, through bridging programs, as international students, or through other integration pathways such as the older adult care sector. In brief, integration has paradoxical effects – it can lead to greater autonomy but also leads to finding employment in precarious sectors of the labour market and to social isolation. Health worker migrants experience deskilling, which can be addressed in part by bridging programs, but these are susceptible to precarity (just like the health worker migrants they are intended to support). Recognizing the challenges experienced by previous cohorts of migrating health workers from their countries, new pathways as international students have been explored, where we know much less about their integration opportunities. Across integration pathways, social exclusion is affected in part by the economic impact of remittance obligations. These findings help to expand upon the elements of the right-hand side of the theoretical framework in figure 5.1– in particular as way to tease apart the micro and meso levels with a specific focus on gender.

There are a number of promising areas to explore. In terms of gendered outcomes of migration for health workers, migration may serve to disrupt traditional gender roles and relations within the family (see A. George, 2007), but research on gender implications of remittances (e.g., Basa et al., 2011) is scarce and offers mixed results. In her book on migration of nurses from Kerala, India, to the United States, S. George (2005) argues that migrant nurses have always been contributing significantly to the economies of the source countries through remittances. She adds that migrating women are more likely to remit to their female relatives in the household, which results in a "'feminized transnational network" that distributes money directly among women. This can result in altered gender relations and family dynamics in those households in countries of origin, as female relatives become head of households who make decisions in terms of household finances and can become more engaged in the community decision-making (S. George, 2005). On the other hand, Gallo (2005) found that Kerala nurses who migrated to Italy pooled their resources to pay for their female relatives' marriage dowries, thus perpetuating a gendered patriarchal system.

Some research also suggests that migration of health care professionals from some patriarchal societies can result in worse matrimonial prospects. In her study with nursing students and faculty in Kerala, Walton-Roberts (2012) found that unlike male migrants who insinuate a sense of modernity, Kerala's female migrants are still subject to deeply embedded negative constructions and perceived as "loose" or "suspicious," which can affect their matrimonial prospects. Thus, future research needs to explore the impact of migration on family dynamics and marriage prospects from a gender perspective.

NOTE

1 While stakeholders involved in our research were concerned about health care brain drain, Adekola's research on Nigerian nurses in Canada suggests that their migration experiences are positive and closely align with the notion of the idea of the "global brain train," as they accumulate more education and professional development over their migratory journey. For more information on this particular study, please see chapter 15 of this volume.

REFERENCES

Atanackovic, J. (2014). *The migration, working, living and integration experiences of immigrant live-in Caregivers in Ontario, Canada* [Doctoral dissertation, McMaster University]. MacSphere Open Access Dissertations and Theses, http://hdl.handle.net/11375/16418.

Atanackovic, J., & Bourgeault, I.L. (2013). The employment and recruitment of immigrant care workers in Canada. *Canadian Public Policy, 39*(2), 335–50. http://dx.doi.org/10.3138/CPP.39.2.335.

Atanackovic, J., & Bourgeault, I.L. (2014). Economic and social integration of immigrant live-in caregivers in Canada. Institute for Research on Public Policy Study. http://irpp.org/research-studies/economic-and-social-integration-of-immigrant-live-in-caregivers-in-canada/.

Atanackovic, J., Bourgeault, I.L., Denton, M., LeBrun, J., McHale, J., Parpia, R., Rashid, A., Toombs, R., & Winkup, J. (2009). *The role of immigrant care workers in an aging society: The Canadian context and experience.* CIHR/Health Canada Research Chair in Health Human Resource Policy. http://tools.hhr-rhs.ca/index.php?option=com_mtree&task=att_download&link_id=6113&cf_id=68&lang=en.

Aure, M. (2013). Highly skilled dependent migrants entering the labour market: Gender and place in skill transfer. *Geoforum,* 45, 275–84. http://dx.doi.org/10.1016/j.geoforum.2012.11.015.

Bach, S. (2003). International migration of health workers: Labour and social Issues. Working Paper, Sectoral Activities Department, International Labour Office. https://www.researchgate.net/publication/242394141_International_Migration_of_Health_Workers_Labour_and_Social_Issues.

Basa, C., Harcourt, W., & Zarro, A. (2011). Remittances and transnational families in Italy and the Philippines: Breaking the global care chain. *Gender and Development, 19*(1), 11–22. https://doi.org/10.1080/ 13552074.2011.554196.

Batnitzky, A., & McDowell, L. (2011). Migration, nursing, institutional discrimination and emotional/affective labour: Ethnicity and labour stratification in the UK National Health Service. *Social and Cultural Geography, 12*(1), 181–201. doi: 10.1080/14649365.2011.545142.

Baumann, A., Blythe, J., & Ross, D. (2010). Internationally educated health professionals: Workforce integration and retention. *Healthcare Papers, 10*(2), 8–20. http://dx.doi.org/10.12927/hcpap.2010.21795.

Bernstein, J., & Shuval, J.T. (1998). The occupational integration of former Soviet physicians in Israel. *Social Science and Medicine, 47*(6), 809–19. https://doi.org/10.1016/S0277-9536(98)00139-7.

Bourgeault, I.L. (2015). The double isolation of immigrants undertaking older adult care work. In C.L. Stacey, A. Armenia, & M. Duffy (Eds.), *Caring on the clock: The complexities and contradictions of paid care work* (pp. 117–25). Rutgers University Press.

Bourgeault, I.L. (2018). Women's work across every aspect of healthcare is largely invisible. Make Evidence Matter. https://evidencenetwork.ca/womens-work-across-every-aspect-of-healthcare-is-largely-invisible/.

Bourgeault, I., Parpia, R. & Atanackovic, J. (2010). Canada's live-in caregiver program: Is it the answer to the growing demand for elderly care? *Journal of Population Aging, 3*(1), 83–102. http://dx.doi.org/10.1007/s12062-010-9032-2.

Bourgeault, I.L., Wrede, S., Benoit, C., & Neiterman, E. (2016). Professions and the migration of expert labour: Towards an intersectional analysis of transnational mobility patterns and integration pathways of health professionals. In M. Dent, I.L. Bourgeault, E. Kuhlmann, & J.L. Denis (Eds.), *The Routledge handbook on professions and professionalism* (pp. 295–312). Routledge.

Boyd, M., & Grieco, E. (2003). Women and migration: Incorporation into international migration theory. Migration Policy Institute. http://www.migrationinformation.org/Feature/ display.cfm?ID=106.

Camlin, C.S., Snow, R.C., & Hosegood, V. (2014). Gendered patterns of migration in rural South Africa. *Population, Space and Place, 20*(6), 528–51. http://dx.doi.org/10.1002/psp.1794.

Canadian Bureau for International Education. (2015). *A world of learning: Canada's performance and potential in international education*. https://cbie.ca/wp-content/uploads/2019/01/A-World-of-Learning-HI-RES-2015.pdf.

Canadian Bureau for International Education. (2016). *A world of learning: Canada's performance and potential in international education*. http://cbie.ca/wp-content/uploads/2017/07/A-World-of-Learning-HI-RES-2016.pdf.

Cardu, H. (2013). Resilience strategies used by immigrant women facing professional deskilling in Quebec: A literature review and a small-scale study. In *Crushed hopes: Underemployment and deskilling among skilled migrant women* (pp. 137–61). International Organization for Migration. https://publications.iom.int/system/files/pdf/crushed_hopes_3jan2013.pdf.

Covell, C, Neiterman, E, Atanackovic, J., Owusu, Y., & Bourgeault, I.L. (2015). The study-migration pathway: Understanding the factors that influence the employment and retention of international students as regulated health professionals in Canada. Pathways to Prosperity Project. http://p2pcanada.ca/wp-content/blogs.dir/1/files/2016/02/Study-Migration-Pathway.pdf.

Covell, C.L., Neiterman, E., & Bourgeault, I.L. (2016). Scoping review about the professional integration of internationally educated health professionals. *Human Resources for Health, 14*(1), 38. https://human-resources-health.biomedcentral.com/articles/10.1186/s12960-016-0135-6.

Cuban, S. (2013). "I don't want to be stuck as a carer": The effects of deskilling on the livelihoods and opportunities of migrant care workers in England. In B. Mollard & S. Umar (Eds.), *Crushed hopes: Underemployment and deskilling among skilled migrant women* (pp. 37–76). International Organization for Migration. https://dx.doi.org/10.18356/7eff69cf-en.

Dicicco-Bloom, B. (2004). The racial and gendered experiences of immigrant nurses from Kerala, India. *Journal of Transcultural Nursing, 15*(1), 26–33. doi: 10.1177/1043659603260029.

Drori, G.S., Höllerer, M.A., & Walgenbach, P. (2014). Unpacking the glocalization of organization: From term, to theory, to analysis. *European Journal of Cultural and Political Sociology, 1*(1), 85–99. https://doi.org/10.1017/jmo.2016.10.

Eckenwiler, L. (2014). Care worker migration, global health equity, and ethical place-making. *Women's Studies International Forum 47*, 213–22. http://dx.doi.org/10.1016/j.wsif.2014.04.003.

Factourovich, A., Ritsner, M., Maoz, B., Levin, K., Mirsky, J., Ginath, Y., ... & Natan, E.B. (1996). Psychological adjustment among Soviet immigrant physicians: distress and self-assessments of its sources. *The Israel Journal of Psychiatry and Related Sciences, 33*(1), 32–9. https://pubmed.ncbi.nlm.nih.gov/8163361/.

Fossland, T. (2013). Crossing borders – getting work: Skilled migrants' gendered labour market participation in Norway. *Norsk Geografisk Tidsskrift–Norwegian Journal of Geography 67*(1), 276–83. http://dx.doi.org/10.1080/00291951.2013.847854.

Gallo, E. (2005). Unorthodox sisters: Gender relations and generational change among Malayali migrants in Italy. *Indian Journal of Gender Studies, 12*(2-3), 217–52. http://dx.doi.org/10.1177/097152150501200204.

George, A. (2007). Human resources for health: A gender analysis. Background paper prepared for the Women and Gender Equity Knowledge Network and the Health Systems Knowledge Network of the WHO Commission on Social Determinants of Health. http://www.who.int/social_determinants/resources/human_resources_for_health_wgkn_2007.pdf.

George, S. (2005). *When women come first: Gender and class in transnational migration.* University of California Press.

Hennebry, J., Williams, K., & Walton-Roberts, M. (2016). *Women working worldwide: A situational analysis of women migrant workers.* UN Women.

Hussein, S., Manthorpe, J., & Stevens, M. (2011). The experiences of migrant social work and social care practitioners in the UK: Findings from an online survey. *European Journal of Social Work 14*(4), 479–96. http://dx.doi.org/10.1080/13691457.2010.513962%7D.

Iredale, R. (2005). Gender immigration policies and accreditation: Valuing the skills of professional women migrants. *Geoforum, 36*(2), 155–66. http://dx.doi.org/10.1016/j.geoforum.2004.04.002.

Kapur, D., & McHale, J. (2005). *Give us your best and brightest: The global hunt for talent and its impact on the developing world.* Center for Global Development.

Kingma, M. (2006). *Nurses on the move: Migration and the global health care economy.* ILR Press.

Mahler, S.J., & Pessar, P.R. (2006). Gender matters: Ethnographers bring gender from the periphery toward the core of migration studies. *International Migration Review, 40*(1), 27–63. http://dx.doi org/10.1111/j.1747-7379.2006.00002.x.

Mathews, M., Kandar, R., Slade, S., Yi, Y., Beardall, S., Bourgeault, I., & Buske, L. (2017). Credentialing and retention of visa trainees in post-graduate medical education programs in Canada. *Human Resources for Health, 15*(38), 1–8. https://human-resources-health.biomedcentral.com/articles/10.1186/s12960-017-0211-6.

Meares C. (2010). A fine balance: Women, work and skilled migration. *Women's Studies International Forum, 5*, 473–81. http://dx.doi.org/10.1016/j.wsif.2010.06.001.

Neiterman, E., & Bourgeault, I.L. (2013). Cultural competence of internationally educated nurses: assessing problems and finding solutions. *Canadian Journal of Nursing Research, 45*(4), 88–107. doi: 10.1177/084456211304500408.

Neiterman, E., & Bourgeault, I.L. (2015). The shield of professional status: Comparing discriminatory experiences of IENs and IMGs in Canada. *Health,* 19(6), 1–20. doi: 10.1177/1363459314567788.

Neiterman, E., Atanackovic, J., Covell, C., & Bourgeault, I.L. (2018). "We want to be seen as partners, not vultures of the world": Perspectives of Canadian stakeholders on migration of international students studying in health professions in Canada. *Globalisation, Societies and Education, 16*(4), 395–408. doi: 10.1080/14767724.2018.1440350.

Neiterman, E., Bourgeault, I., Peters, J., Esses, V., Dever, E., Gropper, R., Nielson, C., Kelland, J., & Sattler, P. (2018). Best practices in bridging education: Multiple case study evaluation of postsecondary bridging programs for internationally educated health professionals. *Journal of Allied Health, 47*(1), 23E–28E. https://www.ingentaconnect.com/contentone/asahp/jah/2018/00000047/00000001/art00015.

O'Brien, T. (2007). Overseas nurses in the National Health Service: A process of deskilling. *Journal of Clinical Nursing, 16(12)*, 2229–36. doi: 10.1111/j.1365-2702.2007.02096.x.

Oikelome, F., & Healy, J. (2013). Gender, migration and place of qualification of doctors in the UK: Perceptions of inequality, morale and career aspiration. *Journal of Ethnic and Migration Studies, 39*(4), 557–77. doi: 10.1080/1369183X.2013.745233.

Piper, N. (2005). Gender and migration: A paper prepared for the policy analysis and research programme of the global commission on international migration. Global Commission on International Migration. https://www.iom.int/jahia/webdav/site/myjahiasite/shared/shared/mainsite/policy_and_research/g cim/tp/TP10.pdf.

Pratt, G. (2001). Filipino domestic workers and geographies of rights in Canada. Norma Wilkinson Endowment Lecture, Geographical Paper No. 156. Department of Geography, University of Reading.

Remennick, L.I., & Ottenstein-Eisen, N. (1998). Reaction of new Soviet immigrants to primary health care services in Israel. *International Journal of Health Services, 28*(3), 555–74. doi: 10.2190/JL9E-XHH9-XC5Y-5NA4.

Ribeiro, J.S. (2008). Gendering migration flows: Physicians and nurses in Portugal. *Equal Opportunities International, 27*(1), 77–87. http://dx.doi.org/10.1108/02610150810844956.

Runnels, V., Packer, C., & Labonté, R. (2016). International health worker migration: Issues of ethics, human rights and health equity. In F. Thomas (Ed.), *Handbook of migration and health* (pp. 117–31). Edward Elgar Publishing.

Ryan, J. (2002). Chinese women as transnational migrants: Gender and class in global migration narratives. *International Migration, 40*(2), 93–116. http://dx.doi.org/10.1111/1468-2435.00192.

Salami, B., Meherali, S., & Covell, C. (2018). Downward occupational mobility of baccalaureate-prepared, internationally educated nurses to licensed practical nurses. *International Nursing Review 65*(2), 173–81. doi: 10.1111/inr.12400.

Salami, B., & Nelson, S. (2014). The downward occupational mobility of internationally educated nurses to domestic workers. *Nursing Inquiry, 21*(2), 153–61. http://dx.doi.org/10.1111/nin.12029.

Salami, B., Oluwakemi, A., & Okeke-Ihejirika, P. (2016). Migrant nurses and care workers' rights in Canada. Working Paper 2016-9. United Nations Research Institute for Social Development. http://www.unrisd.org/80256B3C005BCCF9/httpNetITFramePDF?ReadForm&parentunid=7EEBA6386981AC8FC125800B00295658&parentdoctype=paper&netitpath=80256B3C005BCCF9V)/7EEBA6386981AC8FC125800B00295658/$file/Salami%20et%20al.pdf.

Sattler, P., Peters, J., Bourgeault, I.L., Esses, V., Neilsen, C, Dever, E., Neiterman, E., Gropper, & Kelland, J. (2015). *Multiple case study evaluation of postsecondary bridging programs for internationally educated health professionals*. Higher Education Quality Council of Ontario. http://www.heqco.ca/SiteCollectionDocuments/IEHPs_ENG.pdf.

Spitzer, D.L. (2016). Engendered movements: migration, gender and health in a globalized world. In J. Gideon (Ed.), *Handbook on gender and health* (pp. 251–67). Edward Elgar Publishing.

Spitzer, D., & Torres, S. (2008). Gender-based barriers to settlement and integration for live-in-caregivers: A review of the literature. (CERIS Working Paper Series. No. 71). Centre of Excellence for Research on Immigration and Settlement.

Taylor, A., & Foster, J. (2015). Migrant workers and the problem of social cohesion in Canada. *Journal of International Migration and Integration*, *16*(1), 153–72. http://dx.doi.org/10.1007/s12134-014-0323-y.

Walton-Roberts, M. (2012). Contextualizing the global nurse care chain: international migration and the status of nursing in Kerala, India. *Global Networks*, *12*(2), 175–94. http://dx.doi.org/10.1111/j.1471-0374.2012.00346.x.

Walton-Roberts, M. (2015). Femininity, mobility and family fears: Indian international students. *Journal of Cultural Geography*, *32*(1), 68–82. http://dx.doi.org/10.1080/08873631.2014.1000561.

Walton-Roberts, M. (2019). Asymmetrical therapeutic mobilities: Masculine advantage in nurse migration from India. *Mobilities 14*(1), 20–37. http://dx.doi.org/10.1080/17450101.2018.1544404.

Wojczewski, S., Pentz, S., Blacklock, C. Hoffmann, K., Peersman, W., Nkomazana, O., & Kutalek, R. (2015). African female physicians and nurses in the global care chain: Qualitative explorations from five destination countries. *PLoS ONE 10*(6), e0129464. doi: 10.1371/journal.pone.0129464.

World Health Organization. (2017). *Women on the move: Migration, care work and health*. World Health Organization Press.

6 The Global Intimate Workforce

CAITLIN HENRY

In this chapter I propose a new framework for analysing the intersection of work, skill, migration, and scale: the global intimate workforce. A global intimate workforce is a labour force or profession essential to the reproduction of everyday life that is transnational in its constitution and in its historical formation. The new concept builds on the work on the global and intimate by Pratt and Rosner (2006, 2012) and is a way to bring together the scalar nuances of the global intimate and the study of workforce production and migration. Therefore, the global intimate workforce demands attention to the intimacies of care, work, unconventional migration patterns, the microgeographies of care deficits, and it draws attention to understudied scales and scalar intersections.

In developing the framework of the global intimate workforce, I draw on research on the migration of nurses from the Global South to the United States. Nurses are a powerful case for developing this concept because of their unique position as skilled care workers. They bring together institutions at multiple scales – for example, health care systems, domestic and international regulatory bodies – as well as highlighting the contradictions of commodified care work, contributing to a broader and more nuanced understanding of geographies of care, migration, and social reproduction. Furthermore, the profession and workforce has an always-already global quality (see Choy, 2003; Reddy, 2015). Also, following Kofman and Raghuram's (2006) call to better incorporate skilled workers in care and migration literature, nurses' different positions from domestic workers and nannies adds nuance to the study of care, labour, and migration. In fact, as I will show using the case of nurses, bringing together the global intimate and studies of transnational workforces reveals a co-production of value that is embodied, scalar, and spatial – and struggled over. The juxtaposition of multiple relationalities, including nurse and patient, global and intimate, worker and employee, reveals where processes of (de)valuation are at work across scales and spaces.

After reviewing the global and the intimate, I will explain the framework of the global intimate workforce. Then I will draw on three cases to develop the framework and to illustrate the usefulness of such an approach for studying work, intimacies, and globalization in a more nuanced way. These case studies draw on interviews conducted with migrant nurses in St. Louis, Missouri, and New York City in the United States, as well as on textual analyses of federal legislation and congressional debates on nurse-specific visa programs and media coverage of nursing shortages.

Global Intimate Workforce: A Framework for Analysis

The global and the intimate, as developed by Pratt and Rosner (2006, 2012), is a pairing of concepts that aims to avoid binaries that pairings like global and local perpetuate. Instead, pairing the global with the intimate rather than the local means using terms that "are not defined against one another but rather draw their meaning from more elliptically related domains … thus potentially and productively disruptive of the geographical binaries and hierarchies that often structure our thinking … the intimate forces our attention on a materialized understanding of the body when we theorize on a global scale" (Pratt & Rosner, 2012, p. 2). For Pratt and Rosner (2012), the intimate, much more than just the private or personal, provides a material and experiential basis for exploring the unevenness of globalization, because "[intimacy] is infused with worldliness" (p. 3).

The global intimate emphasizes both the mutual constitution of macro and micro processes and relations and the spaces, places, and scales in between, as it offers a new scale on a new map. It is necessarily both a material and emotional sphere. The intimate is not the global's opposite but "its corrective, its supplement, or its undoing" (Pratt & Rosner, 2006, p. 17). Thus, not only does the intimate shine light on the mutual constitution of the global and the local but also tries to disrupt any possibility of seeing a hierarchy or boundary between the two. It "takes us onto a different map or perhaps entirely beyond the visual register of map reading" (p. 17).

As I will discuss more fully later in this section, nursing work engages directly with the intimate and intimacy, as well as macro scale actors, institutions, and regulatory bodies. In interviews, nurses discussed the fatigue of commuting long distances, efforts at making their work schedules compatible with those of their partners, desires for stimulating work, and the stress that follows them home after working with a tough patient. They discussed the intimate monitoring of patients, such as watching vital signs in surgery or intake interviews in the Emergency Room (ER). They explained the very intimate body work of nursing such as changing bandages on wounds, lifting and moving heavy bodies, or moisturizing a diabetic patient's vulnerable skin. They talked of the mundane and small, but ever important, tasks involved in non-direct care, such as sterilizing surgical tools or coordinating discharge plans for patients and their families. Each of these tasks is intimate, personal, and relational for both patient and nurse, a combination of physical, mental, and emotional labour. Each task also reflects the broader political economic context that influences who does what work and how, such as in the delicate counselling conversations they had with victims of domestic abuse or drug addicts who came to the ER not for medical care but for a meal. The intimate reads the macro policies that are written into bodies and materials, as well as the personal stories and actions that drive, shape, and resist macro-scale politics. While my focus here is on nurses, that does not mean we should understand nurses without the wide range of other workers that they engage with, from orderlies, finance department workers, and surgeons in the everyday to the politicians, state administrators, and border officers that also shape health care work's value and conditions.

The global intimate is an analytic pairing that reveals more than the mutual constitution of global processes and intimacies since the intimate operates on a different spatial logic from traditional scales. As Pratt and Rosner (2012) explain, "the stain of what we call the global complicates and compromises intimacy in productive ways by opening it to histories of imperialism, national formation, global economic development, systematic humiliation and deprivation, and gender and sexuality inequality. Joining the global and the intimate … forces us to question what is big and what is small, what is important and what is inconsequential" (p. 22).

Since the global intimate is attuned to the political and economic, it can be used well for understanding the constitution of labour, hence the global intimate workforce. The global and intimate complements the work done using global care chains or GCC (Murphy, 2014; Williams, 2010; Yeates, 2012). Hochschild (2000) and Parreñas (2001) both emphasize how care duties are transferred from one person in a household to another when the former emigrates to provide care to people in the Global North. The result is a metaphorical chain of caring and care workers (paid and unpaid) that connects the families "left behind" in the Global South with the families in the receiving countries, which are generally in the Global North. GCC means to capture not only the ways that care needs shape the geographies of care within and among families but also the global political economic processes at play, shaping where, when, and how care is given and received. For example, as Yeates (2004) explains, each station on the care chain offers a moment to understand where, how, and by whom care is delivered. Such an investigation provides opportunities to explore global inequalities. Yeates's attempt to add a commodity chain-style analysis to global care chains is a way of reinforcing the political economic strengths of the latter concept. Parreñas (2012) goes further with her international division of reproductive labour, which focuses on reproduction rather than care. The global intimate broadens this even further by addressing the sphere of the intimate.

Furthermore, using the global and intimate avoids the linearity of GCC (see Williams, 2010; Yeates, 2004), as well as the scalar primacy of the household. This risks missing, as others have warned (see Kofman & Raghuram, 2006; Raghuram, 2012), the important roles the state plays in shaping global reproductive migration, something that, in throwing out the map entirely, the global intimate strives to avoid. The global intimate workforce also attempts to rectify this, for example with a focus on state immigration programs directed at nurses, which I will discuss later. Finally, because the care chain focuses mostly on care moving from the Global South to the Global North, there is, respectively, an implicit assumption of a binary care deficit and surplus. Such a model misses the drastic unevenness of care in the West and that there is an uneven *global* care deficit, one that manifests around the world in uneven micro and mezzo geographies (Jackson & Henry, 2017).

Thus, I propose applying a framework of the global intimate workforce onto the intersection of work, migration, and value across space and scale. As the global intimate helps to disrupt ideas of "national" or local labour markets, it requires historicizing the ways in which embodied experiences and global uneven development impinge on each other. Thus, I juxtapose individual nurses' voices with macro-scale policies and migration flows. The concept creates a new scale through which the nursing workforce can be analysed; a scale that accounts for the iterative relationship between local conditions, global power dynamics, the privatization of social reproduction, and policies at intermediary scales.

Because the global intimate entails the entanglements and interconnections rooted in everyday intimacies as well as those made across time and place (Mountz & Hyndman, 2006, p. 447), it suits the nursing workforce well. This framework will help us understand the complex, multi-scalar factors that influence the everyday practice of nursing and its transnational constitution: the Kenyan nurse who migrates, in part, so she does not have to wash and reuse her latex gloves; the Filipina nurse caring for the aging Jewish woman in Brooklyn; and the nurses left jobless when the hospital up the street from the World Trade Center goes bankrupt. The global intimate workforce is a framework for not just understanding everyday work but doing so in a way that situates such work and workers historically. The framework allows us to see workers in relation to political economic forces big and small, from how those forces shape workers' jobs and the reproduction of everyday life to how they resist such power.

Approaching the nursing profession as a global intimate workforce means situating its transnational history in its present, bringing in formal and informal agents of nurse recruitment as well as state and extra-state attempts to manage (ethically and not) the migration and recruitment of nurses, and, finally, letting nurses' stories guide our inquiries. Global, national, and very local politics of race and colonialism shape the federal efforts to fill staffing vacancies with temporary foreign workers. The transnational and local scales collide. Nursing history is helpful for parsing out this intersection because the nursing workforce is inherently global in its constitution (Choy, 2003; Glenn, 2010; Reddy, 2015). But more than this, the profession developed transnationally, with "foreign" trained nurses having been integral to the development of contemporary (read: Western) nursing as it is known today. Reddy's (2015) history of nursing and migration between the US and India during the nineteenth and twentieth centuries demonstrates how US and British imperialism combined with the localized gendering of nurses to produce a nursing profession that has always been transnational.

In following the global intimate's interest in creating a new map of inquiry, the global intimate workforce is a concept attuned to the scalar complexity of nursing shortages and care deficits, as well as overlooked scales and sites in global care chain analyses. Furthermore, as Raghuram et al. (2011) demonstrate, there is a co-production of value between the caregiver and recipient of care, as two marked and devalued bodies are also "both recuperated through this association." Taking this further, understanding the nursing workforce as a global intimate workforce reveals a co-production of value not only between patient and nurse but also between scales and across spaces – a recuperation of the intimate spaces, of spaces that are remote from investment.

Feminist geography literature emphasizes the ways in which people are agents of globalization and global political economic processes rather than passive or puppets or repositories of impacts (Chang & Ling, 2000; Freeman, 2001; Nagar et al., 2002). Following Lawson's (2000) call to find the theoretical potential in migrants' stories, I draw on both migrant nurses' stories and various attempts to manage nurse migration to develop the concept of the global intimate workforce. The following three cases are attempts to bring together nursing history, various actors involved in the global migration of nurses, and understudied scales of analysis in the care-migration literatures. As a way to illustrate and develop the analytic framework of the global intimate workforce, I use the framework to analyse the three cases together. Such an approach to nursing work, I will show, enables a fuller picture of nursing full of subtleties intimate, mezzo, and global. For example, placing acts of solidarity amongst nurses adjacent to state disputes about the spatiality and scale of a nurse shortage and the care needs of aging populations, means seeing these not as distinct but as part and parcel of the same issue – the co-production of devalued bodies, spaces, and scales, and the subsequent revaluation or salvation of these identities through practices of solidarity, political advocacy, and care.

Intimacies of Transnational Migration

Grace, Sophie, and Mary are three friends from Kenya. They now all work as nurses in St. Louis, Missouri. Their story is one of collective solidarity. Grace migrated first in 1993, and then Sophie and Mary each followed. Grace had a cousin living in St. Louis who helped her get settled, and then she helped

Sophie settle in 1997, and they both helped Mary get set up in 1998. They all trained as nurses first at home in Nairobi, but then came to the US first as Bachelors of Science and Nursing (BSN) students as a means of eventually securing work in the US. Only some of their classes and credentials transferred, so they all chose to migrate to the US to complete their four-year degrees.

Through their mutual support, each was able to migrate and settle in St. Louis fairly smoothly. They also found camaraderie with other migrant nurses, helping each other to adjust to the nuances of working in US health care. Furthermore, Grace's and Sophie's stories are ones of literal rehabilitation of bodies. They both work in physical rehabilitation units providing care for people with serious injuries. The body work of rehabilitating and caring for their patients can be taken as a metaphor for the *revaluation* of bodies of both patient and nurse, echoing Raghuram et al.'s (2011) discussion of geriatricians' interests in improving the lives of older adults.

Sophie, Mary, and Grace became friends in nursing school in Nairobi. Grace, the first of the three to immigrate, explains: "You tell your friends, and then those who are capable of coming – and those were the ones who were not committed, who did not have families, and it was easy for them to travel." Sophie followed Grace's advice and came to St. Louis in 1997. She lived with Grace during her first year in the city, as she got settled, started her bachelor's program and found work in a nursing home. For Mary, coming to St. Louis was an easy decision: "Because Grace was here. Grace and Sophie were here. That's why." Their transnational friendship facilitated their migration.

Skills transfer demonstrates the internationality of nursing, its variegated regulation, and its histories of colonialism. Though she did not come to the US on a temporary nurse work visa, Grace did immigrate just as Congress was debating a nurse-specific visa, which I will discuss in the next section. After finishing her BSN, which was her second nursing degree, she began working in urban hospitals in St. Louis, serving the same underserved populations these visa programs targeted. Her knowledge and skills opened the immigration door for her, but were limiting professionally, as only some of her credentials transferred.

American licensing departments did not accept her work experience or her diploma from a Kenyan nursing school, which was one of many established by the British. The intimacies of colonialism, international and subnational nursing regulations, localized gender expectations, and systemic racism all shape Grace's migration story. The global intimate workforce, therefore, brings greater attention to the scalar diversity of care, work, migration, and gender.

Related is the role of immigration intermediaries, such as recruitment companies. As with Grace, Mary, and Sophie, few of the nurses I interviewed used a formal recruitment agency to facilitate their migration to the US. Most used informal means. Glenda's sister had already immigrated, so she followed her to the Midwest. Grace's manager in Kenya overheard her talking about migrating and connected her to someone in the US. Reena followed her nurse husband to New York, and then found none of her credentials transferred from India. After coming to the US for political asylum from Nigeria, Helen decided to retrain as a nurse after her bachelor's biology degree did not transfer to pharmacy programs in the US. Olga had basic nursing training in the Soviet Union but trained as a registered nurse when she arrived in the US as a refugee in the early 1990s. Each of their stories points to the importance of accounting for the diversity of intimate experiences and relationship shaping nurses' migration patterns as well as their work and family experiences and needs. Approaching nurses' migration stories this way reveals the unique and varied combinations of institutions, informal networks, and casual interactions that guide migration stories as well as global uneven development and social reproductive labour.

A focus on home life reveals the transnational solidarity that the women and their families practise. All of the women are married, with Grace having met a Kenyan man in St. Louis, while Mary immigrated with her husband, and Sophie's husband joined her in St. Louis three years after she arrived.

Eventually they all had children. Though all are married, they are, in the everyday, functionally three single mothers, as all of their husbands have returned to Kenya, more or less full-time for work. Each woman sends money home to extended family still living in Kenya. And they've formed their own support group of sorts, helping each other with childcare and supporting each other as good friends do. For example, Mary explains how they get support at home from each other, as well as mothers and grandmothers who come to St. Louis to visit. She explains that "there's always a Kenyan mom who's here for a couple of months and they'll come in and help and babysit … all of us having young children at the same time has been supportive in the fact that you all can relate to what the other person is going through."

The combination of a regular rotation of mothers and grandmothers coming to town and a supportive friend network creates a complex care network. This is all the more important as the nuclear family's care is stretched transnationally with the husbands' return to Kenya.

The husbands' return to Kenya points to the complicated intersection of gender, work, and immigration. Mary, for example, is adamant that her husband is incompatible with gender expectations of life in the US. She explains,

> Each one of us really hasn't had a husband living in the home permanently because they are either working away from the home or they're in Kenya … I'm not dependent on my husband. I was always very independent. So it's continuation of life. He works. I work. He doesn't like it here. I couldn't possibly go back as yet because … the current lifestyle that I'm able to give my child is not the kind of lifestyle I would – and it's really not even about lifestyle; it's about access to opportunities.

At work, everyday differences reveal the challenges of fitting in to the US workplace. Sophie explained that the different names for medications and supplies repeatedly slowed her down when she began working in the US. American nurses called acetaminophen by its brand name or referred to gauze by its measurements, rather than simply as gauze. Communication barriers appear in all sorts of mundane ways.

Communication and other barriers are also filtered through the politics of race and immigration, revealing intimate, local, national, and international politics and histories. For example, Mary explained how patients receive her and how it has shaped her understanding of both herself and race.

MARY: In St. Louis, the minute I walk into a patient's room I think the expectation is that I will talk like an African American. Then I talk to them and I have a different accent – you can see from their face just the surprise and then the confusion because they don't know. They didn't expect that. The second thing is they are not sure what you are talking about (*laughs*) because they didn't expect the accent. And then the next question is "Where are you from?" So I never know, should I say I'm from exactly where in St. Louis or [ask] what do you mean "where am I from"?

CAITLIN: Do you notice that people treat you differently than African Americans?

MARY: Yes. I have to say yes. They're more friendly [to me]. They are more – more interested, and they wanna know about you. What do you think? They want to know what your opinion is on a couple of things … I think for the first time, for the first time in my life, I actually looked at myself through the colour of my skin … So I think thinking of myself in terms of colour has been the hardest thing for me.

Mary's experiences of race and racism when she speaks to patients highlight the global intimacy of nursing work and care migration. Her accent is a very intimate and personal part of her; it is also a reflection of global colonial and neocolonial legacies. Her accent also marks her as different in relation to whiteness and blackness in the St. Louis context, and her experiences working in this context also have influenced her own understanding of herself. Her patients' reactions to her when she speaks shows a schism between their expectations of who a nurse is and who an immigrant might be, as well as the local politics of a segregated city.

Caring for the Nation: Visas for the Global Intimate Workforce

The global intimate workforce foregrounds the intimacies of everyday work, life, and migration. It also necessitates macro-scale analyses alongside these intimate engagements. This section is a scalar shift, investigating the complexity of national policy-making and what spatial, scalar, and value assumptions are embedded in (and fought over) in creating nurse workforce legislation.

Furthermore, these were US programs whose effects ricocheted around the globe, as well as rested upon historical colonial policies.

The visa program drew nurses from around the world, and especially the Global South, magnifying pre-existing immigration flows from, for example, the US's former colony, the Philippines. In the mid-1980s, when the US was experiencing one of its most severe nurse shortages, federal lawmakers created the H1-A visa program. This was a non-immigrant visa program specifically for nurses to fill staffing vacancies in chronically understaffed facilities for up to seven years. Congress created the program with the recognition that the *nation* was in a care deficit. The visa program lasted six years and brought in thousands of nurses from around the world to the US. For the US health care system, these nurses filled important roles, providing care to patients and communities, predominantly in facilities that were in underserved areas.

The program ended in 1995, but Congress instituted a similar program in 1997, the H1-C visa.

This new program was much more restricted than the earlier H1-A visa and was created in a decade of US hostility towards immigrants overall. By mobilizing rhetoric of immigrant labour stealing jobs from Americans, Congress included such strict requirements that only fourteen facilities in the entire US qualified to hire H1-C nurses. Central to accomplishing this was the lawmakers' refusal to acknowledge a national care deficit.

In debating the new program, experts and members of Congress explicitly denied the US had a shortage of nurses, claiming that staffing issues were persistent only in many urban and rural areas (U.S. House of Representatives, 1999a). Politicians repeatedly invoked the idea that the nurse shortage of the 1980s was over, something the American Nursing Association echoed along with calls to protect American jobs. With the H1-C visa, politicians did not discuss the care deficit as a national issue but as a problem faced by "a few special acute care hospitals that, because of geography and demographics, have a very difficult time attracting health care professionals" (U.S. House of Representatives, 1999a). It is important to identify these special hospitals. As evidenced by (1) the areas facing the most chronic staffing shortages and (2) the limitations placed on which facilities qualified to use the H1-C visa program, these lawmakers were discussing hospitals that served poor populations. These included communities that were rural as well as ones within urban centres but which had been rendered remote because of a chronic underinvestment in health and social services (see Henry, 2015; Jackson & Henry, 2017). These hospitals predominantly served patients who were racialized, who used public insurance for the poor, and who often had lower health literacy. For nurses, these facilities had harder and more stressful working conditions and a patient population that required much more care than more affluent patient populations. Hospital representatives opined of the inability to attract and retain nurses to these underserved and under-resourced facilities, refraining from speaking to the scale of the shortage and focusing instead on their particular situations. Such a complicated scalar debate, and one dominated by a dismissal of a national shortage, greatly shaped the specifics of which employers could qualify for visas.

Importantly, this scalar debate also represented a struggle over the value of care in terms of where care is offered, who receives sufficient care, and who provides the care. Thus, the connections between location of hospital, patients receiving care, and nurses providing care spatializes the value of all three.

Particular Bodies and Spaces Are Devalued Simultaneously and in Relation to Each Other

When a fellow lawmaker asked Representative Bobby Rush, the sponsor of the legislation behind the H1-C visa, why, if American nurses would not work in these facilities, would an internationally trained nurse want to, Rush responded by saying, "Probably [internationally trained nurses] can earn more or do earn more working in American hospitals, and if this is an opportunity to work in a hospital in this Nation, then they would jump at that opportunity" (U.S. House of Representatives, 1999b). Rush's statement points to an understanding of immigrant nurses as "cheaper labour." This racialized exploitation of global uneven development is an easy way around the problem of nurse retention in underfunded facilities, as Rush continues, explaining that he was "not sure whether or not there is adequate political resolve … to force us to deal with additional incentives for nurses" to work in such hospitals (U.S. House of Representatives, 1999c). Devalued workers fill care deficits without the state, politicians, or health care administrators having to address root causes of the staffing issues in the first place: an overall devaluation and lack of respect for the important care work nurses provide.

Lawmakers denied a national crisis of care, instead referring to the nursing problem as a "spot shortage." Congresspersons' reframing of the nursing crisis – from a national crisis to one manifesting only in specific but numerous localities – is an important intersection of scales. The intimacies of racial politics and community dynamics interface with global political economic conditions that enable the uneven impacts of austerity governance. Federal lawmakers and policies obscure national care deficits; they depend on global legacies of colonialism; they devalue immigrant nurse labour to maintain a care deficit locally. Through the framework of the global intimate workforce, these nurse visa programs call for a rethinking of the geographies of care deficits as the very scale and spatiality of the shortage was in debate. This debate is also a contestation over the value of patient, nurse, and place.

Furthermore, approaching nurses as a global intimate workforce draws attention to not only how value is spatial and scalar. It also enables a richer understanding of federal US policies in relation to the policies outside of the US. For example, Blouin's (2005) analysis of the influx of nurses from Canada to the US in the 1990s demonstrated that NAFTA was very likely not the central driver of Canadian nurse emigration. Rather, as most Canadian nurses migrating to the US were coming from Ontario, the austerity policies of the Mike Harris administration implemented in the mid-to-late-1990s were just as or more important. As Harris's administration fired nurses and closed hospitals, Ontario nurses moved south for more potentially lucrative jobs in American hospitals. Therefore, not only was the H1-C visa being debated during a time when the US was hostile to immigrants, it was also a time of outright attacks on nurses and health care across the border in Ontario, the most populous Canadian province. A cross-border struggle over the value of nurses and patients was intensifying at this time.

Changing Care Needs for Aging Populations

As the aging of populations in the Global North presents new and changing care needs for individuals, families, populations, and nation states, nursing and other forms of intimate care work are receiving increased attention. In the US, proposed and enacted restrictions on and changes to immigration will potentially greatly impact elder care (Bailey, 2018; Jordan, 2018). Immigrants and workers of colour are disproportionately represented in elder care, particularly in home care and long-term care (Espinoza, 2017; Stone & Wiener, 2001). While the US has been suffering from a chronic nursing shortage (in varying degrees of severity over the past three decades), immigrant workers have often filled gaps in care needs. When former US President Donald Trump announced cancelling various immigrant programs, such as the temporary protected status for Haitians, many families, care recipients, and health care specialists were concerned about the impacts of cancelling such programs (Bailey, 2018). Though many of these programs are ultimately intact through legal challenges to Trump's cancellation and then through their extension under Biden's administration (Lind, 2021), concerns about access to care work remain for both care recipients and immigrants. Many immigrants earn their living by providing important care to elderly people who need it. Importantly, many of these workers are employed in lower ranks of nursing and care, including personal support workers, home care aides, nursing assistants, and licensed practical nursing. This growing sector of health and nursing care must be thought of in relation to, not in distinction from, registered nursing. Registered nurses (RNs) work most commonly in hospitals, while lower ranks of nurses and care workers find employment in clinics, long-term care, and home care. For this chapter, I approach each of these jobs as falling under the umbrella of nursing, with workers fulfilling different tasks, to different levels of respect and remuneration, but all doing nursing work.

This is not necessarily how these different jobs are understood within the nursing profession, which has, over the past 150 years, struggled to isolate registered nursing as the highest skilled, most powerful (and whitest) rank in the profession (Glenn, 1992; Hine, 1989; Reverby, 1987). For example, as nursing has professionalized, RNs have become responsible for less of the direct care work that once characterized their jobs. Instead, lower ranks of nurses and aids have picked up these tasks. For elder care, this means that while RNs do provide elder care, most of this intimate care work, which is one of the least respected sectors of health care, falls to those in the lower ranks of the profession. Employment sites support this, as hospitals are where most RNs work, while nurses in the lower ranks work in long-term care facilities and home care (Dyck et al., 2005; Stone & Wiener, 2001). Adding elder care to the analysis introduces a way to think about both the lower and upper ranks of health care, as well as introducing a more global understanding of the nursing hierarchy and a way of thinking across different types of care work.

Additionally, elder care forces an explicitly transnational analysis because of the scope of care needs. While I have focused on the US in my analysis, this demographic phenomenon is global in that populations around the world are getting older (Dugarova, 2017), which is bringing new challenges for meeting care needs and which is seeing migrant care workers fill this care deficit. Attention to the aging of the global population raises awareness not just of the need for elder care but also of how these care skills have been understood – as cheap labour and low-skilled work – which is part of a global stress put on social reproduction (Federici, 2004). Because of the diversity of intimate care needs, the elder care crisis sheds light on the diversity of not only care workers but also of the individuals, families, and populations in need of care. It also highlights the centrality of care to international relationships, such as economic agreements. For example, current debates over what a post-Brexit UK will look like frequently focus on care workers. As the UK relies heavily on care workers from around the world, and especially the European Union, the end of free movement to the UK threatens the reproduction of everyday life, placing thousands of people's care needs in jeopardy.

Raghuram et al. (2011) demonstrate how, by being pushed into relationships with each other, two marginalized groups in the UK – the elderly and migrant South Asian doctors – restore each other's dignity, bringing about a "reconfiguration" of each other's status, as bodies "out of place," as "both black bodies and older bodies are marked by disgust and a distancing from the desirable body" (p. 326). Yet, in producing and reproducing new practices of care for older people, understandings of both bodies are rewritten, reclaimed.

This is a useful theorization for the global intimate workforce in relation to the care needs of the aging US population. People need different degrees of support and care as they age. This means they need support from people with a range of skills. Yet much of their care falls to care workers whose work has been devalued through low wages and an underinvestment in resources. This underinvestment includes the lack of available spaces for older people, the dependence that many families have on cheap care labour, meeting the demands of intergenerational care needs, and the chronic shortage of workers willing to care for the elderly. As activist Ai-jen Poo (2015) explains, "People getting older is not a crisis; it's a blessing. We're living longer; the question is how we should live" (p. 3). That question of how to live is about the relation between care giver and recipient and between the intimacies of care and the macro-scale analyses of populations. In sum, it is the valuation of bodies, scales, and spaces. Perhaps, then, in the face of attacks on elder care, both in elders and in care, there is space for reclamation of both groups, similar to Raghuram et al.'s (2011) study. We only need to turn to the activism of domestic workers, home care workers, and nurses for inspiration. Over the past decade, these groups have launched some of the most public and successful labour rights campaigns (England, 2017; Poo, 2015; Semuels, 2014). And these campaigns have, importantly, foregrounded the relationship between worker and care recipient. This is both a strategic move and an honest one. In interactive service work, meaning is found in the relationship to each other.

Building the Global Intimate Workforce Framework

The global intimate workforce is a useful way to understand these three different cases concerning nursing work. On the surface, they are quite different, dealing with work-life balance, colonial legacies, contemporary immigration policies, the limitations of the state regulations, and changing care needs brought on by demographic shifts. Yet the global intimate workforce provides a framework for understanding not only commonalities across these cases (beyond simply nursing or migrant care work) but also for showing how each of these cases is part of a dynamic and complex workforce, shaped by myriad forces operating at multiple scales. It reveals how this workforce is part and parcel of the struggles over the valuation of intimate scales, spaces of uneven development, and relations of care.

First, the global intimate workforce helps to more fully understand the worker herself. She is a globetrotter taking care of people at their most intimate moments in very intimate ways. She provides emotional support for, and performs bodywork on, people when they are injured and sick; in other words, she is with them in very private, vulnerable, intimate moments and ways. She is also, as a migrant nurse, a pairing of global and intimate, embodying the experience of global migration. This understanding of the migrant nurse follows other scholars of the global and the intimate (Mountz & Hyndman, 2006), but develops the framework further by accounting for the relationality of the global intimate worker's labours: from her patients in the receiving country to her family she attends to near and far to the caring profession to which she belongs.

Second, the global intimate is apparent in the national (even global) health care system dependent on the global migration of nurses. Nurses in the US are a globally constituted workforce of intimate workers, who also experience for themselves the embodied experience of migration. American nursing, then, is a global intimate workforce in that it is composed of a workforce of nurses trained in the US as well as around the world, and the constitution of which is a product of the ongoing relationship between global history and the intimate work of social reproduction: the legacy of British and American colonialism in the Philippines, India, and much of Africa; the devaluation of social reproductive work; global political economic inequalities; the invisibilization of the dirty work of health care; and the global gender division of labour.[1]

By pairing migrant nurses' stories with multiple scales of regulation, new understandings of how nurses migrate and how global processes and intimacies shape their experiences emerge. Furthermore, by considering care and migration from a less linear approach, such as the lens that a global intimate workforce provides, we can see not only the transnational connections of care that global care chains foreground but also a richer and more nuanced analysis of political economic conditions that nurses migrate within, push against, and react to. The global intimate workforce brings colonial legacies, the history of nursing, and neoliberal austerity together with the patient's reaction to a nurse's accent and the inner-city hospital's chronic struggle to staff enough nurses, offering a more nuanced and holistic approach to analysing work, immigration, and care. Ultimately, these are questions of value, of which spaces, scales, bodies, and labour are valued and devalued. A global intimate workforce analysis attempts to bring these relations together, in the case of nurses, highlighting co-productive processes of care, othering, and the struggle for valuation.

NOTE

1 See Raghuram (2008) and Choy (2003) on colonialism; Duffy (2007, 2010), Glenn (1992, 2010), Federici (2004, 2012), Kingma (2006), and Nagar et al. (2002) on global political economic inequalities and the global gender division of labour; and Zuberi (2013), Weinberg (2003), and McDowell (2009) on the dirty work of health care.

REFERENCES

Bailey, M. (2018, March 26). As Trump targets immigrants, elderly brace to lose earegivers. *Kaiser Health News*. https://khn.org/news/trump-immigration-policies-put-immigran-caregivers-and-elderly-patients-at-risk/.

Blouin, C. (2005). NAFTA and the mobility of highly skilled workers: The case of Canadian nurses. *The Estey Centre Journal of International Law and Trade Policy*, *6*(1), 11–22. http://www.nsi-ins.ca/wp-content/uploads/2012/10/2005-NAFTA-and-the-Mobility-of-Highly-Skilled-Workers-The-Case-of-Canadian-Nurses.pdf.

Chang, K.A., & Ling, L.H.M. (2000). Globalization and its intimate other. In V. Marchand & A.S. Runyan (Eds.), *Gender and global restructuring: Sightings, sites, and resistance*. Routledge.

Choy, C.C. (2003). *Empire of care: Nursing and migration in Filipino American history*. Duke University Press.

Duffy, M. (2007). Doing the dirty work: Gender, race, and reproductive labor in historical perspective. *Gender & Society*, *21*(3), 313–36. https://doi.org/10.1177/0891243207300764.

Duffy, M. (2010). *Making care count: A century of gender, race, and paid care work*. Rutgers University Press.

Dugarova, E. (2017). *Ageing, older persons, and the 2030 agenda for sustainable development*. United Nations Development Program. https://www.un.org/development/desa/ageing/wp-content/uploads/sites/24/2017/07/UNDP_AARP_HelpAge_International_AgeingOlderpersons-and-2030-Agenda-2.pdf.

Dyck, I., Kontos, P., Angus, J., & McKeever, P. (2005). The home as a site for long-term care: Meanings and management of bodies and spaces. *Health & Place*, *11*(2), 173–85. https://doi.org/10.1016/j.healthplace.2004.06.001.

England, K. (2017). Home, domestic work and the state: The spatial politics of domestic workers' activism. *Critical Social Policy*, *37*(3), 367–85. https://doi.org/10.1177/0261018317695688.

Espinoza, R. (2017, June 20). New Study: Immigrants are the future of long-term care. *Huffington Post*. https://www.huffpost.com/entry/new-study-immigrants-are-the-future-of-long-term-care_b_5949191de4b04d8767077b98.

Federici, S. (2004). *Caliban and the witch*. Autonomedia.

Federici, S. (2012). *Revolution at point zero: Housework, reproduction, and feminist struggle*. PM Press.

Freeman, C. (2001). Is local: global as feminine: masculine? Rethinking the gender of globalization. *Signs*, *26*(4), 1007–37. https://doi.org/10.1086/495646.

Glenn, E.N. (1992). From servitude to service work: Historical continuities in the racial division of paid reproductive labor. *Signs*, *18*(1), 1–43. https://doi.org/10.1086/494777.

Glenn, E.N. (2010). *Forced to care: Coercion and caregiving in America*. Harvard University Press.

Henry, C. (2015). Hospital closures: The sociospatial restructuring of labor and health care. *Annals of the Association of American Geographers*, *105*(5), 1094–110. https://doi.org/10.1080/00045608.2015.1059169.

Hine, D.C. (1989). *Black women in white: Racial conflict and cooperation in the nursing profession, 1890–1950*. Indiana University Press.

Hochschild, A.R. (2000). Global care chains and emotional surplus value. In W. Hutton & A. Giddens (Eds.), *On the edge: Living with global capitalism* (pp. 130–46). Jonathan Cape.

Jackson, P., & Henry, C. (2017). The needs of the "other" global health: The case of remote area medical. In C. Herrick & D. Reubi (Eds.), *Global health and geographic imaginaries* (pp. 176–94). Routledge.

Jordan, M. (2018, April 6). When the elderly call for help, a "chain" Immigrant often answers. *The New York Times*. https://www.nytimes.com/2018/03/25/us/immigration-labor-trump.html.

Kingma, M. (2006). *Nurses on the move: Migration and the global health care economy*. Cornell University Press.

Kofman, E., & Raghuram, P. (2006). Gender and global labour migrations: Incorporating skilled workers. *Antipode*, *38*(2), 282–303. https://doi.org/10.1111/j.1467-8330.2006.00580.x.

Lawson, V.A. (2000). Arguments within geographies of movement: The theoretical potential of migrants' stories. *Progress in Human Geography*, *24*(2), 173–89. https://doi.org/10.1191/030913200672491184.

Lind, D. (2021, March 16). Biden opened temporary legal status to thousands of immigrants. Here's how they could end up trapped. *ProPublica*. https://www.propublica.org/article/biden-opened-temporary-legal-status-to-thousands-of-immigrants-heres-how-they-could-end-up-trapped.

McDowell, L. (2009). *Working bodies: Interactive service employment and workplace identities*. Wiley-Blackwell.

Mountz, M., & Hyndman, J. (2006). Feminist approaches to the global intimate. *Women's Studies Quarterly*, *34*(1/2), 446–63. https://www.jstor.org/stable/40004773.

Murphy, M.F. (2014). Global care chains, commodity chains, and the valuation of care: A theoretical discussion. *American International Journal of Social Science*, *3*(5), 191–9. https://www.aijssnet.com/journals/Vol_3_No_5_October_2014/19.pdf.

Nagar, R., Lawson, V., McDowell, L., & Hanson, S. (2002). Locating globalization: Feminist (re)readings of the subjects and spaces of globalization. *Economic Geography*, *78*(3), 257–84. https://doi.org/10.2307/4140810.

Parreñas, R.S. (2001). *Servants of globalization: Women, migration and domestic work*. Stanford University Press.

Parreñas, R.S. (2012). The reproductive labour of migrant workers. *Global Networks*, *12*(2), 269–75. https://doi.org/10.1111/j.1471-0374.2012.00351.x.

Poo, A. (2015). *The age of dignity: Preparing for the elder boom in a changing America*. The New Press.

Pratt, G., & Rosner, V. (2006). Introduction: The global and the intimate. *Women's Studies Quarterly*, *34*(1/2), 13–24. doi:10.7312/PRAT15448-INTRO.

Pratt, G., & Rosner, V. (Eds.). (2012). *The global and the intimate: Feminism in our time*. Columbia University Press.

Raghuram, P. (2008). Thinking UK's medical labour market transnationally. In J. Connell (Ed.), *A global health system: The international migration of health workers* (pp. 198–214). Routledge.

Raghuram, P. (2012). Global care, local configurations-challenges to conceptualizations of care. *Global Networks*, *12*(2), 155–74. https://doi.org/10.1111/j.1471-0374.2012.00345.x.

Raghuram, P., Bornat, J., & Henry, L. (2011). The co-marking of aged bodies and migrant bodies: Migrant workers' contributions to geriatric medicine in the UK. *Sociology of Health & Illness, 33*(2), 321–35. https://doi.org/10.1111/j.1467-9566.2010.01290.x.

Reddy, S.K. (2015). *Nursing & empire: Gendered labor and migration from India to the United States.* University of North Carolina Press..

Reverby, S. (1987). *Ordered to care: The dilemma of American nursing, 1850–1945.* Cambridge University Press.

Semuels, A. (2014, November 3). The little union that could. *The Atlantic.* http://www.theatlantic.com/business/archive/2014/11/the-little-union-that-could/382206/.

Stone, R.I., & Wiener, J.M. (2001). *Who will care for us? Addressing the long-term care workforce crisis.* The Urban Institute. http://www.urban.org/research/publication/who-will-care-us-addressing-long-term-care-workforce-crisis.

U.S. House of Representatives. (1999a). Nursing Relief for Disadvantaged Areas Act of 1999: Hearings on H.R. 2759, Before the Subcommittee on Immigration and Claims, Committee of the Judiciary, 105th Cong. 30 (1997) (statement of Ron Campbell of St. Bernard Hospital, Chicago, IL). http://commdocs.house.gov/committees/judiciary/hju53246.000/hju53246_0f.htm.

U.S. House of Representatives. (1999b). Nursing Relief for Disadvantaged Areas Act of 1999: Hearings on H.R. 2759, Before the Subcommittee on Immigration and Claims, Committee of the Judiciary, 105th Cong. 15 (1997) (statements of Rep Lamar Smith and Rep Bobby Rush, respectively). http://commdocs.house.gov/committees/judiciary/hju53246.000/hju53246_0f.htm.

U.S. House of Representatives. (1999c). Nursing Relief for Disadvantaged Areas Act of 1999: Hearings on H.R. 2759, Before the Subcommittee on Immigration and Claims, Committee of the Judiciary, 105th Cong. 30 (1997). http://commdocs.house.gov/committees/judiciary/hju53246.000/hju53246_0f.htm.

Weinberg, D.B. (2003). *Code green: Money-driven hospitals and the dismantling of nursing.* ILR Press.

Williams, F. (2010). Migration and care: Themes, concepts and challenges. *Social Policy and Society, 9*(03), 385–96. https://doi.org/10.1017/S1474746410000102.

Yeates, N. (2004). A dialogue with "global care chain" analysis: Nurse migration in the Irish context. *Feminist Review, 77*(1), 79–95. https://doi.org/10.1057/palgrave.fr.9400157.

Yeates, N. (2012). Global care chains: A state-of-the-art review and future directions in care transnationalization research. *Global Networks, 12*(2), 135–54. https://doi.org/10.1111/j.1471-0374.2012.00344.x.

Zuberi, D. (2013). *Cleaning up: How hospital outsourcing is hurting workers and endangering patients.* ILR Press

SECTION 3

Transnational Health Mobilities: Networks, Regulation, and Intermediaries

7 Networking through *Kafala*: Skilled Workers and Transnational Networks in the Governance of Health Care Migration in the Gulf

CRYSTAL A. ENNIS

Introduction

"We bestow our last hope in you." So signs off a collective memorandum from members of the Indian nursing community in the Sultanate of Oman to the Indian ambassador on 17 August 2016. The letter was seeking intervention after more than 300 expatriate medical professionals in Oman lost their jobs with the Ministry of Health (MOH) due to a contracting state budget and an intensification of workforce nationalization policies – known locally as Omanization. As part of the public sector, the MOH is usually considered the best choice for health care work in Oman. Yet with this series of dismissals, the MOH was reinterpreting its contractual obligations to pay end-of-service gratuities in full. According to the letter addressed to the Indian ambassador, there were large discrepancies between how certain individuals were having their contracts interpreted and honoured (or not), and how the state was interpreting civil service articles. As skilled migrants, nursing professionals are able to advocate on their own behalf, navigate complex legal and governance situations, and find appropriate networks and organizations to advance their interests. While their personal agency is empowered by their education level and place in global networks, it is also constrained by their lack of connection to elite networks from which knowledge workers or those with "expert" status may more easily benefit. The nursing occupation is situated in a high-skilled limbo, existing between nationalization processes, between expert and labour categories, and between financial remuneration levels. Examining this positional limbo offers valuable insights into how the lines of authority and governance interact around professions and, in this case, the avenues nurses may pursue to network through the employment sponsorship system and navigate the temporality and precarity of their work.

Nurses, doctors, and other medical professionals are in-demand occupations globally, no less than in the labour-hungry Gulf countries of the Arabian

Peninsula. Yet growing populations and an increasingly educated national workforce means there is more local supply of skilled labour. In Oman, the government has been responding to a boiling unemployment crisis among citizens with piecemeal reforms to calm the waters. In the ebb and flow of economic pressures, the government ramps up workforce nationalization enforcement in various sectors. This chapter demonstrates how intersections of economic conditions (oil price lows), nationalization, sponsorship systems, and sending-country selection and recruitment restrictions impacts the governance of migration corridors and how migrants navigate through these overlapping bodies and conditions. It illuminates how precarity lives alongside the sense of privilege and empowerment that education and skill level provide. Through this, the chapter shows how an examination of particular migration skill levels can help untangle the "multiplex" system (Ennis & Walton-Roberts, 2017) in which globalized migration transpires, and inform understandings of multilevel governance patterns from the bottom up. This chapter further explores how nurses navigate various overlapping realms of authority that govern their migration and their work. It argues that migrants themselves form a node in the migration governance complex and utilize networks and knowledge to migrate, to understand migration regimes, and to negotiate through workplace grievances. Building on the findings of Bal (2016) that show how migrant labour politics under different political regimes are shaped by the articulation of production politics with various modes of political participation, this chapter shows how nursing, situated as a high-skilled limbo occupation, generates different forms of advocacy and contestation. This chapter begins with a discussion of migration governance and the positionality of nurses within migrant work spaces. It then addresses the demography and structure of the health care labour market in Oman. This is followed by an overview of visa sponsorship systems in the region, known as *kafala*. Finally, it looks at the levels of governance and networks that exist in the migration governance space.

Governing Migration and Navigating High-Skilled Limbo

In the study of the governance of the migration of health care workers, debates about the nature of care work, the types of regulations that frame care migration and employment, the types of factors that encourage and restrict migration, and the actors involved in the migration process require consideration (Pittman, 2016; F. Williams, 2011; Yeates, 2012). Research shows that focusing on a particular profession, and the nature of that profession, is an important factor in explaining migration flows (Iredale, 2001, p. 20). Studies of migration governance in Gulf countries usually focus on low-skilled workers – particularly in the construction and domestic work sectors. More recently, other studies have focused on the category of expatriate "experts" – those workers who have been recruited to fill positions demanded by a shift towards the knowledge economy (Vora, 2015). These works underline how the intersection of nationality, ethnicity, gender, and perceptions of skill level and occupational value influence the governance of migration at various levels, and also how individuals within these categories negotiate their position and belonging (Ennis & Walton-Roberts, 2017; Ewers & Dicce, 2016; Vora, 2013). The nature of work production and governance of both labour and migration are intricately tied to the kind of workplace politics and procedures that emerge. These become visible through migrant workers' struggles for resolution of grievances over working, pay, visa, and living conditions. As Bal (2016) finds, "how these grievances transform into different forms of contention is a matter of contextual contingency" (p. 228). Focusing on nursing labour markets in Oman as a particular space of production allows us to explore the means and venues migrants exploit to navigate the migration process and work experience.

I consider migration governance a multilevel process that engages a multiplicity of state and non-state actors operating through the state, but also above, below, and beyond it. I use the term "multiplex" to capture the complexity of the governance space and its overlapping realms of authority and power of the governance space. The "migration multiplex system" refers to the overlap of multiple actors' (states, organizations, agencies, non-profits, and migrants) interests that influence how migration and labour market policies are formed (Ennis & Walton-Roberts, 2017). Migration, as an emerging realm of global governance, thus involves "a multitude of actors operating at a number of levels and in a range of spheres" (Yeates, 2002, p. 76). In addition to the influence of elite actors (states, regional organizations, corporate agents) on migration decision-making and processes, these governance spaces are also influenced by the "linkages between global politics and individuals' everyday spaces" (Onuki, 2007, p. 126). Such actors and forms of networks are worth examination within the multiplex migration governance system and yet receive little attention in this context. As Taylor (2016) argues, migrant networks are "small-scale forms of migration governance that are able to fill some of the gaps in the currently fragmented global migration governance system" (p. 351). These small-scale forms of migration governance facilitate nurses' ability to network through the migration process and various regulatory realms of *kafala* systems.

This chapter chooses the case of nursing labour markets in Oman to examine how the nurses network through these overlapping lines of authority and navigate the precarity of their work. In line with Vora's (2015) call for accounts of migration that are "more complete and complex, and thus more humanising" (p. 194), I focus on how the space of transnational nurses in Gulf migration results in articulations of migrant governance that both attempt to control but also respond to various forms of contestation. I conceptualize the nursing occupation as situated in a space of high-skilled limbo. The occupation is viewed, and therefore regulated and monitored, differently than other highly skilled work that contributes to "knowledge work," and is therefore granted expert status. And yet, nursing is separated from images of manual labour and exploitation often associated with the Gulf migration story. Importantly, nursing as a form of care work is a gendered occupation and, in the case of migration, often racialized (see chapter 5, Bourgeault et al. and chapter 14, Peralta in this volume). That is, nursing migrants to the Gulf are overwhelmingly female and from South and Southeast Asia. Nursing is also in-between in terms of salary packages. While remunerated higher than domestic and construction work, wage levels for nurses are far from the salaries of "expert" labour. Nursing's status as a form of gendered care and skilled reproductive labour, combined with these characteristics and external perceptions, influence governmental responses to nursing demands and grievances.[1] Like other occupations, nursing is also influenced by economic conditions, neo-liberal patterns of supply and demand, and evolving policy environments. In the case of Oman, nursing is an occupation both affected by a tightening fiscal situation and by workforce nationalization drives.

Demography and Private Sector Dilemmas in Nationalizing Health Care Work

Like most of the Omani economy, and especially in the private sector, health care provision is heavily dependent on foreign labour. As Oman tries to extract itself from dependence on oil and expatriate workers through economic diversification, it also uses a combination of education and more direct labour market interventions to facilitate jobs for citizens. Such labour force nationalization policies are known locally as "Omanization" and entail reserving a certain percentage of job positions for citizens. Successful implementation of these policies have been patchy and sector dependent. When oil prices are high, there is less overall incentive and demand to push the private sector to hire (more expensive) local labour. When oil prices are low, as during the revenue crunch following the oil price downturn at the end of 2014, Oman faces greater pressure to find jobs for its citizens. The health care sector, too, faces these stresses. As more Omanis graduate as nurses and other medical professionals, there is an expectation of employment. This has reverse implications for the migrant workforce who, when hearing the louder public calls for workforce nationalization, fear the loss of their jobs.

The rate of workforce nationalization in the health care sector is slow. As in other economic sectors, Omanis employed in health care prefer work in government-run institutions (see table 7.1). The percentage of Omani nurses at Ministry of Health (MOH) institutions increased from 49 per cent in 2003 to 59 per cent in 2015. While other public sector institutions also saw an increase, nearly doubling to 30 per cent, the private sector remained rather stagnant and

Table 7.1. Omanization of nurses across type of health care facility

Year	2003				2012				2015			
Category	MOH	Public Sector (Non-MOH)*	Private Sector	Total	MOH	Public Sector (Non-MOH)	Private Sector	Total	MOH	Public Sector (Non-MOH)	Private Sector	Total
Nurses	49%	17%	5%	44%	66%	26%	5%	54%	59%	30%	4%	49%

*Public Sector (Non-MOH) refers to those governmental health care providers not under the auspices of the Ministry of Health such as the Armed Forces medical services, the Royal Oman Police, the Royal Court, Petroleum Development Oman, and Sultan Qaboos University Hospital.
Source: Sultanate of Oman Ministry of Health (2014, p. 126; 2016, p. 41, translation by author).

even declined to 4 per cent. Perhaps peculiarly, the number of male general nurses in Oman's private sector (27.4 per cent) is quite high (see figure 7.1). In comparison, only 9.6 per cent and 11.4 per cent of registered nurses in the US and UK, respectively, are men (Landivar, 2013; R. Williams, 2017). The data on Omani nurses show that more men than women work in private sector establishments, which is a trend also seen in other sectors. Across fields, Omani women have a strong preference for employment in the public sector. Yet the nursing profession is still overwhelmingly female, with the private sector primarily expatriate women.

In fact, the overall Omanization rate in nursing declined between 2012 and 2015 by 5 per cent. This indicates rising international recruitment of nurses and difficulty recruiting and retaining local ones. The demand for foreign nurses seems unlikely to subside, especially in private sector institutions. As figure 7.1 shows, even during the first three years of the oil price crunch, the overall number of expatriate nurses in the private sector rose significantly. In particular, the number of Filipino and Indian nurses climbed while the number of Omanis declined. These data suggest two things. First, even in health care the private sector does not attract or retain Omanis and likewise seems reluctant to hire them. Moreover, the nurse supply from India and the Philippines is high and readily available while the demanded salary is lower and therefore more attractive to employers than Omani salaries. Second, the fear of sudden, sharp job loss due to Omanization has not become reality. There appears to be a steady, rising demand for foreign health care workers.

That said, migrant nurses face a variety of other formal and informal pressures that complicate their presence in Oman. With renewed youth protests in the fall of 2017 for jobs, the Omani government has been increasing pressure on private sector employers to make way for Omanis and abide by Omanization quotas. In the last few years, a series of moves indicate the intensity of these pressures for policymakers. The *Times of Oman* reported in May of 2018

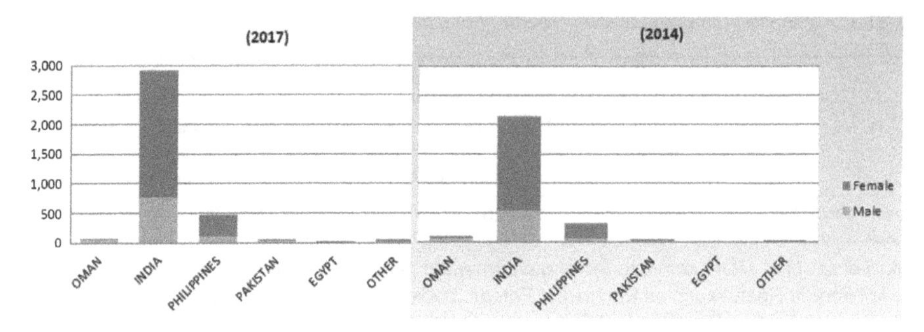

Figure 7.1. General nurses in the private sector by nationality (Oman)
Source: Based on data received from the Ministry of Manpower, Oman (2017).

that the Ministry of Manpower had penalized 161 companies for not hiring even one Omani citizen (Times News Service, 2018b). Health care workers were not among the eighty-seven professions that were banned from foreign recruitment for six months in 2018 ("Scores of Companies Fined," 2018), but the general intensifying atmosphere of nationalization heightens the sense of precarity among expatriate health care employees (Ennis & Walton-Roberts, 2017). Moreover, certain policy moves in the health sector give the impression of either tightening belts or of creating a less welcoming environment. On 10 May 2018, the deputy director general of administration and finance of Khoula Hospital under the Ministry of Health in Muscat, Oman, sent a circular to all directors, heads of department, and superintendents of the hospital with the subject line "legal opinion covering all female non-Omani employees" (Al-Hilali, 2018). It stipulated the following:

1. All female employees to transfer their children's visas to their husband's employer visa both in government and private sectors.
2. This should be done within three months of receiving this notice.
3. Henceforth tickets or financial compensation for tickets and free treatment shall not be provided to children under the MOH visa. (Al-Hilali, 2018)

A few days later, the local newspaper reported that all expatriate women in the medical sector were impacted and would no longer be allowed to serve as sponsors for their children (Times News Service, 2018c). The gendered nature of this new regulation is felt more heavily in a profession like nursing. Even though not all expatriate nurses earn enough above the minimum salary threshold to be eligible to bring their families to accompany them to Oman, the targeted nature of the legal change leaves a heavy, negative impression.

This move, seen alongside a variety of similar decisions, raises very real concerns. The denial of end-of-service gratuities to the nurses whose contracts were not renewed in 2016, mentioned in the opening of this chapter, is one case in point. When such decisions are announced and unfair practices implemented, how nurses respond and undertake action for their benefit reveals a lot about the makeup of labour market regulation and migration governance. Successes as well as defeats in pushing back against such changes also illustrate the power and limits of the occupation, and how the nature of the limited political space in the workplace "impact[s] the extent to which migrant labour regimes can be contested" (Bal, 2016, p. 229).

Contesting *Kafala* and Negotiating Migration, Temporariness, and Status in Gulf Migration

This section challenges the singularity of narratives around temporariness and the *kafala* system while pointing to the effects that interact with national regimes in migration governance. Migration governance in the Gulf is characterized by a sponsorship system known by the Arabic term *kafala*, derived from the Arabic root k-f-l – to guarantee or to care for. The singular noun *kafeel*, denotes the individual sponsor. Officially, migrants to the Gulf are temporary workers (Dito, 2015). When they enter the country, they do so as guest workers on fixed-term contracts and are obligated to leave the country on the end or termination of this contract.

The contractual nature of migrant work in Gulf countries is, in some ways, structurally similar to Western European guest worker programs of the mid-twentieth century. Such programs are created for the purpose of increasing the supply of labour without increasing the number of permanent residents in the country. Removing the possibility of permanence, in theory but rarely in practice, removes the need for governments to think about integration and alleviates concerns about demographic transformation. In most cases, guest workers do not have labour market mobility, meaning they cannot freely compete for other jobs in the labour market. Moreover, guest workers usually have legal restrictions placed upon them that prevent them from applying for permanent residence or naturalization, and from having their families join them. In practice, temporary migration often becomes a permanent feature of labour markets and societies (Castles, 2006; Dauvergne & Marsden, 2014). Host governments, as in the Western European experiences, were unable to restrict overstay and prevent worker and family settlement (see Afonso & Devitt, 2016; Castles, 1986; Ellermann, 2014).

While guest worker systems are not unique to the Gulf, the myth of temporariness persists. Forms of settlement have been occurring across the region even in the absence of paths to citizenship. Many migrants to the Gulf have spent their entire careers living and working in the Gulf. Sometimes their children, born in the region, do the same. As Martin (2006) observes, "there is nothing more permanent than temporary workers" (p. 4). In Oman, as in other Gulf countries, visas may always remain limited and temporary but persistently renewed. Where they are not renewed, there is a continuous and ready supply of labour from migrant-sending regions. Thus, the idea of temporariness is contradicted by the long-term experience of many individual migrants, and by the persistence of this structural feature of the economy.

Gulf Cooperation Council (GCC) countries do not have uniform migration control or labour regimes, despite the element of exoticism and similarity implied by using the Arabic word *kafala* within English-language discourse on the topic. Similarities certainly exist, however. For instance, foreign worker visas are associated with a sponsor – an individual citizen or a company. This aspect of *kafala*, which ties individuals to the particular job and sponsor, is what Gardner (2010) calls "the cornerstone of the orchestration of the extremely unequal relations between foreigners and citizens, and hence a cornerstone in the mechanics of structural violence" (p. 54). Although it is illegal in all Gulf countries, sponsors regularly confiscate the passports of low-skilled workers for "safe-keeping" (International Labour Organization [ILO] and Deutsche Gesellschaft für Internationale Zusammenarbeit [GIZ], 2015, p. 25). Higher-skilled migrants and migrants from wealthier countries are more likely to know their rights and resist such controls.

Differences are wide-ranging. For example, new regulations in the United Arab Emirates (UAE) offer long-term visa possibilities for knowledge workers (Duncan, 2018). In Qatar, expatriates of any skill level require an exit visa in order to leave the country for vacation, travel, or to return home. Receiving an exit visa requires sponsor permission, a stipulation heavily criticized by human rights organizations as one of the "shortcomings in Qatar's legal and regulatory framework … responsible for the abuse and exploitation of Qatar's non-citizen workforce" (Human Rights Watch [HRW], 2013). Reforms introduced in January 2017 were supposed to remove this stipulation, but were quickly rolled back in favour of introducing a grievance committee to hear complaints about sponsors who denied expatriates exit permits (Khatri, 2017). By September 2018, however, the exit visa system was removed for many workers, and by January 2020 it was abolished for all expatriate workers (HRW, 2020).[2] Bahrain's Labour Market Regulatory Authority (LMRA) manages entry and exit processes and employment transfers instead of private sponsors and licenses recruiters. Labour laws in Bahrain have also long applied to both citizens and foreigners, including the right to join trade unions (ILO and GIZ, 2015, pp. 23–4). There are trends to move away from individual sponsorship to larger, centralized, and regulated agencies, like the LMRA, or to a large employment agency model, as has been proposed in Saudi Arabia. *Kafala* shapes a lot of the discourse on migration to the Gulf and frames the legal codes. However, the shape of this system is not as cohesive as suggested in popular discourse, and the law and practices vary by state and often by skill class of labour. This diversity means recruitment agencies, international organizations, and transnational activist communities, among others, need to be constantly aware of evolving migration governance regimes and their shape and interaction with sending-state policies.

At one point, Oman was viewed as being ahead of the game on reforming its migration regime. Despite easing restrictions around the mobility of labour in the 2000s, Oman reintroduced restrictions over the last decade. For example, in 2014 the requirement for a non-objection certificate (NOC) was reinstituted, which requires expatriates to acquire a letter from their employer when they leave a job in order to move to another employer or be faced with a two-year visa ban (Kuttapan, 2016b). Some employers use this power and leverage to keep employees and prevent them from moving on to work for competitors. The rationale from the state is twofold. First, imposing difficulties on hiring foreign labour may encourage companies to hire citizens. Second, the NOC is supposed to help safeguard the investment (visa, relocation, and training costs) of small companies in foreign recruitment. In reality, these restrictions could force employees to put up with unpleasant work conditions and prevent their career progression and learning opportunities. Omani trade unionists have argued that the visa ban not only damages the private sector and worker productivity, but importantly hurts many expatriates by essentially keeping "a worker bonded to the company" (Al-Lawati, 2017).

Despite this, Oman has managed to maintain a friendlier image to expatriates than its GCC neighbours. The impressions of a long-term expatriate and migrant rights activist are illustrative. In a 2018 interview he insisted that Oman was the best for migrants in the region, despite its restrictive regulations:

> Bahrain, Qatar, and maybe the UAE are the best when it comes to [recent] regulatory changes to support migrant issues. Qatar and Bahrain have improved a lot. Bahrain has the LMRA and Qatar now has something similar. Based on the report of a colleague in Qatar at this last Global Compact for Migration meeting, things have been steadily improving … Oman, Kuwait, and Saudi Arabia are the worst [on regulatory change]. Oman brought back the NOC. In Kuwait, Kuwaitis won't even talk to migrants like they're people. Oman, even though maybe in regulation they went back when they were in front, it is still the best place. Maybe because of culture. Here everyone – nationals, government, people – talk to you and deal with you like you're a human. I really like this place. This is my home, and I don't want to move back. Really, if you follow the rules here, everything is fine. Even with all these horrible things I see, I feel this. (Personal interview, 27 March 2018)

As someone who regularly intervenes to rescue migrants in the most dire of circumstances, this activist's perspective may seem surprising. He feels the rules of the labour market and procedures for filing complaints are clear and can be used by migrants and citizens alike. In his view, the problem lies more in the difficulties of regulatory implementation and rights awareness. For instance, because it is an offence for an employer to hold an employee's passport in Oman, such cases will be ruled in favour of the expatriate. However, according to the activist, the court will simply tell the employer to return the passport to the employee without further investigation. The activist explained, "the Court usually rules in the favour of workers, but the problem is with execution. There is no execution. Who is responsible? The court gives it to the police who give it to someone else, who is it?" (personal interview, 27 March 2018). Individuals and migrant networks need to be aware of the rules and labour arbitration possibilities and follow up.

This underlines the importance of migrant communities and informed networks. Certainly regulations like the above needs to be examined within the state while keeping in mind that the factors that influence, and the contestation to, the practices of *kafala* are not bound within national borders. Thus, scholars should pay heed to migrant agency and empowerment in affecting their conditions across skill levels, as well as how migrants "cope and wrest autonomy within a patriarchal system of both the sending and receiving countries" (Mehta, 2017). As Vora (2015) asserts, transnational labour migrants should not be viewed "solely as trapped or duped, but as knowledgeable people with certain skills and networks that allow them to become transnational in the first place" (p. 194). In the case of health care migrant workers, going to work in the GCC may not simply be an employment option of last resort. Rather, it is often viewed as a way to improve training and skills, gain international experience, earn higher income to secure a better future for children and retirement, and the first step in a two-step migration process towards North America or Western Europe. This is a common pattern in the practice of nursing migration in the Gulf. While not offering the possibility for citizenship, the Gulf meets the desire for outward migration to improve family prospects and life conditions at home, and offers the prospect of upgrading experience to make skilled-worker applications to countries in Western Europe or North America more competitive (Percot, 2006, 2016).

It is not only the reasons for migration, but also the ways through which skilled workers like nurses engage networks at global, transnational, and local levels to increase their bargaining power in employer-employee relations that impacts how the *kafala* system and the governance of globalized migration in general functions in different spaces. As unpacked in the next section, the types of bottom-up awareness, forms of available assistance, and empowerment practices influence the process and the agency of the migrant. In this way, the skill-level and type of professional credentials a migrant has influence the avenues and practices for networking through the sponsorship regime within the migration multiplex system.

Networks and Governance in Oman's Health Care Labour Market

Through examining the nursing labour market in Oman, we can obtain a view of how migrants in a particular sector and skill level of migration negotiate through the multiple levels of governance and regulations that affect migration and employment. Migrant nurses in Oman, as educated professionals, are informed, technologically savvy, and connected to family members, to colleagues and nurses working at home and abroad, to embassies and political lobby groups, to domestic civil society groups in their home states, and to migrant support volunteers and activists in their host state, home state, and abroad. The way they make connections and form networks, and the means through which they engage these networks to ensure they are better informed and have access to legal recourse or mechanisms to contest workplace decisions, all form part of the governance multiplex system, operating from below but interacting with other, more formal mechanisms of regulation and governance. Indeed, migrant networks function as a form of governance and "exist within the broader framework of global migration governance, a fragmented system" (Taylor, 2016, p. 353).

Host and home countries are the starting points in global migration governance. The regulations in host states have a direct impact on the hiring, employment, and dismissal practices affecting migrants within their borders. In the case of nurses, the Ministry of Health is a central regulating authority in addition to the Ministry of Manpower, the Ministry of Interior, and the Royal Oman Police – each regulating different aspects of entry and sponsorship rules across occupations. Home countries often have policies or practices that facilitate the export of workers and regulate recruitment, hiring, health, and wage requirements. Increasingly, home countries are also involved in labour market regulation within host states, in some cases determining wage levels for particular skill categories and making demands around workplace conditions. Such practices are especially common from the Philippines, with its labour export policy, active embassies, and the Philippines Overseas Labour Offices (Ennis & Walton-Roberts, 2017; Solomon, 2009).

Host states have been increasingly active in responding to international critique around worker treatment. News from the Oman Human Rights Commission (OHRC) is illustrative. While established in 2008, the OHRC only became active in the last few years. Alongside experts from a variety of government departments, the OHRC recently formed 200 committees focused on workers' rights in order to resolve pay and entitlement issues. In an interview in April

2018, Dr. Obaid Al Shaqsi, the secretary general of the commission, stated that any workers with grievances can approach the OHRC and have their concerns dealt with quickly. He continued:

> Human rights deal with individuals, regardless of creed, ethnic group, language or colour, or any other category that human beings choose to identify themselves ... It is very important to ensure that these rights are safeguarded, and there are national, regional and international bodies that ensure that these rights are really observed. (Times News Service, 2018a)

Host states are sensitive to pressures from the international community, transnational forums, and global governance institutions, but also to pressures from below. The latter have been increasingly interacting with multiple levels of governance to make their grievances heard.

Global institutions like the International Organization for Migration (IOM) and the International Labour Organization (ILO) are becoming louder in the global migration governance space. Processes like those leading to the signing of the Global Compact for Migration transpire, and include rounds of consultation in various regions with national, regional, and civil society actors (Koser, 2010; Phillips & Mieres, 2015). Moreover, individuals in both migrant sending and receiving countries are engaged in workshops, forums, training sessions, and consultative processes with these different organizations. The migration governance multiplex also includes regional organizations like the Association of Southeast Asian Nations and (to a lesser extent) the South Asian Association for Regional Cooperation, which coordinate efforts for unified approaches to migration governance within their organizations and as collectives of labour attachés within host countries (Ennis & Walton-Roberts, 2017). Meanwhile, transnational platforms like the Colombo Process and the Abu Dhabi Dialogue are designed to discuss issues around *kafala* systems and labour mobility across Asian countries.

State-to-state interactions also take the form of memorandums of understanding and various agreements on populations flows. As Koser (2010) notes, the "legal and normative framework affecting international migrants cannot be found in a single document, but is derived from customary law, a variety of binding global and regional legal instruments, nonbinding agreements, and policy understandings reached by states at the global and regional level" (p. 301). This multiplex space of competing regulatory authorities results in a variety of forums to advocate for better migration governance but has limited implementation power when buy-in from all actors, and especially states, is not secured. Some view this combination of national policies; bilateral, regional, and global dialogues; international organizations; and transnational bodies and legal frameworks a type of "soft governance" where the "sum of the parts had not resulted in coherent global governance" (Marchi, 2010, p. 326).

A significant space where national *kafala* systems intersect with transnational systems are with recruitment agencies in migrant sending and receiving states, and with migrant home country governments in point of origin and in embassies abroad. The supply of nurses in sending countries like India and the Philippines outpaces the demand in the Gulf. This imbalance means that migrants are vulnerable to abuses in the recruitment process on both sides, including around contractual promises about wages and working conditions and excessive recruitment fees (International Labour Organization, 2016, p. 3). Indeed, economic incentives create distortions that facilitate and reinforce structures of exploitation. Sending countries continue to provide workers and, in cases like the Philippines, encourage the "export" of migrants. The Indian government claims not to promote the export of workers, but neither does it aim to discourage it. In fact, it relies on it. Moreover, recruiters contribute by facilitating migration. Subagents and unofficial agents in many cases cause and deepen the exploitation by the fees, debts, and other burdens levied (often illegally) on emigrants. These gaps in migration governance are often filled by migrant networks and information-sharing practices through word-of-mouth and also online forums and social media.

Skilled health workers also have to navigate bodies that verify and facilitate professional credentials and their transfer across borders (see van Riemsdijk, chapter 10 in this volume). States in India, for example, validate nursing degrees and credentials before emigration. In most states, educational certificates should be attested at the Ministry of Human Resources and Development. In Kerala, NORKA Roots, a state government agency for emigrant welfare matters, attests certificates. For the second stage, certificates should be attested at the Ministry of External Affairs, and a third, at the embassy of the host country. If a nurse was educated in the state in which she/he is migrating from, it is not difficult to receive the first stage of verification, but it can become more difficult if one has to travel back to their home state to attest the certificate. Because both India and Oman are signatories to the Hague Apostille Convention, which abolished the requirement for legalization for foreign public documents, the process is easier and acceptance of documents that bear this seal is smooth and eliminates the multiple stages. Among GCC countries, only Oman and Bahrain have signed this treaty (Hague Conference on Private International Law [HCCH], 2019). Oman recognizes Indian medical training, however, all health workers to Oman must also pass the Prometric examination.[3]

Attestation and examination processes remain ambiguous to many aspiring migrants and it is common to reach out to others who have gone through the process for guidance. Indeed, there is a general expectation that migrants will help future migrants, and this is realized in practice. Individuals often recommend their family members and friends for positions and readily provide advice to prospective migrants at home. Migrants are viewed as having a certain amount of social capital for prospective migrants, with the expectation that they utilize this capital to help facilitate chain migration as well as to help ease the transition of newcomers to a new work environment (Shah & Menon, 1999; Taylor, 2016, pp. 352–3). In the host country, the reliance on networks and social connections continues, with migrants helping each other "learn geographies, languages, laws and residency processes. They also develop social and affective ties to others, not only from the same nationality, and these ties impact their sense of identity and belonging" (Vora, 2015, p. 195). Most politicization remains connected to the home state. However, migration experiences can instigate political activism in host states as well. This politicization sometimes instigates membership in community associations and transnational human rights and labour rights networks.

Social media including Facebook and WhatsApp, among others, provide ways to stay connected to contacts at home, but also a means to join groups and share information about working in certain countries, at certain institutions, or under certain sponsors. WhatsApp, and its end-to-end encryption, is particularly popular, but other messenger services like Facebook Messenger, Viber, and imo are also used regularly. Multiple online forums exist to share information. Some online social media forums are organized along professional and community lines (e.g., Facebook groups like Filipino Nurses in Oman and Indian Nurses in Oman), while others are organized by sector or profession, or by affiliation with certain linguistic or cultural groups. Some online forums cross these lines. International expatriate networking groups like Expat.com provide an English-language forum for expatriates from and in any country to set up forums and discuss issues. Here one can see nurses and other medical professionals set up threads under the "Oman Forum" (Oman Forum, n.d.). While some members post requests for jobs, there are also numerous queries from prospective and in-country migrants with questions about procedures, norms, and rules. Other members are ready to provide their opinions and answers. (Citations from "Oman Forum" posts reflect their authors' original spelling and punctuation.)

Inquiries about securing health care jobs are often met with advice and sometimes specific advertisements. For example: "I will advice you not to follow any recruitment agency, but contact & apply online to various Private hospitals (e.g-Badr-al sama, stracare, kims, Apollo etc.) of Muscat, Oman … AND Never prefer to work in Interior/Desert part of Oman. Good Luck" (drabhay, 2012). Sometimes nurses ask about particular hospitals:

> Kindly please help me in obtaining information regarding ROP [Royal Oman Police] Hospital ... I applied as staff nurse ER. I would like to know anything that may help me in deciding if im gonna accept the offer or not. Is it a good hospital? Is it big, how many bed capacity? Is the accommodation for Male Nurses are good? What is the acceptable range of offer for nurses? (danjoe03, 2012)

The first response exclaimed, "This is one of the best hospital in Oman. Once you get offer, then you should ask with firm details" (bushrcom7, 2012). It is especially common to ask about salary offers, to check whether received offers are within a reasonable range, and whether it suits the cost of living. "Hi there, I got an offer from a hospital in Oman. May I ask of an average nurse salary in a government hospital in Oman? Thanks!" (rona88, 2018).

Asking questions seems to form an important part of the decision-making process. One Filipino member writes, "Hello, i would like to ask opinions, it is okay and safe in Oman as private Nurse. I have a job offer by my agency but then Im still in doubt if im going accept the offer. thanks" (emilyvon, 2018). The profile associated with the first response states the member is a Pakistani who worked in Saudi Arabia before moving to Oman. "In most of cases it is safe," he answered, "but still do some research for your immediate Employer. Also you should completely know your rights & point of contacts while working here" (Asif_Hassan, 2018). An Indian national chimed in, "Working in Oman is very safe. Safer than any other GCC nation," and then noted that he or she was looking for a new job and seeking assistance (ravibabukondala, 2018).

Pay issues and confusion around rules like the NOC come up repeatedly. An Indian nurse asks, "[C]an i complain in MOH regarding payment issue?" after suggesting that the private hospital she had just transferred to seemed notorious for delays in payment and refused to give NOCs when employees quit and thereby broke their contract (pinky8209, 2017). A quick answer from another Indian expatriate: "Yes indeed. If you have worked and have not been paid for your work, then you have every right as an employee to complain to the authorities concerned. You can also file an online complaint against your faulting employer in the Ministry of Manpower's website" (Sumitran, 2017). How to obtain an NOC, and how to ease the transfer regulations to a new sponsor form common answers. Often these are closely connected to reality, with members sharing specific news articles or labour regulations to clarify issues. Sometimes forum posts represent fears, hearsay, and complaints about a system that can seem ambiguous and difficult to navigate. Through these forums and social networking groups, migrants connect with each other, sharing important information and at times connecting in person.

Such online engagement also facilitates the connection of individual migrants with local and transnational migrant and labour rights advocacy networks. Individuals may join these networks and represent issues at global meetings like those convened by the Migrant Forum for Asia or the Global Compact for Migration consultations, or they may simply inform the process through their information sharing over and across these networks. Migrants may learn about community associations to join in the host states, and they often remain tied to civil society organizations in their home state as well. Since formal associations and civil society organizations require governmental approval in Oman, networking and activism generally occur across spaces or in quieter and often informal ways. This underlines other research that illustrates how the forms of activism among migrant communities take shape and evolve are connected to the type of political regime (Bal, 2016). While labour strikes do occur, they often come at a high cost. Civil society actors on the ground in the Gulf usually advise quieter approaches, pursuing all legal mechanisms as a first step.

In the case involving 300 nurses, introduced at the beginning of this chapter, the nurses referred to their contracts and consulted the legal stipulations of the civil service law. Nurses from Kerala joined together and used this information to write an informed letter to the Indian ambassador. They also consulted the Indian Social Club and volunteers who support migrant rights in Oman. The MOH told these nurses they would receive twelve-year gratuity payments when they lost their jobs. Even those who had spent more than twenty-five years in Oman and expected the gratuity to reflect this were initially denied. The MOH appeared to be reinterpreting contracts in light of government cutbacks and nationalization pressures. In September of 2016, the Indian ambassador met with the MOH officials on the matter and reported that they had promised to review the case (Kuttapan, 2016a). Nurses also approached migrant advocates who wrote to the chief minister of Kerala, and to Sushma Swaraj, the minister of external affairs in India, known for her Twitter engagement on expatriate issues. Activists spoken to in the context of this research reported that the case did not meet a collectively favourable outcome (16 May 2018; 15 May 2018; 27 March 2018). Many nurses ultimately gave up and did not file a legal case. Some found their original paperwork, made a claim, and secured their full end-of-service benefits. Some nurses left the country. Others moved on to other jobs. According to one interlocutor, "Many times our activism and interventions and intermediation with the embassy or officials helps and solves problems. Non-payment of wages cases get solved. But sometimes no. For instance this case" (personal interview, 2018). Thus, these informal networks, while providing assistance and essential support, do not always meet favourable outcomes as they attempt to plug the governance and oversight gaps around migration issues. Rather than a unified outcome, this combination of resistance and acceptance, of successfully securing new jobs or organizing procedures of return, shows how contention and resignation can coexist and oscillate in production politics, and how it influences strategies of navigating through the sponsorship system.

The ways migrants network through this fragmented global governance system is therefore a significant part of the story. Applying a multilevel governance and transnational lens is helpful because the governance of migration and the issues migrants confront occur at both origin and destination countries (Piper, 2009, p. 218) and in the corridors that cross them simultaneously. Migrant networks emerge as instruments of migration governance. The "simultaneity afforded by the space of flows" (Taylor, 2016, p. 352) in migrant lives and the networks in which they are embedded offers the ability to occupy multiple spaces in a transnational migration realm concurrently.

Conclusion

This chapter contributes to scholarly attempts at untangling the multiplex global migration system by examining how political and economic factors and layers of governance interact with a particular profession. Focusing on nursing migration to Oman, a high-skilled limbo occupation, we can begin to unpack the multiple, overlapping realms of authority alongside the ways nurses promote their welfare by mobilizing various sociopolitical techniques that seek to improve their individual and collective welfare as one transnational socio-economic class. Migrants in high-skilled limbo occupations can examine regulations that concern them, or consult individuals about them, and have better chances of drawing on local (home and host country), transnational, and global actors to inform their decisions, empower their bargaining positions, and intervene in their grievances. Their access to technology and ability to read and engage in multiple languages improves their access to information. They can use their networks and social media to become better informed, look for solutions to difficult legal situations, connect with other migrants, and support their transitions to new places of work.

Indeed, the governance of legally temporary (while in practice permanent) migration in the Gulf is shaped by global, transnational, national, and local dynamics. *Kafala* is not a system that emerges and functions in a vacuum, but is rather identified, constructed, and re-framed according to various discourses and in response to evolving sources and structures of power. Skill-level, profession, country of origin, and gender are all embedded in this process. Precarity can coexist with the empowerment offered by education and skill. Nursing migration in the Gulf produces different forms of contestation that, while contextually specific insofar as skilled health workers must navigate particular local sponsorship rules, extends across transnational and global advocacy spaces. Migrants play a role in the migration governance complex and utilize networks and knowledge structures to migrate, to understand migration regimes, and to negotiate through workplace dilemmas.

ACKNOWLEDGMENTS

My deepest thanks to the nurses, activists, scholars, and other contacts who shared their knowledge, expertise, and experiences with me, informing the shape of this chapter. Earlier drafts of this chapter were presented at the Global Migration, Gender, and Professional Credentials workshop in Waterloo, Ontario (25–26 May 2017), and at the Leiden Interdisciplinary Migration

Seminar (31 May 2018). I am grateful for the feedback from the participants of these events, and especially to Salvador Santino Regilme Jr. and Margaret Walton-Roberts for their detailed comments that helped me sharpen the draft.

NOTES

1 To read more about skilled reproductive sectors, see the chapter "Skills and Social Reproductive Work" in Kofman and Raghuram (2015, pp. 100–27).
2 Despite the removal of the exit visa requirement, domestic workers must give their employers seventy-two hours' notice prior to departure. This is supposedly to ensure workers have time to receive due financial benefits prior to leaving, but raises concerns about the potential of abuse and misinterpretation of the notice period (HRW, 2020).
3 Information on procedures shared via correspondence with officials at NORKA Roots in 2018.

REFERENCES

Afonso, A., & Devitt, C. (2016). Comparative political economy and international migration. *Socio-Economic Review 14*(3), 591–613. https://doi.org/10.1093/ser/mww026.

Al-Hilali, H.H. (2018, May 10). Circular: Legal opinion covering all female non-Omani employees. [Unpublished document].

Al-Lawati, H.S. (2017, January 16). Minister of Manpower backs 2-year visa ban in Oman. *Times of Oman*. http://timesofoman.com/article/100707.

Asif_Hassan. (2018, January 7). *Work as private nurse is it safe?* [Online forum post]. Expat.com. https://www.expat.com/forum/viewtopic.php?id=744117.

Bal, C.S. (2016). *Production politics and migrant labour regimes: Guest workers in Asia and the Gulf*. London: Palgrave Macmillan. www.palgrave.com/gp/book/9781137548580.

bushrcom7. (2012, October 15). *Nurse for Royal Oman Police Hospital* [Online forum post]. Expat.com. https://www.expat.com/forum/viewtopic.php?id=204354.

Castles, S. (1986). The guest-worker in Western Europe – an obituary. *International Migration Review 20*(4), 761–78. https://doi.org/10.2307/2545735.

Castles, S. (2006). Guestworkers in Europe: A resurrection? *International Migration Review, 40*(4), 741–66. https://doi.org/10.1111/j.1747-7379.2006.00042.x.

danjoe03. (2012, October 14). *Nurse for Royal Oman Police Hospital* [Online forum post]. Expat.com. https://www.expat.com/forum/viewtopic.php?id=204354.

Dauvergne, C., & Marsden, S. (2014). The ideology of temporary labour migration in the post-global era. *Citizenship Studies, 18*(2), 224–42. https://doi.org/10.1080/13621025.2014.886441.

Dito, M. (2015). Kafala: Foundations of migrant exclusion in GCC labour markets. In A. Khalaf, O. AlShehabi, & A. Hanieh (Eds.), *Transit states: Labour, migration & citizenship in the Gulf* (pp. 79–100). Pluto Press.
drabhay. (2012, September 26). *Doctor jobs in Oman* [Online forum post]. Expat.com. https://www.expat.com/forum/viewtopic.php?id=200164.
Duncan, G. (2018, May 21). New residency visa rules "start of a new era" for UAE's professional workforce. *The National*. https://www.thenational.ae/uae/new-residency-visa-rules-start-of-a-new-era-for-uae-s-professional-workforce-1.732676.
Ellermann, A. (2014). Do policy legacies matter? Past and present guest worker recruitment in Germany. *Journal of Ethnic and Migration Studies*, *41*(8), 1235–53. https://doi.org/10.1080/1369183X.2014.984667.
emilyvon. (2018, January 5). *Work as private nurse is it safe?* [Online forum post]. Expat.com. https://www.expat.com/forum/viewtopic.php?id=744117.
Ennis, C.A., & Walton-Roberts, M. (2017). Labour market regulation as global social policy: The case of nursing labour markets in Oman. *Global Social Policy*, *18*(2), 169–88. https://doi.org/10.1177/1468018117737990.
Ewers, M.C., & Dicce, R. (2016). Expatriate labour markets in rapidly globalising cities: Reproducing the migrant division of labour in Abu Dhabi and Dubai. *Journal of Ethnic and Migration Studies*, *42*(15), 2439–58. https://doi.org/10.1080/1369183X.2016.1175926.
Gardner, A.M. (2010). *City of strangers: Gulf migration and the Indian community in Bahrain*. Cornell University Press.
Hague Conference on Private International Law. (2019). *12: Convention of 5 October 1961 Abolishing the Requirement of Legalisation for Foreign Public Documents*. https://www.hcch.net/en/instruments/conventions/status-table/?cid=41.
Human Rights Watch. (2013). Qatar: Abolish exit visas for migrant workers. https://www.hrw.org/news/2013/05/30/qatar-abolish-exit-visas-migrant-workers.
Human Rights Watch. (2020). Qatar: End of abusive exit permits for most migrant workers. https://www.hrw.org/news/2020/01/20/qatar-end-abusive-exit-permits-most- migrant-workers.
International Labour Organization. (2016). Background note for the information session: "Fair migration with a focus on recruitment." https://www.ilo.org/wcmsp5/groups/public/---ed_norm/---relconf/documents/meetingdocument/wcms_534146.pdf.
International Labour Organization and Deutsche Gesellschaft für Internationale Zusammenarbeit. (2015). *Labour market trends analysis and labour migration from South Asia to Gulf Cooperation Council countries, India and Malaysia*. Eschborn and Geneva: Deutsche Gesellschaft für Internationale Zusammenarbeit GmbH and International Labour Organization. http://www.ilo.org/wcmsp5/groups/public/---ed_protect/---protrav/---migrant/documents/publication/wcms_378239.pdf.
Iredale, R. (2001). The migration of professionals: Theories and typologies. *International Migration*, *39*(5), 7–26. https://doi.org/10.1111/1468-2435.00169.

Khatri, S.S. (2017, January 5). New Qatar law amended to keep employers in charge of exit permit. *Doha News*. https://dohanews.co/new-qatar-law-amended-to-remove-moi-exit-permi-procedure/.

Kofman, E., & Raghuram, P. (2015). *Gendered migrations and global social reproduction*. Palgrave Macmillan.

Koser, K. (2010). Introduction: International migration and global governance. *Global Governance, 16*(3), 301–15. http://dx.doi.org/10.1163/19426720-01603001.

Kuttapan, R. (2016a, September 6). Oman health: Ministry assures review of nurses' gratuity issue, says Indian envoy. *Times of Oman*. http://timesofoman.com/article/91987.

Kuttapan, R. (2016b, November 30). Thumbs up for NOC compromise in Oman. *Times of Oman*. http://timesofoman.com/article/97693.

Landivar, L.C. (2013). *Men in nursing occupations*. United States Census Bureau. https://www.census.gov/content/dam/Census/library/working-papers/2013/acs/2013_Landivar_02.pdf.

Marchi, S. (2010). Global governance: Migration's next frontier. *Global Governance, 16*(3), 323–9. https://doi.org/10.1163/19426720-01603003.

Martin, P. (2006, June 28–30). *Managing labour migration: Temporary worker programmes for the 21st century* [Paper presentation]. International Symposium on International Migration and Development, Turin, Italy. https://www.un.org/en/development/desa/population/events/pdf/other/turin/P07_Martin.pdf.

Mehta, S.R. (2017). Contesting victim narratives: Indian women domestic workers in Oman. *Migration and Development, 6*(3), 395–411. https://doi.org/10.1080/21632324.2017.1303065.

Oman Forum. (n.d.). Expat.com. http://www.expat.com/forum/viewforum.php?id=420.

Onuki, H. (2007). Migration workers as political subjects: Globalization-as-practices, everyday spaces, and global labour migrations. *Refuge, 24*(2), 125–34. http://dx.doi.org/10.25071/1920-7336.21390.

Percot, M. (2006). Indian nurses in the Gulf: Two generations of female migration. *South Asia Research, 26*(1), 41–62. https://doi.org/10.1177/0262728006063198.

Percot, M. (2016). Choosing a profession in order to leave: Migration of Malayali nurses to the Gulf countries. In P.C. Jain & G.Z. Oommen (Eds.), *South Asian migration to Gulf countries: History, policies, development* (pp. 247–63). Routledge.

Phillips, N., & Mieres, F. (2015). The governance of forced labour in the global economy. *Globalizations, 12*(2), 244–60. https://doi.org/10.1080/14747731.2014.932507.

pinky8209. (2017, March 26). *Changing job before completion of contract* [Online forum post]. Expat.com. https://www.expat.com/forum/viewtopic.php?id=660969.

Piper, N. (2009). Temporary migration and political remittances: The role of organisational networks in the transnationalisation of human rights. *European*

Journal of East Asian Studies, 8(2), 215–43. https://doi.org/10.1163/156805809X12553326569678.

Pittman, P. (2016). Alternative approaches to the governance of transnational labor recruitment. *International Migration Review, 50*(2), 269–314. https://doi.org/10.1111/imre.12164.

ravibabukondala. (2018, February 20). *Work as private nurse is it safe?* [Online forum post]. Expat.com. https://www.expat.com/forum/viewtopic.php?id=744117.

rona88. (2018, April 2). *Average salary for nurses* [Online forum post]. Expat.com. https://www.expat.com/forum/viewtopic.php?id=771405.

Scores of companies fined for violating Oman's expat work ban. (2018, May 15). *Arab News*. http://www.arabnews.com/node/1302966/business-economy.

Shah, N.M., & Menon, I. (1999). Chain migration through the social network: Experience of labour migrants in Kuwait. *International Migration, 37*(2), 361–82. https://doi.org/10.1111/1468-2435.00076.

Solomon, M.S. (2009). State-led migration, democratic legitimacy, and deterritorialization: The Philippines' labour export model. *European Journal of East Asian Studies, 8*(2), 275–300. https://doi.org/10.1163/156805809X12553326569759.

Sultanate of Oman Ministry of Health. (2014). *Health vision 2050: The main document* (1st ed.). https://www.moh.gov.om/documents/16506/119833/Health+Vision+2050/7b6f40f3-8f93-4397-9fde-34e04026b829.

Sultanate of Oman Ministry of Health. (2016). *The ninth five-year plan for health development (2016-2020)* [in Arabic]. https://www.moh.gov.om/documents/16506/0/The+Ninth+Five+Year+Plan/622b8a37-5ac9-1740-76c8-25d3010fd6c9.

Sumitran. (2017, March 27). *Changing job before completion of contract* [Online forum post]. Expat.com. https://www.expat.com/forum/viewtopic.php?id=660969.

Taylor, S.R. (2016). The role of migrant networks in global migration governance and development. *Migration and Development, 5*(3), 351–60. https://doi.org/10.1080/21632324.2015.1068504.

Times News Service. (2018a, April 28). You have the right to be treated fairly: Human Rights Commission. *Times of Oman*. http://timesofoman.com/article/133056/Oman/You-have-the-right-to-be-treated-fairly-Human-Rights-Commission.

Times News Service. (2018b, May 14). 161 companies penalised for violating Omanisation law. *Times of Oman*. http://timesofoman.com/article/134058.

Times News Service. (2018c, May 15). Expat women in Oman's medical sector can no longer sponsor their children. *Times of Oman*. http://timesofoman.com/article/134160/Oman/Expat-women-in-Omans-medical-sector-can-no-longer-sponsor-their-children.

Vora, N. (2013). *Impossible citizens: Dubai's Indian diaspora*. Duke University Press.

Vora, N. (2015). Expat/Expert camps: Redefining "labour" within Gulf migration. In A. Khalaf, O. AlShehabi, & A. Hanieh (Eds.), *Transit states: Labour, migration & citizenship in the Gulf* (pp. 170–97). Pluto Press.

Williams, F. (2011). Towards a transnational analysis of the political economy of care. In R. Mahon & F. Robinson (Eds.), *Feminist ethics and social policy: Towards a new global political economy of care* (pp. 21–38). UBC Press.
Williams, R. (2017, March 1). Why are there so few male nurses? *The Guardian*. http://www.theguardian.com/healthcare-network/2017/mar/01/why-so-few-male-nurses.
Yeates, N. (2002). Globalization and social policy: From global neoliberal hegemony to global political pluralism. *Global Social Policy*, *2*(1), 69–91. https://doi.org/10.1177/1468018102002001095.
Yeates, N. (2012). Global care chains: A state-of-the-art review and future directions in care transnationalization research. *Global Networks*, *12*(2), 135–54. https://doi.org/10.1111/j.1471- 0374.2012.00344.x.

8 Migration Intermediaries and the Migration of Health Professionals from the Global South

ABEL CHIKANDA

Introduction

The migration of health professionals from the Global South to the Global North still continues to generate opposing viewpoints among researchers. A large number of researchers from the Global South characterize the movement as a "brain drain" and argue that it is associated with falling health care standards in the sending region (Chikanda, 2006). On the other hand, some researchers from the Global North contend that such movements are in tandem with globalization and argue that health professionals' migration should not be a target of policy intervention measures as it violates their right to free movement (Clemens, 2011; Skeldon, 2009).

This debate is not entirely new. Nearly fifty years ago, Don Patinkin, an economist at Hebrew University of Jerusalem, argued against the movement of skilled professionals, highlighting that the migrant-sending country suffers irreparable damage due to the permanent loss of innovative individuals (Patinkin, 1968). This nationalist viewpoint regards the movement of skilled professionals as a "brain drain" and argues that it leads to increased global disparities: the rich countries become richer while the poor nations become poorer (Mundende, 1989). Ten years after the introduction of the nationalist theory, Mejia (1978) argued that countries in the Global North should be guided by "moral concerns" in developing measures to reduce the active recruitment of health professionals from the Global South. Newer literature in this strand emphasize the impacts of health worker migration in the Global South on the quality of care, particularly on the public sector that plays a crucial role in the delivery of health services (Chikanda, 2011; Crush et al., 2012; Labonte et al., 2015). According to participants at a recent high-level dialogue at the World Health Organization (WHO), the situation is worsened by "predatory practices in the recruitment of health workers" by countries in the Global North (WHO, 2017).

Harry Johnson, an economist at the University of Chicago, presented an alternative "cosmopolitan liberal position" (Ellerman, 2006, p. 21) of viewing the global movement of skilled professionals such as health professionals. Johnson (1968) argued that the international flow of human capital benefits both the migrant sending and recipient countries. The internationalist viewpoint argues that developed countries provide an ideal environment for the advancement of migrants' skills while they pass on the end-products of their investment to all nations that want them as members of one big international community (Johnson, 1967). Recently, some "brain drain revisionists" have argued that countries that lose their health professionals can still reap positive results from their exiled professionals through various transnational activities such as sending remittances to their home country, return migration, skills transfer, and philanthropic activities (Clemens, 2011; Skeldon, 2009).

Caught up in the midst of this polarizing debate are migration intermediaries such as recruitment agencies that play a crucial role in the movement of health professionals from the Global South to the Global North. Migration intermediary operators provide services that assist the mobility, labour market entry, and integration of migrant workers into the destination country (Lindquist et al., 2012; van den Broek et al., 2016). Because of the visible role they play in the migration process, migration intermediaries have become the "public face" of the so-called brain drain of skilled professionals from the Global South. Not surprisingly, migration intermediaries have been referred to in the migration literature as "migration merchants" (Kyle & Liang, 2001), "merchants of medical care" (Connell & Stilwell, 2006), "poachers" (Snyder, 2009), and "global raiders" (Crush, 2002).

As debate on the role of migration intermediaries in facilitating the migration of health professionals rages on, scholars have offered differing viewpoints on their involvement. Connell and Stilwell (2006) have observed that recruitment agencies stimulate migration in labour-supplying countries. Rogerson and Crush (2008), on the other hand, argue that recruitment agencies cannot be blamed for creating a desire among the health professionals to migrate, as this is often the result of an interplay of a host of factors. In some cases, as Sward (2015) notes, the health professionals may initiate contact with recruitment agencies themselves, positioning themselves as active rather than passive agents in the migration process.

This chapter examines the role played by migration intermediaries in the global movement of health professionals. It provides a descriptive account of the operation of migration intermediaries, particularly those that connect professionals from the Global South with the Global North. In addition, it shows how the implementation of the various codes of practice have forced migration intermediaries to reorganize their operations in an effort to comply with the new migration governance system.

Migration Intermediaries and the International Movement of Health Professionals

Even though the movement of health professionals has been a subject of study by researchers for the past five decades, there is scant information available on the infrastructure that support their international mobility (Maybud & Wiskow, 2006; van der Broek at al., 2016). As a result, despite the significance of recruitment agencies in promoting the movement of skilled professionals globally, not much is known about this important industry. Consequently, anecdotal evidence has been presented as facts in various studies about the role played by migration intermediaries in the international mobility of skilled professionals.

Migration intermediaries help potential migrants navigate the complex procedures involved in the migration process (Bludau, 2011). As such, they offer services that assist in the entry of skilled professionals to the labour market of the host country and facilitate their successful integration (van der Broek at al., 2016). Chikanda (2006) notes that the growth of formalized channels of movement has coincided with the increasingly important role of the internet, where information to potential migrants is often "just a mouse click away" (Shrecker & Labonte, 2004, p. 412). In addition, the migration of health professionals is taking place within the broad context of accelerating globalization of the service sector that has occurred over the past three decades (Connell & Stilwell, 2006). Advertisements are regularly posted in an attempt to lure disgruntled health professionals to destinations in the West. As Maybud and Wiskow (2006) note:

> Private agencies specializing in the recruitment of health care professionals for Western markets invite the loggers-on to explore a myriad of opportunities. Go ahead, they entice, just click on this website and you are one step closer to a better life. They advertise hundreds of fabulous hospital and health care nursing jobs in exciting places, and claim that they can make all the difference to health care careers. And it is so easy. Potential candidates just have to register, and they will be helped all along the way. Examination requirements will be demystified, job interviews will be arranged, and visas and permits will be handled. (p. 223)

Migration intermediaries represent the core of the formal migration industry and are key actors in the international mobility of health professionals (Acacio, 2011). Some migration intermediaries offer training programs to help migrants negotiate the different labour and cultural environments, essentially "producing migrants" (Bludau, 2011). Health professionals may be recruited under three basic models of recruitment as identified by Pittman et al. (2007) and illustrated in table 8.1.

Table 8.1. Models of health professionals' recruitment

Model	Description
Employer-led model	Health care organizations use their own resources to recruit health professionals from the source country. They may work with facilitators, such as recruiters or immigration lawyers, but many of the recruitment activities are conducted by the health care organization itself. Research has shown that this model is the most successful recruitment model in terms of the overall number of international hires (Squires, 2008).
Placement model	Health care organizations subcontract all the recruitment and immigration functions to a vendor. Essentially, the vendor serves as a placement agency whose main role is to facilitate the process of placing foreign-trained health professionals with health care organizations.
Staffing/lease model	In this model, the agency carries out most of the recruitment and immigration functions on its behalf. Health professionals hired under this model are contracted directly to the recruitment agency for a period of 18 to 36 months (Sward, 2015). This model is quite profitable for the staffing agencies. Staffing agencies typically reap annual profits of approximately US$50,000 per nurse, making this model a highly profitable business (Pittman et al., 2010).

Source: Pittman et al. (2007).

Recruiting Health Professionals from the Global South

As noted previously, most of the available information on the international recruitment of health professionals remains anecdotal at best, which makes it difficult to gauge the extent to which active recruitment contributes to the international flow of health professionals. However, over the past fifteen years, a large number of studies have sought to document the scale of recruitment of health professionals from the Global South, as discussed below.

EVIDENCE FROM THE GLOBAL SOUTH

Studies in South Africa by Rogerson and Crush (2008) and by Dambisya and Mamabolo (2012) illustrate the scale of medical recruiting that took place in South Africa in the 2000s. Over the period 2000–4, Western recruiters placed a total of 2,522 recruitment advertisements in the *South African Medical Journal* (SAMJ). A follow-up study by Dambisya and Mamabolo (2012) yielded a total of 1,176 recruitment advertisements over the period 2006–10 (figure 8.1). As Dambisya and Mamabolo further note, even though advertisements are not

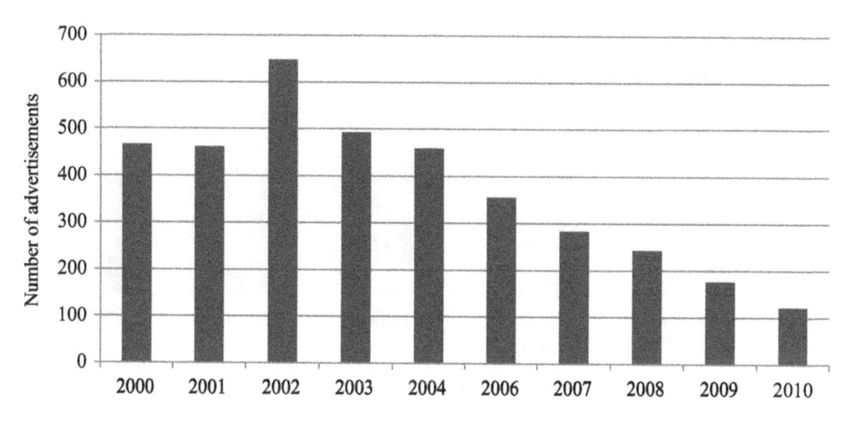

Figure 8.1. Recruitment advertisements in SAMJ by year, 2000–2010
Note: Data for 2005 were not available.
Sources: Rogerson & Crush (2008); Dambisya & Mamabolo (2012).

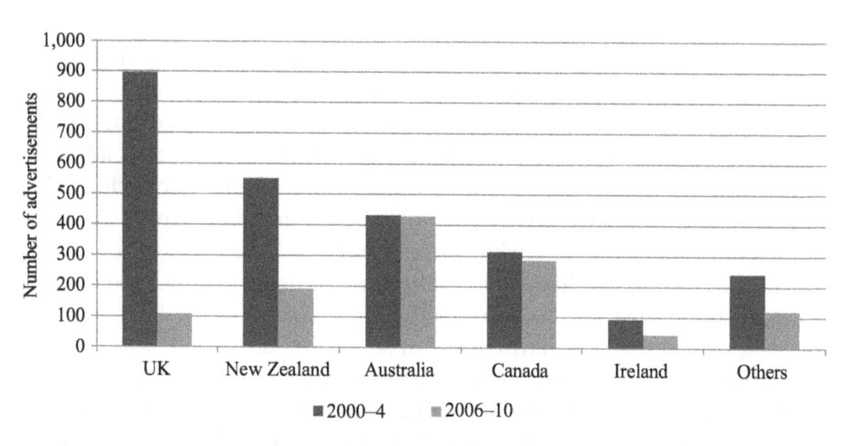

Figure 8.2. Recruitment advertisements in SAMJ by originating country

indicative of success of recruitment efforts, recruiters place advertisements with an expectation of a return on investment. Again, the fall in recruitment numbers may also suggest that the advent of other advertising channels has made it less worthwhile for recruiters to place advertisements for doctors in the SAMJ.

An analysis of originating countries of the advertisements reveals an interesting trend in medical recruiting in South Africa (figure 8.2). Even though there was a 53 per cent drop in the overall number of medical advertisements that were placed in the SAMJ between 2000–4 and 2006–10, the decrease was not evenly distributed among the originating countries of the advertisements. The largest decline was experienced in advertisements originating from the

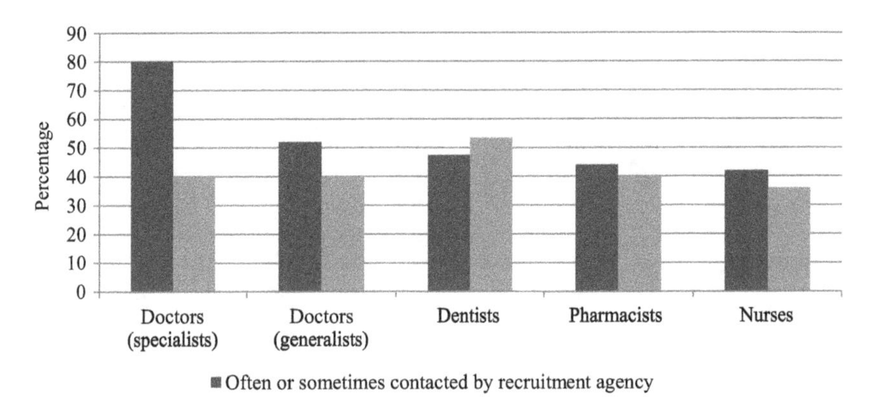

Figure 8.3. Contacted by recruitment agency and migration consideration
Source: Adapted from Labonte et al. (2015).

UK, which declined from 896 to only 108 (90 per cent) for the period 2000–4 and 2006–10, respectively. The number of advertisements originating from New Zealand also declined sharply over the same time period, dropping from 551 to 191, a decline of about 65 per cent. However, the number of advertisements from Australia and Canada have not witnessed a similar decline. Thus, the number of advertisements from Australia fell marginally from 431 to 428 during the same time frame, while those from Canada also fell from 312 to 286. Clearly, there is some evidence of continued recruitment of health professionals from Australia and Canada on the South African market. The US is not a major originating country of advertisements in South Africa.

Recent research sheds more evidence on the continued recruitment of health professionals from South Africa. Labonte et al. (2015) showed that nearly half of their sample of health professionals had been contacted "often or sometimes" by recruitment agencies operating in South Africa (figure 8.3). Contact by recruitment agencies was highest among specialist doctors with as many as 80 per cent indicating that they were contacted "often or sometimes" by recruitment agencies. Contact by recruitment agencies was lowest among nurses, with slightly more than 40 per cent reporting having been contacted by recruitment agencies. Labonte et al. showed that there is a strong correlation between contact by recruitment agencies and consideration to migrate, suggesting that the contact can hasten the migration process.

The movement of health professionals from the Global South has been associated with negative outcomes on the health care systems of the countries involved. In Zimbabwe, for instance, the out-migration of medical doctors has led to increased wait time for patients and overworking of the health professionals

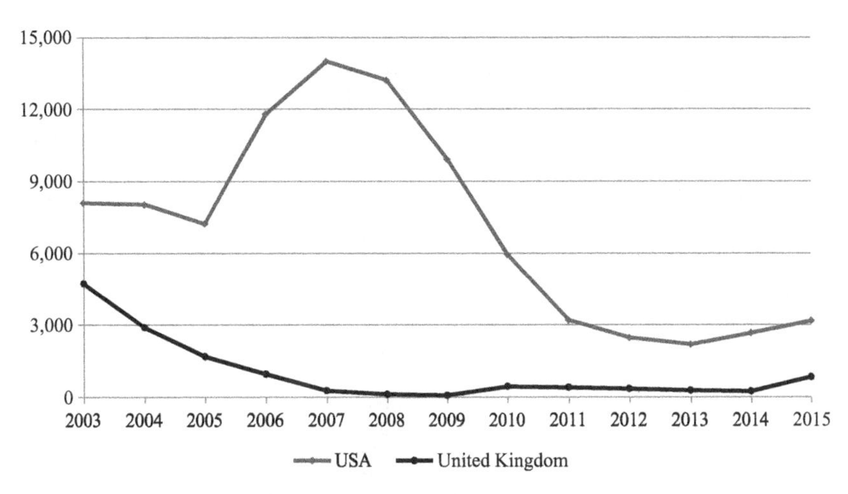

Figure 8.4. Annual flow of Filipino nurses to the US and UK, 2003–2015
Source: Data from OECD (2019).

who choose to remain (Chikanda, 2006). In fact, it has been highlighted that Sub-Saharan Africa accounts for 25 per cent of the global disease burden but only 1.3 per cent of the world's health workers (Naicker et al., 2009). However, not all active recruitment of health professionals is bad for the health system or associated with negative outcomes. India and the Philippines have historically encouraged the migration of health professionals to overseas destinations (Calenda, 2016). Since 1974, the Philippines has led an aggressive labour export campaign under its "overseas employment programme" (Acacio, 2011). The state-run program was eventually replaced by a private sector program with the state playing a regulatory role. The US and the UK are two of the most important destinations of nurses from the Philippines. The Organisation for Economic and Co-operative Development (OECD, 2019) has shown that more than 90,000 Filipino nurses were registered to practise in the US between 2003 and 2015 while a further 13,000 were registered to practise in the UK over the same time period (figure 8.4). However, the number of Filipino nurses migrating to the UK fell significantly after the adoption of the Commonwealth Code of Practice for the International Recruitment of Health Workers in 2003. Likewise, the number of Filipino nurses practising in the US fell significantly after 2007, partly due to an increase in the domestic production of nurses.

EVIDENCE FROM THE GLOBAL NORTH

An examination of the workforce composition in some countries in the Global North provides evidence on the extent of active recruitment of health professionals from the Global South. According to the OECD (2019), about 460,000

Table 8.2. Medical practitioners in Australia by country/region of training, 2017

Country or Region of Training	Medical Practitioners	%
Australia	58,521	64.7
Global North		
United Kingdom	5,577	6.2
New Zealand	2,054	2.3
Ireland	1,179	1.3
United States	168	0.2
Canada	124	0.1
Global South		
India	5,182	5.7
South Africa	1,816	2.0
Sri Lanka	1,607	1.8
Pakistan	925	1.0
Bangladesh	676	0.7
China	643	0.7
Philippines	596	0.7
Malaysia	328	0.4
Zimbabwe	196	0.2
Other foreign country	7,929	8.8
Not stated	2,896	3.2
Total	90,417	

Source: Data from OECD (2019).

foreign-trained doctors and 570,000 foreign-trained nurses were working in OECD countries in 2013–14. Thus, foreign-trained health professionals account for 17 per cent of all doctors and 6 per cent of all nurses on average in OECD countries. While as many as a third of these health professionals come from other OECD countries, the remaining two-thirds originate from countries in the Global South that are experiencing critical shortages of health professionals. This section reviews evidence of the recruitment of health professionals in three countries in the Global North: Australia (doctors), the United States (doctors), and the United Kingdom (nurses).

Australia is a good example of a country that has traditionally relied on foreign-trained medical doctors to meet its labour needs (Hawthorne, 2012). For instance, foreign-trained doctors made up 29,000 of the country's 90,417 (32.1 per cent) medical doctors practising in the country in 2017 (table 8.2). The majority of foreign-trained medical professionals were trained in the Global South. For instance, of the 15,168 medical doctors that were added to the Australian workforce between 2006 and 2011, 10,452 (or 68.9 per cent) were foreign born, with significant increases from countries such as India, Nepal, the Philippines, and Zimbabwe (Negin et al., 2013).

Migration intermediaries play a huge role in the movement of health professionals to Australia. Since the late 1990s, the Department of Health and Ageing (DoHA) has directed funding to rural workforce agencies and has introduced legislation to encourage the hiring of foreign graduates to work in the rural parts of the country (Negin et al., 2013; Pillinger, 2012). The situation is not expected to change significantly in the near future with the Health Workforce Australia report noting that the country will continue to experience shortages of health professionals due to "continued reliance on poorly coordinated skilled migration to meet essential workforce requirements – with Australia having a high level of dependence on internationally recruited health professionals relative to most other OECD countries" (Health Workforce Australia, 2012, p. iii). Interestingly, Australia's recruitment of foreign health professionals is guided by the 2003 Commonwealth Code of Practice, which "seeks to encourage the establishment of a framework of responsibilities between governments – and the agencies accountable to them – and the recruits. This framework would balance the responsibilities of health workers to the countries in which they were trained – whether of a legal kind, such as fulfilling contractual obligations, or of a moral kind, such as providing service to the country which had provided their training opportunities – and the right of health professionals to seek employment in other countries" (Commonwealth Secretariat, 2003, p. 5). It also urges recruiting governments to consider measures of reciprocating for the advantages gained by recruiting foreign health professionals. Current evidence suggests that Australia is not adhering to the ethical recruitment guidelines detailed above.

The US is one of the major destinations of health professionals globally. International recruitment plays a big role in the movement of professionals to the country. In 2016, there were more than 215,000 foreign-trained doctors working in the country (OECD, 2019). Hagopian et al. (2004) note that nearly two-thirds of the doctors who move to the US come from lower-income or lower-middle-income countries and most of them migrate soon after completing their training. India is a major source country of foreign-trained doctors in the US and recent figures suggest a declining trend in the number of doctors choosing to migrate there. The post-2008 decline has been fuelled in part by a rapid increase in the number of domestic production of nurses and doctors, which might, in turn, force potential migrants to seek replacement markets. It also shows the impact of retrogression, a procedural delay in the processing of employment-based visas due to oversubscription, which started in 2006 (Shaffer et al., 2016). As shown in figure 8.5, foreign-trained medical doctors made up 28 per cent of the new doctors who registered to practise in the country in 2017. From 2002 to 2017, a total of 114,254 foreign-trained doctors registered to practise in the US, at an average rate of 7,140 per year during that timeframe. The number of foreign-trained doctors has fallen over the past decade,

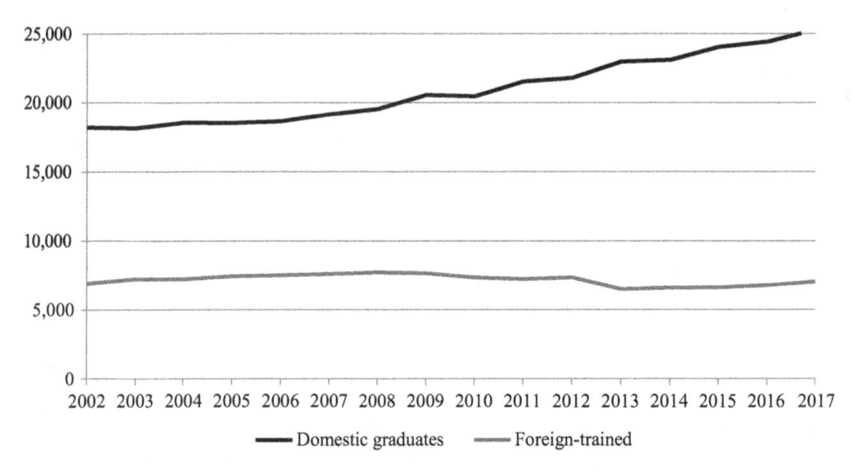

Figure 8.5. Changes in the number of medical graduates and inflow of foreign-trained medical doctors, United States, 2002–2017
Source: Data from OECD (2019).

declining from 7,678 new registrations in 2008 to 7,033 new entrants in 2017. The declining share of foreign-trained medical doctors is partly explained by the rapid rise in domestic output of doctors since 2007. Thus, the number of domestically trained doctors rose from 18,214 of the new doctor registrations in 2002 (67 per cent of the total) to 25,274 of the new registrations in 2017 (78 per cent of the total).

So what is the current state of medical recruitment in the US? Two main data sources shed light on the issue. First, data on the number of recruitment agencies that facilitate the inward flow of health professionals to the US show a declining trend. A study by Pittman et al. (2007) showed that there were 273 US-based international recruitment companies. Out of these, 147 were actively involved in the recruitment of health professionals from 74 countries. More worryingly, 74 companies acknowledged that they actively recruit from 11 of the 57 countries that have been identified by WHO as experiencing critical shortages of health professionals. In another study by Pittman et al. (2014), the research showed that the number of companies that actively recruit health professionals had declined to just 97 in 2012, which suggests either a consolidation of operations of migration intermediaries or demonstrates an overall decline in the number of health professionals migrating to the US.

Second, an analysis of data on new registration of health professionals from certain geographic regions may be indicative of the extent to which health professionals are being actively recruited into the US. Tankwanchi et al. (2015)

have examined the trends in migration of Sub-Saharan African-trained medical doctors to the US. Their goal was to determine whether the adoption of the 2010 WHO Code of Practice had reduced the recruitment of medical doctors from Sub-Saharan African countries. Using data from the 2013 American Medical Association Physician Masterfile, the authors estimated that there were 12,847 Sub-Saharan African-trained medical doctors working in the US in 2015, representing an increase of 2,908 (or 22.7 per cent) over the number recorded when the WHO Code of Practice was adopted in 2010. This suggests the continuation of recruitment of health professionals in the US, at least from regions such as Sub-Saharan Africa. Even more worrying is that fact that an estimated 80 per cent of migrant doctors from Sub-Saharan Africa are entering the US by age thirty-five (Tankwanchi et al., 2015).

The Voluntary Code of Ethical Conduct for the Recruitment of Foreign-Educated Health Professionals to the United States (Alliance Code) was launched in 2008 and lays down "standards of practice" to ensure ethical recruitment of health professionals in the US (Shaffer et al., 2016). Its stakeholders include representatives of unions, hospitals, nursing organizations, regulatory bodies, credentials evaluators, recruiters, staffing agencies, and immigration attorneys (Nichols et al., 2011). It establishes minimum standards for employers and recruiters in the US (e.g., adherence to particular laws, adequate cultural and clinical training of foreign-trained health professionals). It also establishes some best practices "to protect the right to labour autonomy of foreign-educated nurses while acknowledging and mitigating the harms to source countries healthcare systems caused by the migration of nurses" (Shaffer et al., 2016, p. 115). The Alliance Code has since evolved into the Alliance for Ethical International Recruitment Practices. However, the main focus is on curbing the abusive practices of recruitment agencies to health professionals working in the US (Nichols et al., 2011). As it stands, the Alliance Code, though local and voluntary, is a positive step in regulating the activities of recruitment agencies operating within the US. The commitment of the US in establishing ethical recruitment practices of health professionals from the Global South is yet to be seen.

The UK has traditionally relied on foreign-trained health professionals to fill vacancies in the country's health sector. The Colonial Nursing Service was created in 1940 to create an imperial labour market and to assist in the recruitment of foreign-trained nurses (Rafferty & Solano, 2007). Thus, active recruitment has played an important role in filling positions in the UK's health sector. A study by Winkelmann-Gleed (2006) showed that as many as 41 per cent of overseas-trained nurses had migrated due to active recruitment. Buchan and Sochalski (2004) note that, for the first time, between 2001 and 2002, there were more foreign-educated nurses added to the UK nurse register than domestically trained ones. By 2003, foreign-educated nurses made up 15,358 of the 33,452 (or 45.9 per cent) of the new registrations in that year, and falling even

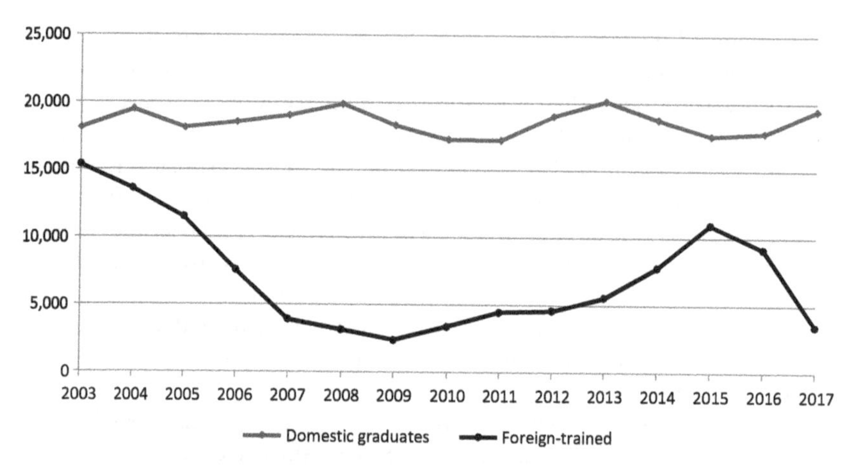

Figure 8.6. Changes in the supply of domestic and foreign-trained nurses in the UK, 2003–2017
Source: Data from OECD (2019).

further to just 3,462 of the 22,910 (or 15.1 per cent) of the new registrations in 2017 (figure 8.6). At the same time, the domestic production of nurses has risen from 18,094 in 2003 to 19,448 in 2017. This trend suggests an overall decline in importance of foreign health professionals' recruitment in the UK.

The fall in foreign-educated nurse recruitment in the UK can be attributed to the adoption of the Code of Practice for international recruitment for National Health Service (NHS) employers in 2001 (Sward, 2015). The Code of Practice banned the active recruitment of health professionals from countries in the Global South, unless an agreement was in place with the government in question (Buchan et al., 2009). The Code of Practice was updated in 2004. The adoption of the Code of Practice followed a well-publicized case in which the former South African president, Nelson Mandela, castigated the UK for "poaching" South African nurses and doctors (Pittman et al., 2010; Bach, 2015). Clearly, international recruitment was a cost-effective strategy of addressing the problem of the shortage of health professionals in the UK before 2004 (Sward, 2015). A study by Mills et al. (2011), for instance, showed that the UK saved at least US$2.7 billion by hiring health professionals from Sub-Saharan Africa.

Readjustment and Realignment of Migration Intermediaries

The turn towards ethical recruitment of health professionals has caused migration intermediaries to realign their activities with the new policy environment. At the global level, WHO introduced the Global Code of Practice on the International Recruitment of Health Personnel in May 2010 to provide an ethical and human rights framework guiding destination and source countries to mitigate the migration of health professionals through international agreements and national policies (Dambisya & Mamabolo, 2012). However, a major weakness of the Global Code of Practice is that it is voluntary. Promising, though, are initiatives by individual countries such as the UK, which introduced a Code of Practice in 2004 that led to a significant reduction in international recruitment of health professionals. It still remains to be seen whether other major destinations of health professionals such as Australia will adopt measures to reduce active recruitment from countries in the Global South with low rates of supply of health professionals.

In an environment marked by reduced agencies that facilitate health professional migration to the Global North, are migration intermediaries still relevant? There is no simple answer to this question. First, with a number of traditional destination countries tightening work and entry conditions, new destinations have emerged within the Global South (WHO, 2017). As Labonte et al. (2015) note, the decline in active recruitment from Anglo-American destination countries has led to a growth in migration to countries in the Gulf States such as Saudi Arabia, the United Arab Emirates, and Oman. South African nurses in particular are reportedly being actively recruited by Gulf States (Labonte et al., 2015). In another case, Connell and Stilwell (2006) have shown that the implementation of the Code of Practice caused Caribbean nurses, who had previously been migrating to the UK, to start moving to the US. Migration intermediary agencies, therefore, will continue to play an active role in the identification of, and building links with, new markets for health professionals globally. Thus, the implementation of the WHO Code of Practice will not be a success without its global adoption and enforcement.

A second observation is that migration intermediary agencies are now channelling health professionals into the less visible sectors of the economy. As Sward (2015) observes, the UK's code of practice prohibits active recruitment in the public sector but this does not apply to the private sector (i.e., private hospitals, nursing homes, and clinics). What this means is that it is possible that active recruitment from the Global South to the UK is still occurring. For instance, Buchan et al. (2006) in a study of 400 London-based foreign-trained nurses from both OECD countries and those in the Global South, found that nurse recruitment was still occurring "through the back door." The Code of Practice does not prevent private hospitals and care homes from recruiting from the Global South and the health professionals can switch from these institutions to the NHS at a later date (Connell & Stilwell, 2006). In fact, Buchan et al. (2006) found in their study that as many as 57 per cent of the nurses who had been hired in the private sector had switched jobs since coming to the UK, with most of them moving to the NHS. The research also found that two-thirds of the 400 nurses interviewed had their move to the UK facilitated by a recruitment agency. This provides further evidence that recruitment agencies can still manage to find a way of channelling health professionals to markets in the Global North, even in the presence of considerable barriers to movement.

Third, there has been major changes in the overall structure of the migration intermediary industry. In an effort to realign their activities with the new policy framework, migration intermediaries have been forced to adopt a number of coping strategies. In the US, consolidation has become commonplace as the shrinking market has left only a small space for the operation of intermediary agencies. For instance, the number of US companies recruiting foreign nurses fell from 273 in 2007 to only 97 in 2012 (Pittman et al., 2010, 2014). Even in the UK, there is evidence of a drop in the number of intermediary agencies in the wake of the implementation of the Code of Practice. The NHS maintains a list of approved agencies that employers should cooperate with that apply ethical recruitment strategies set out in the Code of Practice (Maybud & Wiskow, 2006). Not surprisingly, there were 11 per cent fewer recruitment agencies operating in the UK in 2013 compared with 2002 (Calenda, 2016). While part of the reason can be attributed to the decline in the overall recruitment of nurses to the UK, there is evidence of recruitment agencies coming together to form procurement consortia, which now account for 80 per cent of what the NHS spends on recruitment (Calenda, 2016).

The introduction of the 2004 Code of Practice in the UK forced recruitment agencies to change their approach to international recruiting. Drake Medox UK, for instance, has specialized in the international recruitment of nurses since 1997 for the NHS and some private sector organizations (Gamwell, 2016). In line with ethical recruitment guidelines, the company operates a "no-fee policy" and only works with organizations that do not charge fees to candidates. Working together with its recruitment agency, ASC Global Recruitment, Drake Medox UK recruits nurses only from the Philippines, which is described as an "ethical source country" (Gamwell, 2016). In order to avoid excessive recruitment of health professionals from a single area or hospital, Drake Medox UK "ensures that recruitment campaigns are geographically dispersed, avoiding areas that are experiencing shortages of nursing staff" (Gamwell, 2016, p. 32). Such an approach does not only serve to facilitate the flow of nurses to the UK from a country where active recruitment is permissible under the Code of Practice, but it also ensures that the right to health of the people in the Philippines is respected. Only time will tell whether other agencies that are involved in the international recruitment will follow suit and implement measures that help ensure that health professionals are recruited from source countries that have an excess supply and also takes into consideration the geographical distribution of the health professionals in the source country at various scales. In India, recruitment firms are allowed to charge fees for their services but firms such as IFANglobal, a private recruitment agency, have decided to forego charging such fees and instead rely on profits from their foreign clients, that is, the employers (Calenda et al., 2016).

Evidence from the Philippines show that recruitment agencies in the Global South are positioning themselves as development champions that are more than profit-seeking enterprises. This is in line with the Philippines government portrayal of overseas contract workers as "heroes of national development" (Gibson et al., 2001, p. 365). Recruitment and placement agencies that help nurses move to overseas destinations view their work as a service and vocation to the Filipino people who face limited opportunities at home (Acacio, 2011). One co-owner of a recruitment agency in the Philippines who was interviewed by Acacio (2011) observed, "The vision/mission of the agency is to help nurses find job opportunities in the States. This is a big help to the Philippines. You don't just help one nurse, you help the whole community. The nurse you help will help their families. In ten years, a Filipino nurse will earn P840,000; whereas in the U.S., a nurse will earn the equivalent of P3 million in one year" (p. 113).

Thus, the recruitment and placement of health professionals has clearly become more than a business in some countries in the Global South. Some recruitment agencies in the Global South have adopted altruism as a framework to guide their operations. As the case in the Philippines shows, three important elements hold the key to future ethical recruitment practices. First, co-operation between recruitment agencies in the Global North and Global South can yield a win-win scenario for both parties. Second, governments in the Global South need to play a proactive role in regulating the role of recruitment agencies in their countries. Abuse and exploitation of migrant workers by recruitment agencies and foreign clients can be minimized when governments support state agencies in regulating the operation of the migration intermediary industry. Third, migration intermediaries operating in the Global South are increasingly focusing on the success of individuals they place in the destination country. Consequently, they have expanded their services to offer packages that might include loans, transportation services, housing advice, employment contract negotiation, and assistance with visa documents (van der Broek et al., 2016). This "paradigmatic shift" by the private sector recruiters in the provision of migrant services and information has been labelled as a game changer (Nyberg Sorensen & Gammeltoft-Hansen, 2013).

Concluding Remarks

The movement of health professionals in this increasingly globalizing world has been enhanced by the activities of migration intermediary operators. The current influence of migration intermediaries partly reflects the impact of market liberalization, which has seen the decentring of state mobilized services since the late 1980s (Groutsis et al., 2015). This transfer of authority from state-owned and managed institutions to market-based interests means that there is "no one centre but multiple centres; there is no sovereign authority because networks have considerable autonomy" (Rhodes, 1997, p. 109).

Migration intermediaries not only assist health professionals in their mobility and entry to the new labour market but also play an important role in their successful integration in the destination country. Consequently, in regions that are experiencing shortages of health professionals, migration intermediaries have attracted a lot of bad publicity. In a bid to counter this negative view, the UK has led the charge towards ethical recruitment of health professionals, as shown by its adherence to the Commonwealth 2003 and its own 2004 Code of Practice on ethical recruitment of health professionals. There is evidence of a considerable decline in active recruitment of health professionals from the Global South in the UK, with a few notable exceptions such as India and the Philippines, which have historically exported health professionals globally. Future studies could also seek to uncover the role played by migration intermediaries in the temporary movement of health professionals as described by Jonathan Crush (chapter 16 in this volume). There is scant evidence to suggest that the adoption of the WHO Global Code of Practice has slowed down the recruitment of health professionals in Australia and the US, which represent two of the most important destinations of health professionals globally. On the other hand, the movement of health professionals within the Global South (South-South migration) continues to gather pace, and there is need for urgent research to understand the role played by migration intermediaries in this South-South movement of health professionals.

REFERENCES

Acacio, K. (2011). *Getting nurses here: Migration industry and the business of connecting Philippine-educated nurses with United States employers* [Unpublished doctoral dissertation]. University of California Berkeley.

Bach, S. (2015). Nurses across borders: The international migration of health professionals. In B. Parry, B. Greenhough, T. Brown, & I. Dyck (Eds.), *Bodies across borders: The global circulation of body parts, medical tourists and professionals* (pp. 155–68). Ashgate.

Bludau, H. (2011). Producing transnational nurses: Agency and subjectivity in global health care labor migration recruitment practices. *Anthropology of East Europe Review* 29(1), 94–108.

Buchan, J., & Sochalski, J. (2004). The migration of nurses: Trends and policies. *Bulletin of the World Health Organization*, *82*(8), 587–94. https://doi.org/10.1016/j.jmwh.2005.03.011.

Buchan, J., Jobanputra, R., Gough, P., & Hutt, R. (2006). Internationally recruited nurses in London: A survey of career paths and plans. *Human Resources for Health, 4*(14), 4–14. HTTPS://DOI.ORG/10.1186/1478-4491.

Buchan, J., Mcpake, B., Mensah, K., & Rae, G. (2009). Does a code make a difference –Assessing the English code of practice on international recruitment. *Human Resources for Health, 7*(1), 4. https://doi.org/10.1186/1478-4491-7-33.

Calenda, D. (2016). Main empirical findings. In D. Calenda (Ed.), *Case studies in the international recruitment of nurses: Promising practices in recruitment among agencies in the United Kingdom, India, and the Philippines* (pp. 14–21). International Labour Organization.

Calenda, D., Joshi, B., & Sharma, S. (2016). Promising practices emerging from the recruitment industry in India. In D. Calenda (Ed.), *Case studies in the international recruitment of nurses: Promising practices in recruitment among agencies in the United Kingdom, India, and the Philippines* (pp. 43–62). International Labour Organization.

Chikanda, A. (2006). Skilled health professionals' migration and its impact on health delivery in Zimbabwe. *Journal of Ethnic and Migration Studies, 32*(4), 667–80. https://doi.org/10.1080/13691830600610064.

Chikanda, A. (2011). The changing patterns of physician migration from Zimbabwe. *Journal of International Migration, Health and Social Care, 7*(2), 77–92. https://doi.org/10.1108/17479891111180057.

Clemens, M. (2011). The financial consequences of high-skill emigration: Lessons from African doctors abroad. In S. Plaza, & D. Ratha (Eds.), *Diasporas for development in Africa* (pp. 165–82). World Bank.

Commonwealth Secretariat. (2003). *Commonwealth code of practice for international recruitment of health workers*. Commonwealth Secretariat.

Connell, J., & Stilwell, B. (2006). Merchants of medical care: Recruiting agencies in the global health care chain. In C. Kuptsch (Ed.), *Merchants of labour* (pp. 239–53). International Institute for Labour Studies.

Crush, J. (2002). The global raiders: Nationalism, globalisation and the South African brain drain. *Journal of International Affairs, 56*(1), 147–72. https://doi.org/10.1080/03768350601165769.

Crush, J., Chikanda, A., & Pendleton, W. (2012). The disengagement of the South African medical diaspora in Canada. *Journal of Southern African Studies, 38*(4), 927–49. https://doi.org/10.1080/03057070.2012.741811.

Dambisya, Y.M., & Mamabolo, M.H. (2012). Foreign advertisements for doctors in the SAMJ 2006–2010. *South African Medical Journal, 102*(8), 669–72. https://doi.org/10.7196/samj.5803.

Ellerman, D. (2006). The dynamics of migration of the highly skilled: A survey of the literature. In Y. Kuznetsov (Ed.), *Diaspora networks and the international migration of skills: How countries can draw on their talent abroad* (pp. 21–57). The World Bank.

Gamwell, S. (2016). Promising practices emerging from the recruitment industry in the United Kingdom. In D. Calenda (Ed.), *Case studies in the international recruitment of nurses: Promising practices in recruitment among agencies in the United Kingdom, India, and the Philippines* (pp. 22–42). International Labour Organization.

Gibson, K., Law, L., & McKay, D. (2001). Beyond heroes and victims: Filipina contract migrants, economic activism and class transformations. *International Feminist Journal of Politics, 3*(3), 365–86. https://doi.org/10.1080/14616740110078185.

Groutsis, D., van den Broek, D., & Harvey, W.S. (2015). Transformations in network governance: The case of migration intermediaries. *Journal of Ethnic and Migration Studies, 41*(10), 1558–76. https://doi.org/10.1080/1369183X.2014.1003803.

Hagopian, A., Thompson, M.J., Fordyce, M., Johnson, K.E., & Hart, L.G. (2004). The migration of physicians from sub-Saharan Africa to the United States of America: Measures of the African brain drain. *Human Resources for Health*, 2(17). https://doi.org/10.1186/1478-4491-2-17.

Hawthorne, L. (2012). International medical migration: What is the future for Australia? *Medical Journal of Australia Open*, 1(Supplement 3), 18–21. https://doi.org/10.5694/mjao12.10088.

Health Workforce Australia. (2012). *Health workforce 2025: Doctors, nurses and midwives* (Vol. 1). Health Workforce Australia. https://apo.org.au/sites/default/files/resource-files/2012-01/apo-nid154456.pdf.

Johnson, H.G. (1967). Some economic aspects of brain drain. *The Pakistan Development Review*, 7, 379–411.

Johnson, H.G. (1968). An internationalist model. In W. Adams (Ed.), *The brain drain* (pp. 69–91). Macmillan.

Kyle, D., & Liang, Z. (2001). Migration merchants: Human smuggling from Ecuador and China to the United States. In V. Guiraudon & C. Joppke (Eds.), *Controlling a new migration world* (pp. 200–21). Routledge.

Labonte, R., Sanders, D., Mathole, T., Crush, J., Chikanda, A., Dambisya, Y., Runnels, V., Packer, C., MacKenzie, A., Tomblin Murphy, G., & Bourgeault, I.L. (2015). Health worker migration from South Africa: Causes, consequences and policy responses. *Human Resources for Health, 13*(92). https://doi.org/10.1186/s12960-015-0093-4.

Lindquist, J., Xiang, B., & Yeoh, B.S. (2012). Opening the black box of migration: Brokers, the organization of transnational mobility and the changing political economy in Asia. *Pacific Affairs, 85*(1), 7–19. https://doi.org/10.5509/20128517.

Maybud, S., & Wiskow, C. (2006). "Care trade": The international brokering of health care professionals. In C. Kuptsch (Ed.), *Merchants of labour* (pp. 223–38). International Institute for Labour Studies and International Labour Office.

Mejia, A. (1978). Migration of physicians and nurses. *International Journal of Epidemiology, 7*(3), 207–15. https://doi.org/10.1093/ije/7.3.207.

Mills, E.J., Kanters, S., Hagopian, A., Bansback, N., Nachega, J., Alberton, M., Au-Yeung, C.G., Mtambo, A., Bourgeault, I.L., Luboga, S., Hogg, R.S., & Ford, N. (2011). The financial cost of doctors emigrating from sub-Saharan Africa: Human capital analysis. *British Medical Journal, 343.* https://doi.org/10.1136/bmj.d7031.

Mundende, D.C. (1989). The brain drain and developing countries. In R. Appleyard (Ed.), *The impact of international migration on developing countries* (pp. 183–95). OECD.
Naicker, S., Plange-Rhule, J., Tutt, R.C., & Eastwood, J.B. (2009). Shortage of healthcare workers in developing countries – Africa. *Ethnicity & Disease, 19*(S1), 60–4. https://doi.org/10.5414/CNP74S129.
Negin, J., Rozea, A., Cloyd, B., & Martiniuk, A.L.C. (2013). Foreign-born health workers in Australia: An analysis of census data. *Human Resources for Health, 11*(1), 69. http://www.human-resources-health.com/content/11/1/69.
Nichols, B.L., Davis, C.R., & Richardson, D.R. (2011). *The future of nursing: Leading change, advancing health.* National Academies Press.
Nyberg Sorensen, N., & Gammeltoft-Hansen, T. (2013). Introduction. In T. Gammeltoft-Hansen & N. Nyberg Sorensen (Eds.), *The migration industry and the commercialization of international migration* (pp. 1–23). Routledge.
Organisation for Economic Co-operation and Development. (2019). *OECD health statistics 2018.* OECD Publishing. http://www.oecd.org/els/health-systems/health-data.htm.
Patinkin, D. (1968). An internationalist model. In W. Adams (Ed.), *The brain drain* (pp. 92–108). Macmillan.
Pillinger, J. (2012). *International migration and women health and social care workers program, quality healthcare and workers on the move-Australia national report.* Public Services International.
Pittman, P., Folsom, A., Bass, E., & Leonhardy, K. (2007). *U.S.-based international nurse recruitment: Structure and practices of a burgeoning industry.* Academy Health.
Pittman, P.M., Folsom, A.J., & Bass, E. (2010). U.S.-based recruitment of foreign-educated nurses: Implications of an emerging industry. *American Journal of Nursing, 110*(6), 38–48. https://doi.org/10.1097/01.NAJ.0000377689.49232.06.
Pittman, P., Frogner, B., Bass, E., & Dunham, C. (2014). International recruitment of allied health professionals to the United States: Piecing together the picture with imperfect data. *Journal of Allied Health, 43*(2), 79–87. https://doi.org/10.1177/1077558711432890.
Rafferty, A.M., & Solano, D. (2007). The rise and demise of the Colonial Nursing Service: British nurses in the colonies, 1896–1966. *Nursing History Review, 15*(1), 147–54. https://doi.org/10.1891/1062-8061.15.147.
Rhodes, R.A.W. (1997). *Understanding governance.* Open University Press.
Rogerson, C., & Crush, J. (2008). The recruiting of South African health care professionals. In J. Connell (Ed.), *The international migration of health workers* (pp. 199–224). Routledge.
Shaffer, F.A., Bakhshi, M., Dutka, J.T., & Phillips, J. (2016). Code for ethical international recruitment practices: The CGFNS alliance case study. *Human Resources for Health, 14*(31). https://doi.org/10.1186/s12960-016-0127-6.
Shrecker, Y., & Labonte, R. (2004). Taming the brain drain: A challenge for public health systems in Southern Africa. *International Journal of Occupational and Environmental Health, 10*(4), 409–15. https://doi.org/10.1179/oeh.2004.10.4.409.

Skeldon, R. (2009). Of skilled migration, brain drains and policy responses. *International Migration*, 47(4), 3–29. https://doi.org/10.1111/j.1468-2435.2008.00484.x.

Snyder, J. (2009). Is health worker migration a case of poaching? *American Journal of Bioethics*, *9*(3), 3–7. https://doi.org/10.1080/15265160802654152.

Squires, A. (2008). Ethical international recruitment: Many faces, one goal – part 1. *Nursing Management*, *39*(9), 16–21. https://doi.org/10.1097%2F01.NUMA.0000335253.23789.98.

Sward, J. (2015) International recruitment: Current trends and their implications for small states. In W.H. Konje (Ed.), *Migration and development: Perspectives from small states* (pp. 325–52). The Commonwealth Secretariat.

Tankwanchi, A.B.S., Vermund, S., & Perkins, D. (2015). Monitoring sub-Saharan African physician migration and recruitment post-adoption of the WHO Code of Practice: Temporal and geographic patterns in the United States. *PLoS ONE*, *10*(4), e0124734. https://doi.org/10.1371/journal.pone.0124734.

van den Broek, D., Harvey, W., & Groutsis, D. (2016). Commercial migration intermediaries and the segmentation of skilled migrant employment. *Work, Employment and Society*, *30*(3), 523–34. https://doi.org/10.1177/0950017015594969.

Winkelmann-Gleed, A. (2006). *Migrant nurses: Motivation, integration and contribution*. Radcliffe.

World Health Organization. (2017, November 14). *International health worker migration: A high level dialogue*, Herbert Park Hotel, Dublin, Ireland. http://www.who.int/hrh/news/2017/high_level-dialogue-int-health-worker-migration-meeting-summary.pdf?ua=1.

9 The Ethical Recruitment of Internationally Educated Nurses? An Examination of the Devaluing of Nursing in the Philippines, a Sending Region

MADDY THOMPSON

Introduction

The Philippines is one of the largest exporters of nurses, and it is estimated that around 25 per cent of the world's migrant nurses are Filipino born (Lorenzo et al., 2007). The stories – the vulnerabilities, forms of exploitation, and successes – of Filipino nurse migrants, and Filipino migrants more widely, have been well-documented (Dalgas, 2014; Guevarra, 2010; Pratt, 1999; Terry, 2014; Tyner, 2004). Yet, in the overwhelming majority of cases, these are the stories of those nurses who have already migrated, those who have reached "greener pastures" (Ronquillo et al., 2011). I therefore orient attention to nurses living and working in the Philippines to draw attention to the various ways nursing is devalued in the Philippines as a consequence of the global migrations of Internationally Educated Nurses (IENs).

In this chapter, I explore how the construction of the profession of nursing in the Philippines is rooted in the complexities involved in the transfer of internationally educated healthcare professionals (IEHPs). By turning to a sending region, I draw attention to the ways that the global migration of nurses causes nurses' skills and training to be devalued in their home country. This is important, as both the existing body of academic literature and the national and international policies, codes, and practices concerning ethical recruitment of health care professionals are oriented towards practices in places that recruit IENs. There is a gap in understanding how the skills and credentials of IEHPs are valued and exploited in sending regions. I argue that a deeper understanding of the complexity of issues surrounding the migration of IEHPs in sending regions is necessary to ensure ethical codes and practices are truly ethical.

I begin by contextualizing the wider issues surrounding the migration of IENs and international credential recognition before discussing existing ethical recruitment practices and their limitations when it comes to aiding sending regions. I then turn to the case of the Philippines and the ways in which the recruitment practices of those hiring IENs gives rise to a culture of "volunteerism" in the national nursing system, where nurse graduates are expected to pay hospitals in order to gain the experience needed to migrate and/or receive permanent domestic employment as a nurse (Ronquillo et al., 2011; Thompson & Walton-Roberts, 2018). Following a brief overview of my methodology, I draw on empirical data to demonstrate how the lack of a clear international system of nursing credential recognition has contributed to the normalization of severe forms of exploitation for Filipino nurses in the Philippines. Here I show how Filipino nurses frame experiences of volunteerism as "luck" and present themselves as hardworking, subservient, and uncomplaining in order to compete in both domestic and international job markets.

Ultimately, I argue that the work of Filipino nurses in the Philippines and overseas is devalued through the process of volunteerism, which is intrinsically connected to the practices of overseas recruiters. This demonstrates a clear need for a more effective system of hiring IENs.

IENs and Global Credential Recognition

The global migration of nurses and health care workers is significant, and the percentage of IENs employed "in Australia, Canada, the United Kingdom, and the United States is reported to be between 21 and 33 percent" of their national health care workforces (Kingma, 2007, p. 1282). Existing literature regarding IENs largely focuses on experiences of adaptation and integration into the host society. Issues centre on differing standards and practices of nursing (Al-Hamdan et al., 2015; Choi & Lyons, 2012); discrimination and racism in caring institutions (Al-Hamdan et al., 2015; Ball, 2004; Nichols & Campbell, 2010; Smith & Mackintosh, 2007); and processes of deskilling (see Baumann et al., chapter 4 in this volume for a detailed account of this process in the Canadian context; see also Nichols & Campbell, 2010; O'Brien, 2007). Much has also been written concerning the wider pressures that led to the contemporary global demand for nurses, largely under the umbrella of global care chain thinking (see Walton-Roberts, 2012; Wojczewski et al., 2015; Yeates, 2004, 2012). In this section, I briefly summarize the key forces driving the demand for migrant nurses and discuss issues concerning standards, practices, and credential recognition (also see Kingma, 2006).

Contemporary nurse shortages are occurring in the Global North as populations (and nursing workforces) age (Ball, 2004; Buchan, 2001; Choi & Lyons, 2012; Kingma, 2006; Matsuno, 2009), and as women are afforded increased labour market choices (Kingma, 2006). Additionally, in the Global North in particular, childcare and elderly care is increasingly provided beyond the home or within the home by paid health care personnel rather than by family members, further increasing the demand for care workers (Bach, 2015; Kingma, 2006).

In the Gulf region, a key importer of migrant nurse labour, restrictive cultural practices limit the ability of women to enter nursing, as nurse-to-patient physical contact may be construed as inappropriate (Ball, 2004), while the availability of locally trained labour fluctuates in line with the oil sector (see Chikanda, chapter 8 in this volume). Furthermore, nurses in most national contexts are severely under-remunerated for their work compared to other occupations with similar educational, skill, or responsibility requirements, a legacy of the historic and contemporary association of nursing with caring, reproductive roles, and womanhood, making nursing an unattractive option for many (Kingma, 2006; Ohlén & Segesten, 1998; Smith & Mackintosh, 2007).

Geopolitical and global economic events and the resulting national governance and decision-making also impact the supply of nurses. The UK is a salient example of this, in part due to the deep involvement of the government in health care decisions through the National Health Service (NHS). Measures of austerity implemented since 2010 have included ending government subsidies of nursing education from 2017 (Department of Health and Social Care, 2017). Removing one of the few advantages of studying nursing has caused the number of applications to plummet (Matthews-King, 2018). Furthermore, the "Brexit" vote has resulted in fewer applications from EU trained nurses, as well as more EU trained nurses leaving the UK. Indeed, in the first year following the vote, there was a reported "96% drop in EU nurses registering to work in Britain" (Siddique, 2017). Beyond the Global North, shortages are caused by health epidemics; an inability to meet minimum numbers of health care personnel stipulated by international bodies and agreements, such as the Millennium and Sustainable Development Goals; and a lack of infrastructure that makes health care difficult to access (Thompson & Walton-Roberts, 2018). Yet in certain places, it also is the emigration of nurses that further exacerbates these problems. This has prompted various bodies including the NHS, the Commonwealth Nurses Federation, and the International Council of Nurses to suggest sets of ethical guidelines for the recruitment of overseas nurses in the last twenty years, although the extent to which these codes can be enforced is debatable (Connell & Walton-Roberts, 2016; Kingma, 2006).

Importing IENs is a common "fix" to nursing shortages, and the international mobility of nurses is a key strategy for many regions and countries suffering from nursing shortages (Kingma, 2006). However, despite constructions of nursing as a global occupation based on the universal notion of care, in reality nursing is a highly differentiated occupation. The professional status of nursing varies internationally, meaning educational requirements differ. For example, in the US, as well as in the Philippines and other countries influenced by American nursing styles, nursing has been professionalized for over a century and has always required a degree (see chapter 14 in this volume; Thompson & Walton-Roberts, 2018). In the UK, conversely, it is only since

September 2013 that nursing has become a fully professionalized occupation, as previously there were both degree and diploma options available (National Health Service [NHS], 2016). In countries where nursing has been influenced by the UK, such as in Singapore, nursing is still stratified into professional/registered/degree-holder nurses and semi-professional/enrolled/diploma holder-nurses (Choi & Lyons, 2012). Furthermore, cultural understandings of what care should be conducted beyond the home, combined with socio-economic restrictions on welfare opportunities, influence the types of caregiving nurses are exposed to (Huang et al., 2012; Raghuram, 2012; Vaittinen, 2014). A key example of this is elderly care, which in many Southeast Asian societies, the Philippines included, is largely conducted in private spheres, but in the Global North and Gulf regions is increasingly outsourced beyond the family unit. Indeed, elderly care tends to suffer most from nursing shortages, the work involved being relatively low-skill and with few opportunities for career development. It is therefore a key sector for migrant nurses (O'Brien, 2007). Finally, it is important to note that in the case of Canada, India, and the US, nursing qualifications and standards can differ on a country/provincial/state level, further adding to the complexity of nursing globally (see also Sweetman, chapter 3; Baumann et al., chapter 4; and Bourgeault et al., chapter 5 in this volume).

Global variety in the professional standards of nursing complicates efforts to develop effective systems of credential recognition. Added to this are difficulties concerning language. Industries with more successful examples of international credential recognition, such as IT or engineering, albeit themselves imperfect (see Bauder's 2003 discussion of the concept of "brain abuse"), are less reliant on oral communication, relying instead on universal codes and formulae. Additionally, these industries have fewer immediate dangers connected to mis- or non-communication than nursing, where not understanding a patient who is explaining their symptoms, or administering incorrect medication or procedures, can have life-threatening implications.

Due to these issues, nurse employers and/or the recruiters they work with must develop, maintain, and review highly detailed information concerning not only the country's training and registration systems and nursing curriculum, but issues of corruption, of the delivery and quality of education and training within certain institutions, of facilities in hospitals and other places of employment, and of wider cultures of nursing. This explains why nurse migration and credential recognition between countries with similar historical nursing contexts and similar contemporary socio-economic standing is easier (consider the UK and Australia, for example). Credential recognition for nurses is even more straightforward within the EU/EEA, where not even language proficiency is required (see Connell and Negin, chapter 12 in this volume). It also explains why there may be more difficulties recruiting nurses from poorer regions where educational and training standards are questionable, and where corruption is more common. For example, in the Philippines, the mushrooming in the number of educational institutions offering nursing since 2000 has led to poor pass rates and a lack of training facilities to ensure undergraduates receive adequate education (Masselink & Lee, 2010).

The difficulties involved in employing internationally trained and educated nurses from the Global South has led recruiters and employers to adopt other tactics to ascertain the potential quality of nursing care that will be "imported." These tactics include requiring bridging courses or additional training in the destination country before migrant nurses are allowed to sit nursing board examinations (see Sweetman, chapter 3 in this volume for a detailed examination of these practices and their implications for IENs in the Canadian context), or requiring language proficiency to be evidenced through expensive tests administered by the destination country. Such tactics impose severe financial and time costs on those planning to migrate as well as those who have recently migrated (see also Connell and Negin, chapter 12 in this volume). However, as I demonstrate later, those without any desire to migrate and who instead seek employment in their home country are also impacted by the international transfer of nurses.

Finally, employers and recruiters of IENs tend to expect and look for different qualities when hiring migrant nurses, typically relying on highly racialized and ethnicized representations. This is because of the difficulties in assessing credentials as discussed above, but also because migrant nurses are commonly hired to fill the biggest deficiencies in nursing care – the dirty or mundane aspects of nursing, which generally require skills of bedside care and the carrying out of intimate tasks such as cleaning, hygiene, and food preparation above clinical tasks (see Crush, chapter 16 in this volume; Batnitzky & McDowell, 2011; O'Brien, 2007).

Employers and recruiters of IENs typically desire nurses who will "know their place" in the nursing hierarchy, and who will be pliable, compliant with, and subservient to the native nurses both to avoid friction and to ensure that promotion opportunities are reserved for native-born and trained nurses (Batnitzky & McDowell, 2011; O'Brien, 2007). Migrant nurses are expected to work without complaint and without disrupting the health care systems they enter (Terry, 2014). Costs associated with hiring IENs means there is also a desire to get the best value for one's money, and migrant nurses are expected to be exceptionally hardworking. Migrant nurses from the Global South have repeatedly been found to work unsociable evening, overnight, and weekend shifts, as well as to take on significantly more overtime than their domestic counterparts (Ball, 2004).

The Effectiveness of Ethical Recruitment Codes in Sending Regions

As noted, since the turn of the century there have been some attempts to work towards binational or international forms of ethical recruitment of IENs and IEHPs more broadly. Here, I briefly discuss the existing practices that attempt to alleviate the issues associated with international value transfers, as well as reflect on some of the critiques and limitations of these approaches.

Discussions and debates concerning the ethical recruitment of IEHPs arose from sending nations at the turn of the century, with Southern African nations concerned with the effects on their already fragile health care systems of large numbers of qualified health care professionals emigrating (Pagett & Padarath, 2007). In 2004, at the 57th World Health Assembly, African ministers of health lobbied for policy changes to mitigate the effects from the emigration of health care workers, and a Resolution urged members to develop their own policies. The First Global Forum on Human Resources for Health was held in Kampala, Uganda, in March 2008, and from there the World Health Organization (WHO) began drafting the Code of Practice on the International Recruitment of Health Personnel in collaboration with the Organisation for Economic Co-operation and Development (OECD), non-governmental organizations (NGOs) working in migration and health, and UN member states. It was formally adopted by all 193 WHO member states at the 63rd World Health Assembly in 2010 (Zurn, 2015). The Code was informed by existing policies and recommendations developed in African nations and in key importers of IEHPs, primarily Australia and the UK.

Table 9.1 includes selected international and national organizations who have developed codes that aim to promote the ethical recruitment of IEHPs.[1] In line with this chapter's focus on a sending region, only key elements of these codes that apply to the sending regions are highlighted. It should be noted that all codes also have a clear focus on the rights of health care workers to seek employment overseas and their rights in overseas health care employment settings, as well as a recognition that developing countries need to do more to "recognize the health workforce as a productive investment rather than an expense to curtail" (Cometto, 2015, p. 2).

Table 9.1 demonstrates two important points. First, for the international codes, adopting recommendations is voluntary and enforcement is therefore non-existent; instead, the WHO simply monitors policies. The WHO in particular has limited authority to demand compliance while also having multiple stakeholders with competing interests to appease. Indeed, this is all too evident in the Code's opening objective: "to establish and promote *voluntary* principles and practices for the ethical international recruitment of health personnel, taking into account the rights, obligations and expectations of source countries, *destination countries* and migrant health personnel [emphasis added]" (WHO 2010, Article 1.1). The focus of the WHO and the International Council of

Table 9.1. Selected ethical recruitment codes and statements for the international recruitment of IENs

Name of code/statement	Scope and legal status	Relevant points for sending regions
2001 International Council of Nurses (ICN) Position Statement: International career mobility and ethical nurse recruitment (revised 2007 and 2019)	Represents over 130 national nursing associations. No legal status. Provides means of lobbying and advising.	Briefly refers to idea of "national sustainability" and "regulation of recruitment." 2019 revision adds need for comprehensive and effective international regulation of nursing registration.
2004 Department of Health (UK) Code of Practice for International Recruitment (commonly referred to as the NHS Code)	Employees and employers of the NHS in the UK and Northern Ireland. Private industry on voluntary sign-up basis (only from 2014 revision by the NHS).	Code "underpinned by the principle that any international recruitment of healthcare professionals should not prejudice the healthcare systems of developing countries" but only one guiding principle (out of seven) relates to sending regions: "Developing countries will not be targeted for recruitment, unless there is an explicit government-to-government agreement with the UK to support recruitment activities." Notably, the Philippines has a Memorandum of Understanding with the UK making it an ethical' source country (Lorenzo et al., 2005), despite it being on the list of countries to avoid recruitment as developed by the Department for International Development and Department of Health and Social Care.
2008 Voluntary Code of Ethical Conduct for the Recruitment of Foreign-Educated Health Professionals to the United States (evolved to Alliance for Ethical International Recruitment Practices)	Various health care stakeholders in USA can voluntarily sign up.	Acknowledges that "the interests and responsibilities of nurses, source countries, and employers in the destination country may conflict" and "affirms that a careful balancing of those individual and collective interests" is the best course of action.

(*Continued*)

Table 9.1. Selected ethical recruitment codes and statements for the international recruitment of IENs (*Continued*)

Name of code/statement	Scope and legal status	Relevant points for sending regions
2008 Voluntary Code of Ethical Conduct for the Recruitment of Foreign-Educated Health Professionals to the United States (evolved to Alliance for Ethical International Recruitment Practices) (*continued*)		Best Practices (but not minimum standards) include: Pursuing health facilities partnership agreements with hospitals and colleges, establishing scholarship funds in sending regions, and allowing IENs to return home to help in critical shortages. Also avoiding active recruitment from places identified with chronic nursing shortages.
2010 World Health Organization Global Code of Practice on the International Recruitment of Health Personnel	All 193 UN member states have agreed, but the Code is voluntary and non-binding.	Article 5 examines "health workforce development and health systems sustainability" (prioritizing bilateral agreements), but without providing specific guidance as to what such agreements may entail. Recommends that developed countries "to the extent possible, provide technical and financial assistance … including health personnel development."

Nurses (ICN) is on advising and guiding nations on the development of national and binational policies, codes, and agreements. There is significant variation in the ways nations approach this, and the examples of codes developed in the UK and US demonstrate this clearly. While the UK's code is legally binding and offers some clear steps to ethical recruitment, it is highly limited and does little to actually mitigate effects where bilateral agreements are in place. I provide evidence of this below when discussing the culture of volunteerism. The US's code, while less prohibitive in regards to places from which nurses can be recruited, provides much clearer recommendations to mitigate some of the issues in sending regions, but is entirely voluntary.

Second, the codes have limited effectiveness. The WHO Code and ICN Position are focused only on providing guidance for members developing policy and bilateral arrangements, rather than on providing a global regulatory framework, and are therefore partial rather than comprehensive in terms of their coverage.

Furthermore, in 2015 the WHO underwent a review of the relevance and effectiveness of the Code, finding just 20 per cent of national policies relating to the Code included forms of financial or technical assistance to sending regions, and only 37 per cent had entered bi- or multi-lateral agreements (Salehi Zalani et al., 2015). Conversely, 90 per cent of national policies provided IEHPs with the same legal rights in terms of employment (Salehi Zalani et al., 2015), demonstrating that the Code's application has been focused on the rights of workers as individuals rather than the rights of nations. This further demonstrates the need to orient attention to conditions in the sending regions, which are a low priority in international and national approaches, despite their centrality in bringing this issue to the global stage.

In addition to these two points is the vagueness of the codes in relation to sending countries, as few specific practices beyond "technical assistance" are proffered. Although the WHO has identified various national policies that provide clearer assistance to sending nations, they are not built into the Code's ethical guidelines. Indeed, the WHO's 2015 review of the Code finds that "nowhere does the Code explicitly point to international cooperation or support from developed countries with the aim of supporting HRH [Human Resources for Health] retention in underserved areas in low/middle income countries" (Dhillon, 2015, p. 4). Additionally, the NHS Code includes a clear "get-out clause" in that so long as a bilateral agreement is made, recruitment can continue.

In the following section, I turn to the culture of volunteerism in the Philippines that I argue has grown as a result of the lack of an effective global system for the ethical recruitment of nurses. The latter part of the chapter turns to the experiences of Filipino nurses in the Philippines to demonstrate the ways that this in turn leads to the severe undervaluing of Filipino nurses both at home and overseas.

Exploitation in Filipino Nursing: The Culture of Volunteerism

In many places with high levels of outward migration of trained and qualified nurses, it is poor employment conditions, including under-remuneration, lack of facilities, lack of opportunities for promotion, and unsafe working conditions, that prompts individuals to seek employment overseas (Connell & Walton-Roberts, 2016). This is the case in the Philippines, and as Ball (2004) highlights, exploitative labour conditions are integral in "pushing" nurses into seeking opportunities overseas. In the Philippines, employment conditions for nurses are particularly problematic due to the "culture of volunteerism" that exists (Pring & Roco, 2012). The culture of volunteerism involves nurses working in hospitals, in some cases for more than two years, with no or very little pay in order to gain experience necessary for obtaining paid employment. The practice is prevalent in both public and private sectors, but where "volunteers" in the public sector may secure some reimbursement for travel and uniform expenses, those in the private sector tend to be charged fees to gain hospital experience.

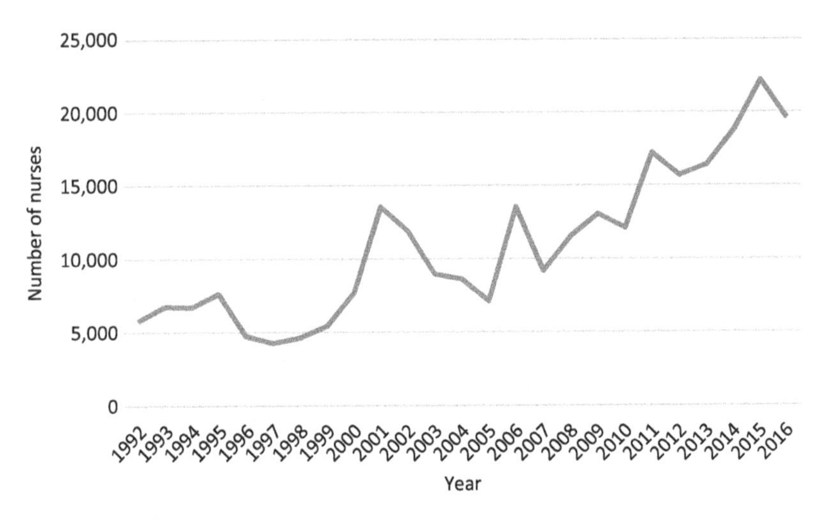

Figure 9.1. Number of deployed Filipino workers hired as nurses, 1992–2016
Source: Data compiled from yearly POEA "OFW Statistics: Deployment per Skill per Sex" (Philippine Overseas Employment Administration [POEA], 2018).

This culture of volunteerism has been prevalent in the Philippine nursing sector since at least the 1990s, correlating with higher numbers of nurses emigrating (see figure 9.1) and requiring hospital-level experience to do so. Many overseas recruiters, unable to easily verify the credentials of IENs, instead stipulate minimum terms of experience, often between one and two years within a hospital setting. The culture of volunteerism has intensified and become more widely spread due to the twenty-first-century expansion of higher education institutions (HEIs) offering nursing education. From the 1970s to 2004 there was a fourfold increase in the number of nursing programs (Lorenzo et al., 2005). In Metro Manila alone there are currently sixty colleges offering nursing courses (FindUniversity.ph, 2020). At minimum there is an estimated 290,000 unemployed or underemployed Filipino nurses in the Philippines competing for limited positions both overseas and at home (Lapeña, 2011).[2]

In 2011, due to ongoing pressure from nursing advocates and educators, the Aquino administration formally outlawed volunteerism in Memorandum 2011-0328. The memorandum states:

> All DOH [Department of Healthcare] hospitals are hereby directed to discontinue all existing programs involving nurses who deliver free services in exchange for work experience/volunteer nurses, volunteer trainings, and all other similar programs. All hospitals-based [*sic*] trainings for nurses should follow a definite career

progression to be defined and accredited by the DOH and Professional Regulatory Commission-Board of Nursing. (DOH, 2011)

The memorandum has a severely limited scope, applying only to DOH hospitals, of which there are about 70 out of some 1,800 (WHO & DOH, 2012). Furthermore, it does not provide any punitive measures for hospitals, DOH or otherwise, for failing to comply. Finally, it does not actually outlaw the system in which nurses pay to train, so long as training provided can be seen to "follow a definite career progression." If anything, this caveat has resulted in the further exacerbation of the culture of volunteerism. Hospitals, DOH ones included, are becoming more and more concerned with nurses being able to evidence their training through certificates, which are rarely valid overseas but increasingly needed to find paid domestic positions. These training courses can be short, ranging from a few days to a month, or an even longer-term placement. They are specialized courses, focusing on clinical skills, such as intravenous (IV) training, or on specialist skills, such as renal dialysis. Nurses must pay for these courses, and costs can vary from a short IV course at around 1,500PHP (CAD$37) to over 10,000PHP (CAD$250) for several months of dialysis training. In most cases, a nurse will receive training in a hospital setting and will essentially carry out the duties of a staff nurse.

Requiring training certificates to have a chance to find paid employment in the Philippines has transformed volunteerism from an informal yet common occurrence to "an institutionalised and formalised practice" (Thompson, 2019, p. 9). Nurses without desires to migrate are now also facing increasing pressures to "volunteer" their time and skills in order to find decent domestic employment in both urban and rural contexts (Pring & Roco, 2012). In this way, the lack of a credential system leads to the significant devaluing of Filipino nurses' work, not just overseas as migrants but in the Philippines as locally employed nurses too.

It is crucial to understand the ways in which this exploitative system of volunteerism and devaluing of work is maintained. Figure 9.2 displays the multiple interconnected and interlocking factors that have led to the development and maintenance of the culture of volunteerism. In the Philippines, perhaps unsurprisingly, the system of volunteerism is highly lucrative for already cash-strapped hospitals that essentially receive free labour from the "volunteering" nurses. Many nurses I spoke with who had experience volunteering in public hospitals reported that while volunteering they would also pay for patients' medicines and bring in supplies. Those in private hospitals, however, described how hospital cleaners were paid more than nursing staff, regardless of employment/volunteer position, and how new building and equipment seemed to suggest that money exists. There is little motivation for further policy changes

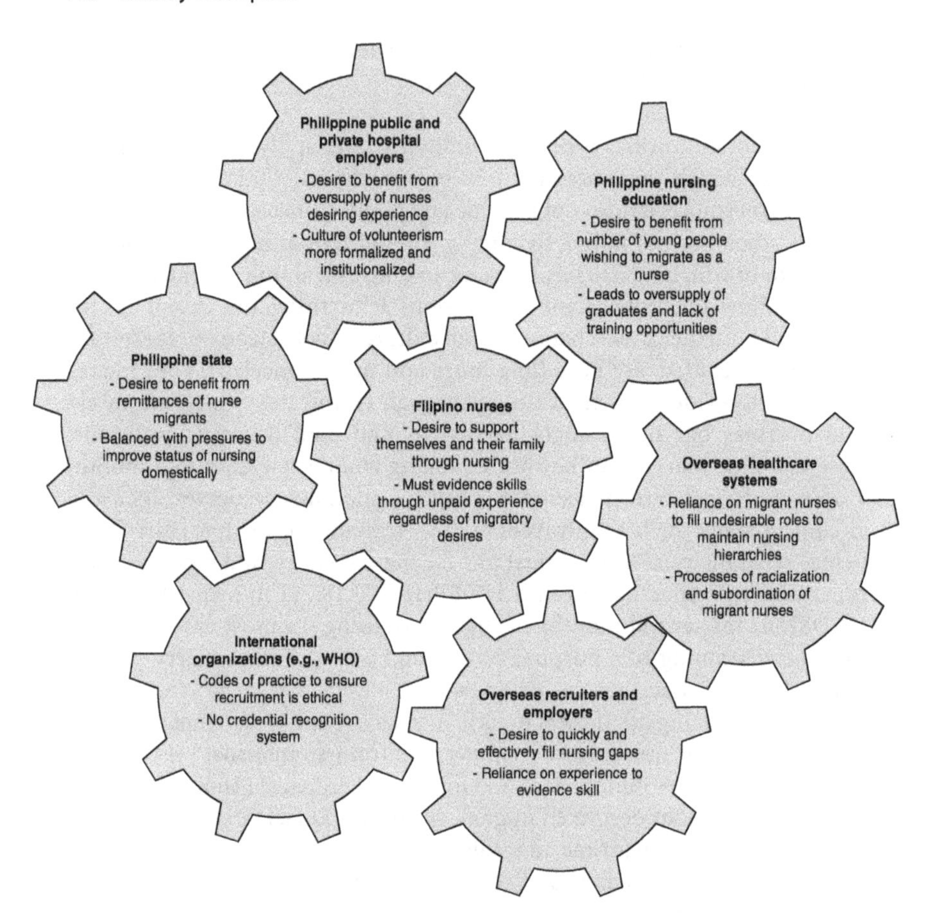

Figure 9.2. Actors involved in the maintenance of the culture of volunteerism and their primary interests
Source: Created by author.

at a governmental level when the export of nurses brings in significant money in the form of remittances and aids bilateral relations.

While it is clear that actors in the Philippines benefit from the lucrative nature of volunteerism and nurse migration, the role of overseas recruiters and states cannot be underestimated. Beyond the Philippines the culture of volunteerism is stimulated through often unethical recruitment practices and requirements. Due to the aforementioned issues in evaluating and recognizing the credentials of IENs, foreign recruiters rely on experience as an indicator of quality. Indeed, some overseas recruiters stipulate in recruitment materials that "volunteer experience" counts towards work experience, demonstrating knowledge of and complicity with the culture of volunteerism. A recent job posting advertised by

Cambridge University Hospitals (UK) (Work Abroad.ph, 2017), targeted only at those of a Filipino nationality (reflecting the "ethical" guidelines followed by the NHS), requires one year of experience but stipulates that "some volunteer experience is acceptable, provided that the work was undertaken full-time and with the responsibilities of a staff nurse position or higher."

Furthermore, this experience is often excessive. One to two years' experience is a standard requirement of IENs, and often, to better ensure quality, experience is restricted to that gained in large hospitals, sometimes with certain unrelated wards listed as essential. An ad displayed on Ingulfjob.com (2016; domain has since expired), a leading recruitment site for migrant positions in the Gulf region, for example, calls for a "well experienced Home Care Registered nurse in Dubai." However, no experience in home care is stipulated; instead, the basic skills requirements required exceed the job role, asking for a minimum of "2 years working experience in Hospital setting / medical, surgical ward, CCU, ICU, P & NICU"[3] ("Female Filipino nurse jobs," 2016). All such wards require an advanced knowledge of medical technologies and require nurses to deliver acute care in often stressful situations. These are the more privileged and desired nursing positions that nations facing severe nursing shortages are generally able to fill with domestic nurses. However, they are by no means necessary experience for a "home care registered nurse," whose duties would be limited to less skilled caring tasks.

This failure of domestic governance and overseas recruiters in effectively addressing the "culture of volunteerism" demonstrates the dominance of global market forces in not only defining the skills needed for Global South nurses to move north, but in creating an environment in which highly exploitative employment practices thrive for all nurses, regardless of migratory desire. Below, I examine the ways nurses experience and respond to these exploitative conditions. As I show, in many cases, to overcome the hypercompetitive nature of nursing as a Filipino, participants engaged in practices of self-exploitation and rewrote their narratives in terms of "luck." This further contributes to the devaluing of skills for these nurses.

Methodology

The following section is based on empirical fieldwork conducted in the Philippines from June–December 2015, as well as two shorter return trips where further materials from recruitment agencies were collected (July 2016 and April 2018). This research was funded by the Economic and Social Research Council, under grant ES/J500082/1. Additionally, online material on recruitment and national health websites has been consulted. The empirical fieldwork consisted of semi-structured interviews with thirty-five nurse graduates and thirteen nurse undergraduate students living in Metro Manila. In all cases, pseudonyms have been assigned to participants to further ensure anonymity. I spoke with both women (thirty-four) and men (fourteen) with a wide range of employment experiences and migratory desires.

The wider project objective is to understand how nurses are drawn into global and international circulations of health care, and nurses were questioned about their decisions to enter nursing, their understandings of what nursing is both in the Philippines and beyond, their experiences of employment in nursing, and their desires of migration (Thompson, 2019). The following section focuses on nurses' experiences and perceptions of employment and volunteerism in order to consider how issues associated with international credential recognition in the nursing sector have profound implications for the lives and opportunities of nurses in the Philippines, regardless of their migratory desires.

Experiencing Exploitation: The Hardworking and Uncomplaining Filipino Nurse

In this section, I examine how the domestic nursing employment sector in the Philippines both produces and undergirds the structural subordination of Global South IENs into nurse receiving regions. I first discuss participants' reflections on their experiences of volunteering and then move to consider how the culture of volunteerism influences how they perceive and represent themselves, and Filipino nurses more broadly, as hardworking, uncomplaining, and subservient. Importantly, although only a few participants are referred to here, this construction of the Filipino nurse was an overwhelming constant throughout the narratives of all forty-eight participants.

Firstly, in relation to the "culture of volunteerism," although participants are aware of the exploitative nature of volunteerism, "think that [it]'s unfair" (Gabriel), and are generally critical of the practice, they simultaneously frame their unpaid experience in relatively positive ways. The sense that one is "lucky" if given a chance to volunteer was common with many responses. For example, Leon states, "I know it's not hard to volunteer 'cause it's really helpful." Ian, who volunteered around fifteen years ago, told me he was "lucky" to "just [pay] 500 pesos" for his volunteering at a public hospital considering costs are now higher. Even though Ian has never desired migration, he was nonetheless trapped within the culture of volunteerism and volunteered for a year to gain "very good experience." He left the profession shortly after his stint volunteering due to burnout and is now an entrepreneur.

Secondly, as discussed earlier, migrant nurses from the global south often have additional and differing expectations than other nurses. Beyond evidencing skill in nursing (through experience levels), the nurses who are employed overseas must also live up to the expectations and requirements of recruiters and employers if they are to keep their jobs and maintain the image of their nation/region as a producer of quality nurses. While this is primarily achieved through the range of material and discursive practices enacted by the Philippine state that seek to push nurses to embody and perform an idealized version of a model nurse and advertise this to the world (for a detailed examination of these practices, see Rodriguez, 2008; see also Guevarra, 2010; Ortiga, 2014; Tyner, 1994, 2004), here I demonstrate how the "culture of volunteerism" is also implicated in producing nurses suited for overseas exploitation and deskilling through normalizing compliance with exploitative practices in the Philippines.

In almost all cases, the participants I spoke with had fully internalized and accepted the imagination of the Filipino nurse as being a hardworking, uncomplaining servant. Hard work, happiness, and subservience are presented as inherent qualities of Filipinos more widely and are advertised as qualities that make Filipinos suited for both nursing and migration (see also Choi & Lyons, 2012). More than this though, participants presented Filipinos as being the embodiment of the ideal nurse. Jessica, for example argues that "we [Filipinos] can do overtime for as long as you want to, but in other countries, you will not do overtime." Sofia, who worked in Singapore, similarly creates binaries between Filipino nurses and "Others," in this case both native Singaporean and non-Filipino migrant nurses she encountered:

> The difference between others and Filipino nurses is, when they [Filipinos] do their jobs, they do it well, and it's like they're not time conscious ... if their work is not yet finished, you know the paperwork, they extend and work more time, and they do it properly ... [But] other citizens from other countries, when the shift ends, whether their work is done or not, they are going [home].

For Sofia, who volunteered for several years in the mid-2000s before finding employment as a nursing assistant in Singapore, unpaid work was nothing new. She made no complaints to her overseas employers and worked all paid overtime they requested of her. Although she was not employed as a registered nurse, Sofia effectively carried out the duties of one – which is not uncommon (Choi & Lyons, 2012) – and again, did not complain to her employers about this. Just as in the Philippines, Sofia's employers in Singapore benefited unduly from her qualifications and labour.

Similarly, Kate, who at the time of interview was dedicating her time to supporting the Red Cross before her imminent deployment to Abu Dhabi, had spent time volunteering in private hospitals, to gain sufficient experience for her overseas role. She clearly identifies that because of the heavy workload in the Philippines, she cannot envisage any reason why as a migrant she would complain about overwork. This naturalizes an uncomplaining and subservient attitude: "Filipino nurses are hardworking. And they do not complain much. Because the workload here, in the public hospitals here, the workload is lighter [than] in other countries. So, why would you complain? ... You can never complain because it's part of your job."

This attitude is also reflected in the accounts of the undergraduate students I spoke with, those without experience in the nursing employment market. Bianca, who plans to remain in the Philippines to work as a nurse rather than migrating, internalizes the notion of the hardworking, and in this case, underpaid Filipino nurse, stating, "Filipinos are hardworking, and they easily get satisfied even with low salary." Again, she is naturalizing future exploitation, in this case, in the Philippines, but this time without direct experience of work/volunteering, or without a desire to migrate.

The sense that one is "lucky" if given a chance to volunteer was common for all those with volunteer experience, and the idea that Filipino nurses are hardworking, uncomplaining, and subservient was a recurring theme noted by all forty-eight participants. The nurses, in their early careers, before migration is a concrete option, even in college, are learning to suffer in silence, to work as hard as they can, to value the work over the remuneration, and to never complain. Above all, they are learning that their skills are not valued, whether as a migrant or in the Philippines. For employers in the Philippines and overseas, however, there is significant value-added (see Ravenhill, chapter 1 in this volume) in hiring hardworking nurses satisfied with low pay.

Concluding Thoughts

Through examining the forces that sustain the culture of volunteerism in the Philippines and the experiences of the nurses whose lives are impacted, it is evident that the lack of a global credential recognition system for nursing has wide-reaching consequences. Nurses in the Philippines, regardless of their desires to migrate or not, are left with little choice but to engage in the culture of volunteerism if they are to remain within the occupation of nursing.[4] Recruiters of IENs desire to source reliable, skilled, hardworking, and compliant workers, and therefore rely on experience rather than credentials to recruit nurses from the Global South. This, combined with an oversupply of nurses in the Philippines, the opportunism of hospital employers, and failures of Philippine governance has created a fertile environment for the highly exploitative culture of volunteerism to thrive. Furthermore, the culture of volunteerism becomes another way Filipino nurses are disciplined into becoming ideal migrant workers who are hardworking, uncomplaining, and pliable, and who not only expect their labour to be devalued and exploited, but have already learnt how to frame their exploitation in positive terms, without complaint.

Solutions to these issues, however, are as complex and multifaceted as the issues themselves (see figure 9.2). Further investment into health care and stricter employment regulations may reduce the plight of nurses in the Philippines, but because nurse migrants are consistently some of the highest remitters of money earned overseas, it is unlikely the Philippine state will seriously attempt to improve domestic conditions, or indeed, financially could (Thompson & Walton-Roberts, 2018). Even if they did, the estimated 290,000+ qualified nurses who are underemployed or unemployed, combined with the lucrative nature of the culture of volunteerism means hospitals would likely attempt to lobby against or circumvent new policy interventions, as they already have.

The cultural embeddedness of volunteerism – it has now existed for at least three decades – means that attempts to influence recruitment practices and national and international codes of conduct are complicated. Outlawing overseas recruiters from accepting experience gained through volunteerism would disadvantage the thousands of nurses who have already dedicated their time and resources to providing unpaid health care in the Philippines. Furthermore, endemic issues of corruption in the Philippines in all levels of governance mean that while regulations concerning the type of experience recruiters can require (hospital versus community-based experience, for example) seems an attractive solution, international recruiters would take on significantly more risk. An international credential recognition system would go some way to towards alleviating these issues, yet the complexities within the sector of nursing globally are significant. With such a wide-ranging number of actors each with distinct and often competing interests, this will require a multi-stakeholder, multi-scalar, international approach with significant financial costs. While nursing continues to be devalued as a profession in most contexts throughout the world, this makes such a system difficult to envisage. While the COVID-19 pandemic may have turned the world's attention to the importance of nurses, including migrant nurses following the case of the UK prime ministers' treatment by Portuguese and New Zealander nurses, it remains to be seen if this will translate into the valuing of nurses through investment.

NOTES

1 Chikanda in chapter 8 of this volume provides a detailed account of the ways in which private recruiters and other migration intermediaries are involved in these "ethical" recruitment practices, while Ennis in chapter 7 considers practices on a subnational scale.

2 Data do not seem to exist concerning the numbers of nurse graduates each year, or is piecemeal at best. Nonetheless, there is a clear oversupply of nurses.

3 CCU is coronary care unit/cardiac intensive care unit. ICU is intensive care unit/ intensive treatment unit/critical care unit. NICU is neonatal intensive care unit. PICU is paediatric care unit.

4 Elsewhere, I demonstrate how many nurses unable to afford the costs of volunteering leave their occupation for other care-related globalized roles such as "call-centre nursing" (Thompson, 2019).

REFERENCES

Al-Hamdan, Z.M., Al-Nawafleh, A.H., Bawadi, H.A., James, V., Matiti, M., & Hagerty, B.M. (2015). Experiencing transformation: The case of Jordanian nurse immigrating to the UK. *Journal of Clinical Nursing, 24*(15–16), 2305–13. https://doi.org/10.1111/jocn.12810.

The Alliance for Ethical International Recruitment Practices. (2008). Health Care Code for Ethical International Recruitment and Employment Practices. https://www.cgfnsalliance.org/wp-content/uploads/2019/03/Health-Care-Code-for-EIREP-Sept- 2017_FINAL.pdf.

Bach, S. (2015). Nurses across borders: The international migration of health professionals. In B. Parry, B. Greenhough, T. Brown, & I. Dyck (Eds.), *Bodies across borders* (pp. 153–69). Ashgate. https://www.researchgate.net/publication/281241849_Nurses_Across_Borders.

Ball, R.E. (2004). Divergent development, racialised rights: Globalised labour markets and the trade of nurses-The case of the Philippines. *Women's Studies International Forum, 27*(2), 119–33. http://dx.doi.org/10.1016/j.wsif.2004.06.003.

Batnitzky, A., & McDowell, L. (2011). Migration, nursing, institutional discrimination and emotional/affective labour: Ethnicity and labour stratification in the UK national health service. *Social and Cultural Geography, 12*(2), 181–201. https://doi.org/10.1080/14649365.2011.545142.

Bauder, H. (2003). "Brain abuse", or the devaluation of immigrant labour in Canada. *Antipode, 35*(4), 699–717. https://doi.org/10.1046/j.1467-8330.2003.00346.x.

Buchan, J. (2001). Nurses moving across borders: "Brain-drain" or freedom of movement. *International Nursing Review, 48*(2), 65–7. http://dx.doi.org/10.1046/j.1466-7657.2001.00073.x.

Choi, S., & Lyons, L. (2012). Gender, citizenship, and women's "unskilled" labour: The experience of Filipino migrant nurses in Singapore. *Canadian Journal of Women and the Law, 24*(1), 1–26. http://dx.doi.org/10.3138/cjwl.24.1.001.

Cometto, G. (2015). Strategic considerations to learn from experience and reinforce implementation. WHO Expert Advisory Group for the Global Code of Practice on the International Recruitment of Health Personnel. Item 7. https://www.who.int/hrh/migration/Item7-HL_WHO_Code_brief_15Apr2015_GiorgioCometto.pdf?ua=1.

Connell, J., & Walton-Roberts, M. (2016). What about the workers? The missing geographies of health care. *Progress in Human Geography, 40*(2), 158–76. https://doi.org/10.1177/0309132515570513.

Dalgas, K.M. (2014). Becoming independent through au pair migration: Self-making and social re-positioning among young Filipinas in Denmark. *Identities, 22*(3), 333–46. https://doi.org/10.1080/1070289X.2014.939185.

Department of Health. (2004). *Code of Practice for the International Recruitment of Healthcare Professionals.* https://www.nhsemployers.org/your-workforce/recruit

/employer-led- recruitment/international-recruitment/uk-code-of-practice -for-international-recruitment.

Department of Health. (2011, August 22). Termination of all "Volunteer Training Programs for Nurses" and all similar or related programs in DOH-retained hospitals. Memorandum 0238, August 22. [Unpublished].

Department of Health and Social Care. (2017, January 27). *NHS bursary reform*. Gov. UK. https://www.gov.uk/government/publications/nhs-bursary-reform/nhs -bursary-reform.

Dhillon, I. (2015). WHO Global COP Recommendations & Policy Drivers Relevance Analysis. WHO Expert Advisory Group for the Global Code of Practice on the International Recruitment of Health Personnel. Item 11. https://www.who.int/hrh /migration/Item11-Tabular_summary-CodeArticlesPolicyDrivers-IbadatDhillon .pdf?ua=1.

Female Filipino nurse jobs in United Arab Emirates. (2016). Retrieved 18 July 2016, http://ae.ingulfjob.com/emirates-jobs-query-female-filipino-nurse.html [site no longer exists].

FindUniversity.ph. (n.d.). *Schools offering nursing courses in Metro Manila*. Retrieved 18 July 2016, from https://www.finduniversity.ph/search.aspx?sch=1&ca=23 ®ion=national-capital-region.

Guevarra, A.R. (2010). *Marketing dreams, manufacturing heroes: The transnational labor brokering of Filipino workers*. Rutgers University Press.

Huang, S., Thang, L.L., & Toyota, M. (2012). Transnational mobilities for care: Rethinking the dynamics of care in Asia. *Global Networks*, *12*(2), 129–34. https://doi.org/10.1111/j.1471-0374.2012.00343.x.

International Council of Nurses. (2019). Position statement: International career mobility and ethical nurse recruitment. https://www.icn.ch/system/files /documents/2019-11/PS_C_International%20career%20mobility%20and %20ethical%20nurse%20recruitment_En.pdf.

Kingma, M. (2006). *Nurses on the move: Migration and the global health care economy*. Cornell University Press.

Kingma, M. (2007). Nurses on the move: A global overview. *Health Services Research*, *42*(3.2), 1281–98. https://doi.org/10.1111/j.1475-6773.2007.00711.x.

Lapeña, C.G. (2011, May 20). PRC: Nearly 290,000 nursing grads underemployed. *GMA News Online*. http://www.gmanetwork.com/news/news/nation/221126 /prc-nearly-290-000-nursing-grads-underemployed/story/.

Lorenzo, F.M.E., Galvez-Tan, J., Icamina, K., & Javier, L. (2007). Nurse migration from a source country perspective: Philippine country case study. *Health Services Research*, *42*(3.2), 1406–18. http://dx.doi.org/10.1111/j.1475-6773 .2007.00716.x.

Masselink, L.E., & Lee, S.-Y.D. (2010). Nurses, Inc.: Expansion and commercialization of nursing education in the Philippines. *Social Science & Medicine*, *71*(1), 166–72. https://doi.org/10.1016/j.socscimed.2009.11.043.

Matsuno, A. (2009). *Nurse migration: The Asian perspective*. International Labour Organization. http://www.ilo.org/wcmsp5/groups/public/---asia/---ro-bangkok/documents/publication/wcms_160629.pdf.

Matthews-King, A. (2018, February 5). Applications to study nursing fall for second year after removal of training bursary. *The Independent*. http://www.independent.co.uk/news/health/nursing-applications-ucas-course-drop-nhs-grants-funding-debt-tuition-fees-costs-a8191546.html.

National Health Service. (2016). How to become a nurse. http://nursing.nhscareers.nhs.uk/skills/pre_registration/fulltime.

Nichols, J., & Campbell, J. (2010). Experiences of overseas nurses recruited to the NHS. *Nursing Management*, *17*(5), 30–5. https://doi.org/10.7748/nm2010.09.17.5.30.c7963.

O'Brien, T. (2007). Overseas nurses in the national health service: A Process of deskilling. *Journal of Clinical Nursing*, *16*(12), 2229–36. http://dx.doi.org/10.1111/j.1365-2702.2007.02096.x.

Ohlén, J., & Segesten, K. (1998). The professional identity of the nurse: Concept analysis and development. *Journal of Advanced Nursing*, *28*(4), 720–7. https://doi.org/10.1046/j.1365-2648.1998.00704.x.

Ortiga, Y.Y. (2014). Professional problems: The burden of producing the "global" Filipino nurse. *Social Science and Medicine*, *115*, 64–71. https://doi.org/10.1016/j.socscimed.2014.06.012.

Pagett, C., & Padarath, A. (2007). *A review of codes and protocols for the migration of health workers*. Regional Network for Equity in Health in east and southern Africa (EQUINET) with the Health Systems Trust (HST) and the East, Central and Southern African Health Community (ECSA-HC), Discussion Paper 50. https://www.aspeninstitute.org/wp-content/uploads/files/content/images/pagett.pdf.

Philippine Overseas Employment Administration. (2018). OFW statistics. http://www.poea.gov.ph/ofwstat/ofwstat.html.

Pratt, G. (1999). From registered nurse to registered nanny: Discursive geographies of Filipina domestic workers in Vancouver, BC. *Economic Geography*, *75*(3), 215–36. http://dx.doi.org/10.1002/9780470755716.ch24.

Pring, C.C., & Roco, I. (2012). The volunteer phenomenon of nurses in the Philippines. *Asian Journal of Health*, *2*(1), 95–110. https://doi.org/10.7828/ajoh.v2i1.120.

Raghuram, P. (2012). Global care, local configurations: Challenges to conceptualizations of care. *Global Networks*, *12*(2), 155–74. https://doi.org/10.1111/j.1471-0374.2012.00345.x.

Rodriguez, R. (2008). The labor brokerage state and the globalization of Filipina care workers. *Signs: Journal of Women in Culture and Society*, *33*(4), 794–800. http://dx.doi.org/10.1086/528743.

Ronquillo, C., Boschma, G., Wong, S.T., & Quiney, L. (2011). Beyond greener pastures: Exploring contexts surrounding Filipino nurse migration in Canada through oral history. *Nursing Inquiry*, *18*(3), 262–75. https://doi.org/10.1111/j.1440-1800.2011.00545.x.

Salehi Zalani, G., Bayat, M., Mirbahaeddin, S.E., Shokri, A., Alirezaei, S., Manafi, F., & Bahmanziaei, N. (2015). Reviewing the relevance and effectiveness of the WHO Global Code of Practice on the International Recruitment of Health Personnel. WHO Expert Advisory Group for the Global Code of Practice on the International Recruitment of Health Personnel. Item 10. https://www.who.int/hrh/migration/Item10-Evidence1stround_reporting-JAHWPF-DrSalehiZalani.pdf?ua=1.

Siddique, H. (2017, June 12). 96% drop in EU nurses registering to work in Britain since Brexit vote. *The Guardian*. http://www.theguardian.com/society/2017/jun/12/96-drop-in-eu-nurses-registering-to-work-in-britain-since-brexit-vote.

Smith, P., & Mackintosh, M. (2007). Profession, market and class: Nurse migration and the remaking of division and disadvantage. *Journal of Clinical Nursing, 16*(12), 2213–20. http://dx.doi.org/10.1111/j.1365-2702.2007.01984.x.

Terry, W.C. (2014). The perfect worker: Discursive makings of Filipinos in the workplace hierarchy of the globalized cruise industry. *Social and Cultural Geography, 15*(1), 73–93. https://doi.org/10.1080/14649365.2013.864781.

Thompson, M. (2019). Everything changes to stay the same: Persistent global health inequalities amidst new therapeutic opportunities and mobilities for Filipino nurses. *Mobilities, 14*(1) 38–53. https://doi.org/10.1080/17450101.2018.1518841.

Thompson, M., & Walton-Roberts, M. (2018). International nurse migration from India and the Philippines: The challenge of meeting the sustainable development goals in training, orderly migration and healthcare worker retention. *Journal of Ethnic and Migration Studies, 45*(14) 2583–99. https://doi.org/10.1080/1369183X.2018.1456748.

Tyner, J.A. (1994). The social construction of gendered migration from the Philippines. *Asian and Pacific Migration Journal, 3*(4), 589–618. http://dx.doi.org/10.1177/011719689400300404.

Tyner, J.A. (2004). *Made in the Philippines: Gendered discourses and the making of migrants*. RoutledgeCurzon.

Vaittinen, T. (2014). Reading global care chains as migrant trajectories: A theoretical framework for the understanding of structural change. *Women's Studies International Forum, 47*(Part B), 191–202. https://doi.org/10.1016/j.wsif.2014.01.009.

Walton-Roberts, M. (2012). Contextualizing the global nursing care chain: International migration and the status of nursing in Kerala, India. *Global Networks, 12*(2), 175–94. https://doi.org/10.1111/j.1471-0374.2012.00346.x.

Wojczewski, S., Pentz, S., Blacklock, C., Hoffmann, K., Peersman, W., Nkomazana, O., & Kutalek, R. (2015). African female physicians and nurses in the global care chain: Qualitative explorations from five destination countries. *PLOS ONE, 10*(6), e0129464. https://doi.org/10.1371/journal.pone.0129464.

Work Abroad.ph. (2017). Nurses Uk-Cambridge University Hospitals-Skype Jobs in United Kingdom. Retrieved 23 June 2017; cite inactive 13 September 2021. https://www.workabroad.ph/job-openings/183a6e-nurses-uk-cambridge-university-hospitals-skype.html.

World Health Organization. (2010). The WHO Global CODE of Practice on the International Recruitment of Health Personnel. Sixty-third World Health Assembly-WHA63. 16 May. https://www.who.int/hrh/migration/code/code_en.pdf?ua=1.
World Health Organization & Department of Healthcare. (2012). Health service delivery profile: Philippines 2012. WHO Western Pacific Region. https://pdf4pro.com/view/health-service-delivery-profile-philippines-3119ea.html.
Yeates, N. (2004). A dialogue with "global care chain" analysis: Nurse migration in the Irish context. *Feminist Review*, *77*(1), 79–95. https://doi.org/10.1057/palgrave.fr.9400157.
Yeates, N. (2012). Global care chains: A state-of-the-art review and future directions in care transnationalization research. *Global Networks*, *12*(2), 135–54. https://doi.org/10.1111/j.1471-0374.2012.00344.x.
Zurn, P. (2015). Developing the WHO Global Code of Practice on the International Recruitment of Health Personnel "The Code" – A history of dialogue to adoption. WHO Expert Advisory Group for the Global Code of Practice on the International Recruitment of Health Personnel. Item 6. https://www.who.int/workforcealliance/15.pdf.

SECTION 4

Domestic Policies in Receiving Countries: Value Transfer, Integration, and Regulation

10 Recognition of Professional Qualifications of Foreign-Born Nurses: Gender, Migration, and Geographic Valuations of Skill

MICHELINE VAN RIEMSDIJK

Although not a problem exclusive to health professions, immigration for health professionals … is far more costly, administratively complex and fraught with risk than is the case for the typical worker.

Grignon, Owusu, and Sweetman (2012), "The International Migration of Health Professionals," 19

Introduction

The World Health Organization (WHO) has predicted a global shortage of 12.9 million nurses, midwives, and physicians by the year 2035 (WHO, 2014). Factors that contribute to the demand for health care workers are advances in medical technologies, increasing complexity of care, and an overall rise in the use of medical services (Ogilvie et al., 2007). In addition, as populations are aging and living longer, more health care workers are needed to provide care. The rising demand for these workers has focused attention on licensure requirements and concerns about patient health and safety (Ogilvie et al., 2007).

Three factors contribute to long-term nursing shortages: First, many nurses have part-time jobs. They may choose to work part-time to balance their professional life with childrearing responsibilities, while others are not able to work full time due to a lack of these positions (Seeberg, 2012). Second, many nurses are nearing retirement age (Organisation for Economic Co-operation and Development [OECD], 2015). Third, some nurses take on management or teaching roles, leave the profession early due to high levels of stress, physical toll, or better opportunities in other sectors (Hasselhorn et al., 2005).

Government agencies can address nursing shortages in various ways: they can redefine the duties of nurses, increase the number of seats in nursing schools, and develop initiatives to recruit nurses back to the profession. Another option is for government agencies and employers to recruit nurses from abroad.

This recruitment has been critiqued as a short-term solution to an institutional problem. Rather than creating more permanent, full-time positions, governments and employers tap into an abundant international skilled labour supply. In addition, the recruitment of nurses from low- and middle-income countries raises ethical questions. These countries often have low nurse-to-patient ratios and lose their investment in the education of nurses who leave. In order to promote the ethical recruitment of internationally trained nurses, the WHO adopted a Global Code of Practice on the International Recruitment of Health Personnel (WHO, 2010). While these guidelines have increased awareness of the ethics of nurse recruitment, the migration of nurses from low- and middle-income to high-income countries continues. This migration is partially fuelled by the education of nurses for "export labour" in the Philippines, the Caribbean, South Africa, Ghana, India, Korea, and China (Li et al., 2014).

Nursing is a registered profession, governed by strict authorization guidelines that aim to maintain quality of care and protect the health and safety of patients. For internationally trained nurses, authorization agencies determine the equivalency of international training and professional experience to educational standards in the country of destination. When deficiencies are identified, bridging courses are often required. Internationally trained nurses may also have to pass a language proficiency exam. These requirements and their implications for the recognition of professional qualifications will be discussed in more detail in this chapter. International migration and labour regulations also shape the migration of nurses (Bach, 2007).

Most high-income countries have liberalized their labour migration policies to attract more skilled migrants, arguing that these workers contribute to economic growth and innovation. Despite the desire to attract more skilled migrants, governments continue to control the admission of these migrants to their territories (van Riemsdijk & Wang, 2017). This is especially the case in registered professions, where stringent requirements for professional training and language proficiency are posing obstacles to the labour market integration of foreign-born skilled workers (Guo, 2009; Sweetman et al., 2015).

This chapter examines the valuation and recognition of professional qualifications of internationally trained nurses.[1] In particular, the chapter studies the actors that are involved in the assessment of qualifications and skills, and explores the intersections of international migration, gender, skill, and geography in the valuation and recognition of nursing qualifications. It also discusses how internationally trained nurses tackle institutional hurdles. These issues are examined through a case study of the authorization of professional qualifications of internationally trained nurses in Norway.

Like many high-income states, Norway is experiencing a nursing shortage that will increase significantly in the next twenty years. Norway is projected to have a shortage of 28,000 nurses by 2035 (Norwegian Broadcasting Corporation [NRK], 2015), and many of these positions are likely to be filled by internationally trained nurses. Norway is an attractive destination due to relatively high salaries, a good work-life balance, and family-friendly labour policies. Most nursing positions, however, require Norwegian language proficiency, demanding considerable time investment from non-Norwegian speakers.

The findings in this chapter are based on articles and reports about the nursing workforce in Norway, authorization data from the Norwegian Department of Health, websites regarding nursing authorization requirements, newspaper articles about institutional obstacles to the authorization of nurses, and semi-structured interviews with foreign-born nurses and other stakeholders involved in international nurse migration. The chapter makes two contributions to literatures on international migration, gender, and skill. First, it focuses on the *geographic* valuation of skill, examining the ease of authorization related to the region where the nursing degree was completed. Second, the chapter examines how internationally trained nurses navigate the (de)valuation of their qualifications, adding an agency perspective to the institutional barriers that skilled migrants encounter.

The chapter first discusses international nurse migration and the ways in which gender and geography shape the valuation of skill. Then, it examines the demand for nurses in Norway and the country's authorization requirements, and institutional obstacles to the international recognition of qualifications. It then examines how internationally trained nurses navigate these institutional and personal challenges, and the conclusion discusses the implications of the findings for literatures on international skilled migration.

International Nurse Migration

The international migration of nurses is governed by national migration regulations, (inter)national qualification standards, labour regulations, and employer preferences. Bach (2007) importantly notes that nurse migration is international rather than global, as national policies and employer preferences shape international migration flows from particular regions. Within the European Union/European Economic Area (EU/EEA), for example, it is easier to recognize professional qualifications obtained in the EU/EEA compared to training programs completed outside the EU/EEA. In addition, employers often recruit from familiar regions to reduce risk. They may ask current employees to recommend co-workers from their countries of origin and use recruitment agencies that have ties with particular regions or countries. Thus, nurses are often recruited from specific regions due to ease of recognition of qualifications, social networks, and employer preferences.

Some source countries have capitalized on the international demand for nurses, educating a large number of nurses to be "exported" to high-income countries. The Philippines is leading this trend, where a government agency assists with the job placement of Filipino nurses abroad. The government benefits from this out-migration through remittances and an alleviation of pressure on the domestic labour market. Filipino nurses are in high demand due to the high quality of their education, their English language proficiency, and their "naturally caring" capacities (Choy, 2003). The migration of nurses from the Philippines is a remnant of its colonial heritage, as it sends many nurses to the United States (for a critical perspective on the "nurse trade" from the Philippines, see Ball, 2004).

The migration patterns of nurses can partially be explained by colonial ties between sending and receiving countries and commonality of languages (OECD, 2016). "Push factors" in the country of origin also contribute to the out-migration of nurses, including low salaries, lack of career progression, unsafe work environments, and political unrest (Buchan et al., 2005). In addition, transnational social networks contribute to international nurse migration, as nurses who have migrated encourage former co-workers and friends to join them (Buchan et al., 2005). The international migration of nurses, however, is constrained by authorization regulations, which will be discussed in more detail later in this chapter.

Gender and the Valuation of Skill

Since most studies of female migrants have focused on low-skilled labour such as domestic work, au pairs, and sex work, we know less about the work experiences of female *skilled* migrants. Acknowledging the increased share of female skilled migrants, scholars have started to pay more attention to the contributions of these migrants to the labour market, economy, and society at large (Iredale, 2005; Kofman, 2000; Kofman & Raghuram, 2006). This research has raised awareness about differences in the experiences of male and female skilled migrants, gendered assumptions in migration policies, and gendered valuations of skill and competences.

Internationally trained nurses encounter several difficulties as their profession and skills are devalued. The nursing profession has lower social status than medical specializations, and the merits of nursing are not adequately recognized in society (Allan et al., 2008). More female than male migrants are employed in regulated professions, while the latter are more prevalent in finance, information technology, and management (Kofman & Raghuram, 2015). In these male-dominated sectors, professional qualifications are more easily transferred across international borders.

The devaluation of the nursing profession operates in three ways: First, nursing labour consists of emotional and affective labour, which is often considered a feminine "innate" quality (Batnitzky & McDowell, 2011). These stereotypical assumptions about caring qualities inform hiring decisions, the assignment of tasks, and career opportunities. Second, nursing is a service profession that does not produce material goods. The latter is valued and remunerated more than emotional labour (Batnitzky & McDowell, 2011). The economic sectors are seen as driving the economy, and thus more highly valued than reproductive labour (Iredale, 2005). Third, caring work is often invisible. It is only visible when it fails to meet quality standards. Thus, it is difficult to show the results of caring work and command a higher salary. Historically, caring labour was performed by women in the home "for love" or by minority women for low compensation (Iredale, 2005). The representation of care as women's work, emotional labour, and "naturally caring" work devalues nursing in terms of professional esteem, wages, and career prospects.

Foreign-born nurses are often considered to be lower-skilled than their native-born colleagues by their managers, co-workers, and patients. The devaluation of their skills can be augmented by a lack of proficiency in the language of the host country, when foreign-born nurses cannot clearly communicate their knowledge and expertise. These assumptions shape the everyday lives of foreign-born nurses and can pose barriers to their career advancement in the long term (van Riemsdijk, 2010; Batnitzky & McDowell, 2011). Female foreign-born nurses often face a double challenge as they experience discrimination in the workplace based on gendered and stereotypical assumptions about their country of origin. The latter is also the case for foreign-born male nurses (Whittock & Leonard, 2003), but they are more likely to have a nursing specialty and tend to earn more than female nurses (Evans, 1997).

Female migrants can also experience deskilling after migration. Man (2004) has found that Chinese skilled immigrant women in Canada were taking jobs below their skill level or taking insecure jobs, or they left the labour force altogether because of stringent state policies, accreditation requirements, and demands for Canadian work experience. In a study of skilled women who migrated from South Africa to New Zealand, Meares (2010) found that migration compromised their careers, and some women took on more domestic duties after migration. Men's skills are also devalued after migration, exemplified by highly skilled immigrant taxi drivers in Toronto (Xu, 2012). These experiences, however, are exacerbated for female migrants through gender discrimination.

These gendered differences in access to employment and valuations of skill have prompted some scholars to argue that accreditation agencies should measure skills in addition to formal educational qualifications (Iredale, 2005). For example, communication skills are undervalued as "natural" female attributes, but crucial on the job (Iredale, 2005). Scholars have noted the importance of "soft skills" in the workplace, including interpersonal understanding, commitment, persuasiveness, compassion, and comforting others (Zhang et al., 2001).

These "innate" skills, however, are not formally recognized and rewarded (Windsor et al., 2012).

Skilled female migrants can also be disadvantaged by standardized assessments of their qualifications and by requirements to take formal exams in the country of destination. Women who have been out of the workforce for an extended period of time – to care for children, for example – are not accustomed to taking exams when they return to the workforce (Iredale, 2005). Authorization agencies do not take these leaves of absence into account when they require applicants to pass competency exams. These requirements produce gendered experiences among highly skilled migrants (Iredale, 2005).

Based on the insights above, it is evident that foreign-born female skilled migrants can experience challenges in the labour market due to gendered and stereotypical assumptions about their skills, language difficulties, and childcare duties. These factors are exacerbated by difficulties with the recognition of professional qualifications, which this chapter addresses.

Nursing Shortages in Norway

Compared to most OECD countries, Norway has a high coverage of nurses. Norway's nurse-to-patient ratio of 12.9 per 1,000 persons is considerably higher than the OECD average of 8.6 per 1,000 persons (OECD, 2015), and the fifth highest ratio in OECD countries after Switzerland, Denmark, Belgium, and Iceland. Despite this good coverage, municipalities experience difficulties with filling positions (OECD, 2014). In order to meet this need, foreign-trained nurses are recruited to Norway.

The demand for nurses in Norway is driven by increased use of health care services, medical technologies, and an aging population. This demand is reflected in an increase of almost 50 per cent of employed nurses in a fourteen-year period, from 55,700 in the year 2000 to 83,800 nurses in 2014. The largest increase in health care positions has been in elder care and disability services (Stølen et al., 2016). Employment in elder care is physically and emotionally demanding labour that is disproportionally performed by foreign-born nurses. In Norway, over 70 per cent of foreign-born nurses work in elder care, 5 per cent work in hospitals, and the remainder work elsewhere (Dahlen & Dahl, 2015).

In 2014, foreign-born nurses made up 9.1 per cent of the nursing workforce in Norway (OECD, 2015). This number is considerably lower than the OECD average of 14 per cent, which may be explained by linguistic differences and stringent authorization requirements. It takes considerable investment of non-Norwegian speakers to learn the language, as Norwegian language proficiency is required in most nursing-related jobs. The next sections will discuss authorization requirements for nurses in general, and in Norway in particular.

Authorization of Professional Qualifications

Nursing curricula differ considerably by degree program and country of origin, with variations in length of training, hours of in-class education, course content, and hours of practical training. Thus, it is a complex process to determine the equivalency of nursing programs among countries. The European Commission started an initiative in 1999 to harmonize the standards and quality of education in institutions of higher education, and to make it easier to recognize qualifications across countries in Europe (European Higher Education Area, n.d.). These efforts resulted in the Bologna Process, which introduced three-year bachelor and two-year master programs with clear guidelines for course content and number of hours of training (Davies, 2008). To date, forty-eight countries have implemented the Bologna Process standards.

The EU started its efforts to streamline the recognition of professional qualifications in Europe in the 1970s. These rules were eventually consolidated in Directive 2005/36/EC of the European Parliament on the recognition of professional qualifications. This directive covers the minimum requirements for the mutual recognition of qualifications for a variety of professions (Baeten & Jorens, 2006). For nurses who complete their training in an EU/EEA member state, the principle of mutual recognition applies. This means that any EU/EEA member state automatically recognizes nursing degrees obtained in another member state (for a more in-depth explanation of EU sectoral directives, see van Riemsdijk, 2013). Applicants who completed their training outside the EU/EEA have to fulfill additional requirements before they are allowed to work as registered nurses.

Nursing Authorization in Norway

In Norway, all health personnel have to seek authorization, including nurses who completed their education in Norway. In January 2016, the responsible agency for authorization was shifted from the Norwegian Registration Authority for Health Personnel (SAK) to the Norwegian Directorate of Health. The authorization agency determines if an applicant's educational and professional qualifications are equivalent to nursing qualifications in Norway. A person who is granted authorization has the full right to work independently as a nurse. If the qualifications are deemed not sufficient, the Directorate can provide a licence for a limited duration, and may require the applicant to work under supervision until all requirements have been met (Norwegian Directorate of Health, 2016).

Between 2008 and 2015, the office denied 3,253 applications for nursing authorizations. Most of these applicants had completed their nursing training outside the EU/EEA. In 2014, almost one in five applicants who had completed their education outside the EU/EEA were denied authorization (NRK, 2015). With these denials, the recognition system contributes to stratification in the labour market as trained nurses often work as nursing assistants.

The Norwegian media has reported on several cases of differential assessments of similar educational backgrounds. For example, three Norwegian citizens completed their nursing degrees at Queensland University of Technology in Australia in 2009. Two of them returned to Norway and were authorized to work as registered nurses. The third remained in Australia for one more year to train as a surgery nurse and was denied authorization when she returned to Norway (Derwing & Waugh, 2012). The authorization office argued that she did not take theoretical courses and that she did not have enough relevant clinical and practical training. During the year when the nurse remained in Australia, decisions made by the appeals committee had created precedents that had to be applied by the authorization agency.[2]

Other nurses who completed their education in Australia faced similar authorization challenges in Norway. The main reasons why authorization was denied were related to the number of hours of training and the content of the training program. The Norwegian curriculum requires six weeks of practical training, which the Australian program does not offer. In addition, nurses educated in Australia often lack training in mental health care and caring for the elderly (Dahl et al., 2018).

The authorization of nursing qualifications in Norway depends on the region where an applicant has completed her/his nursing education. The system reflects a geographical hierarchy where it is easiest for applicants in closest proximity to Norway to have their professional qualifications recognized. It is also relatively easy for applicants with EU/EEA degrees to get their qualifications approved, thanks to the EU Directives on the Mutual Recognition of Professional Qualifications. For professional qualifications obtained outside the EU/EEA, an individual assessment is made. The latter takes more effort and time, and applications are more likely to be denied. Table 10.1 shows the authorization regulations for each region.

Between 2007 and 2016, Norway issued 58,775 authorizations for registered nurses. Sixty per cent of these were issued to persons who completed their nursing education in Norway. Thirty per cent of authorizations were granted to nurses who completed their education in the Nordic countries, and the remainder to persons who completed their education in an EU/EEA member state or outside the EU/EEA. The top twelve countries of education and number of authorizations issued are listed in table 10.2.

Of the 58,775 authorizations granted, 89 per cent of applicants were female and the remainder male. The high proportion of females is common in nursing.

Table 10.1. Recognition agreements for regions where the nursing degree was completed

Region where nursing degree was completed	Recognition agreement
Nordic region	Agreement concerning a common Nordic labour market
EU/EEA member state	EU directives for the mutual recognition of professional qualifications
Outside the EU/EEA	Individual assessment

Table 10.2. Authorizations issued to nurses by country of education, 2007–2016

Country of education	Number of authorizations issued
Norway	35,242
Sweden	12,726
Denmark	4,278
Lithuania	898
Finland	731
Poland	651
Iceland	579
Germany	481
Latvia	416
Spain	414
Philippines	382
Estonia	264

Source: Norwegian Directorate of Health, sent to author by email, 2018.

The number of authorizations granted by country of completed education correlates with geographic proximity to Norway and related ease of recognition of qualifications. Citizens of Sweden, Denmark, and Finland are entitled to automatic recognition according to the agreement concerning a common Nordic labour market. Danes, Swedes, and Swedish-speaking Finns can communicate relatively easily with Norwegians, and they share a similar cultural background.

The listing of Lithuania, Poland, Latvia, and Estonia in the top twelve can partially be explained by the membership of these states in the EU and the high wage differentials between Central and Eastern Europe and Norway. When these states joined the EU in 2004, Norway had implemented transitional regulations to restrict free movement of labour from the new member states. These migrants had to obtain residence permits to be allowed to work in Norway. Permits were granted to applicants with a full-time employment offer who would be paid a wage and whose working conditions were commensurate with a similar job in Norway.

When the restrictions to the Norwegian labour market were lifted in 2011, it became easier for citizens from the EU8 to take up employment in Norway. At that time, nurses from these member states qualified for automatic recognition of professional qualifications in Norway. In addition, recruitment agencies were bringing nurses from the new EU member states to Norway. The pitfalls of recruitment agencies will be discussed in more detail below.

Authorization Requirements: The Nordic Agreement

Norway honours the Nordic Agreement on the mutual recognition of authorization that automatically recognizes degrees obtained in other Nordic states and credentials that have been approved in these states. This agreement has produced surprising results, as reported by the Norwegian Nurses Association and the public broadcasting company NRK. Ingvild Hilling Elsbak, a Norwegian citizen, completed her bachelor's degree in nursing in Australia in 2014 and applied for authorization upon her return to Norway. After reviewing her coursework, the Norwegian authorization agency concluded that the Australian curriculum did not meet the minimum requirements and ruled that she had to complete a nursing degree in Norway before she was allowed to practise as a registered nurse.

Ms. Elsbak then submitted her qualifications to the National Board of Health and Welfare, the authorization agency in Sweden. The board decided that she would have to take an exam, a course in Swedish health services and health legislation, practical training, and a language test to become authorized. Upon authorization in Sweden, her qualifications were recognized by the Norwegian authorization agency under the Nordic agreement on the mutual recognition of authorization (Sykepleien, 2014). Several Norwegian citizens have used this "Swedish trick" (NRK, 2015) to receive authorization to work in Norway.

The divergent valuations of professional qualifications in Norway and Sweden reflect differences in nursing education systems. While applicants in Sweden can take individual courses to make up deficiencies, the Norwegian educational system requires applicants to be admitted to a nursing school to take courses in the subjects that are deemed insufficient. Thus, it usually takes longer for applicants with training from non-EU/EEA countries to be authorized in Norway than in Sweden (NRK, 2015).

The authorization office has been criticized for counting the number of training and practical hours too stringently. The use of instruction hours as a criterion for authorization was adopted in 2011 by the Health Personnel Board (*helsepersonell nemnd*) (Svarstad, 2015). Now, the authorization agency in Norway applies the European ECTS standard, which allocates a number of credits to a course instead of hours of study. Applicants who were denied authorization because they did not meet the required minimum number of hours could reapply for authorization after the rule was changed, but it is unknown how many of these re-evaluations resulted in authorization (Svarstad, 2015).

Authorization Requirements for EU/EEA Nursing Training

Applicants who completed their nursing degrees in an EU/EEA member state have to submit their diploma, transcript, authorization, and a work certificate from the country of graduation to seek authorization as a nurse in Norway (Norwegian Directorate of Health, 2016). These nurses are not required to pass a language exam and do not have to take a course in safe handling of medicine to receive authorization. These exemptions are regulated under EU law that does not allow language tests and professional exams for EU/EEA citizens. These prohibitions were implemented to avoid potential discrimination by language and nationality (Inoue, 2001, p. 8).

It is the employer's responsibility to ensure that an applicant who has completed her/his education in an EU/EEA member state has the required level of language proficiency and professional training and competences. While the level of language proficiency can be documented with transcripts, it is more difficult for employers to assess professional skills and competences in a job interview. When a nursing shortage is particularly pressing, employers may take shortcuts in the vetting process to hire someone as quickly as possible.

Authorization Requirements for Non-EU/EEA Nurse Training

The authorization requirements for nurses who completed their education outside the EU/EEA are the most stringent. In addition to the documents required for nurses who completed their training in the EU/EEA, applicants have to provide a detailed overview of the courses taken, including course content and the duration of training. Applicants also have to submit a work certificate, listing the place of employment and the work that they performed.

Applicants who completed their education outside the EU/EEA are required to pass the written and oral exam of a Norwegian language test called the Bergenstest, which requires a level of Norwegian language proficiency at B2 upper-intermediate level. They have to take a course in Norwegian health services, health legislation, and society (*nasjonale fag*). This course teaches about cultural understanding and national priorities in health services. Non-EU/EEA-trained nurses also have to take a course in safe handling of medicine and they have to pass a test to prove their professional skills. The test consists of written, oral, and practical components and assesses applicants' "theoretical knowledge and practical skills necessary to work within their profession in the Norwegian health services" (Norwegian Directorate of Health, 2016). The course helps shape foreign-born nurses into "ideal migrants" who internalize the norms and values of the host society, and socializes them into the Norwegian health system and the professional habitus (Girard & Bauder, 2007).

If the Norwegian Directorate of Health deems a nurse's training not to be equivalent, applicants are required to take bridging courses. These four-week courses are offered twice a year at Ullevål Nursing School and at Oslo and Akershus University College. The study plan for the bridging courses was approved in 2004, and the course was initially only available to nurses. The course was revised in 2013 and open to nurses, doctors, dentists, pharmacists, and health workers. The main goal of the course is to "provide participants knowledge about the Norwegian health system to ensure the safety of patients, ensure quality in health services and trust in the Norwegian health system, according to paragraph 1 in the health personnel law" (Dahlen & Dahl, 2015, p. 53).

The authorization of nurses who completed their education outside the EU/EEA can take a long time, and many applicants work as auxiliary nurses in nursing homes while they complete the authorization requirements. Thus, the stringent authorization requirements help produce a flexible labour force that fills labour needs. Once foreign-born nurses have obtained authorization, perhaps the largest obstacles are the job search and integration into the workplace. The hiring decisions of employers are often informed by stereotypical and normative assumptions about applicants' gender, race/ethnicity, nationality, citizenship, and other personal characteristics (J.A. Larsen, 2007), and co-workers may have these assumptions as well (van Riemsdijk, 2010). These beliefs can channel applicants into particular positions, with an over-representation of foreign-born nurses in nursing homes and disability care (Stølen et al., 2016).

Recruitment Agencies

The international recruitment of nurses is a lucrative trade, especially in times of acute nursing shortages. The high demand for internationally trained nurses has given rise to unscrupulous agents who take advantage of foreign-born nurses and health care workers. Periodically, Norwegian newspapers report on these abuses. In 2001, a Norwegian businessman, Finn Radmann, had recruited health care workers from Estonia and Latvia to Norway. After arrival in Norway, several migrants did not qualify for authorization to work as nurses. For example, Radmann had brought two midwives from Estonia who were lacking a nursing education (Strømman, 2001). In Estonia, it is not required to complete a nursing degree to become a midwife, but this is required in Norway (Strømman, 2001).

These schemes continue today as language schools and recruitment agencies capitalize on applicants' need for Norwegian language proficiency and employment. The website of the Philippine Embassy in Oslo warns Filipino nurses about abusive practices of "consultancy firms" and language schools that promise students well-paying jobs in Norway (Embassy of the Philippines, 2013).

These practices are likely to continue as long as the demand for foreign-born nurses exists. This is still the case in Norway, where recruitment agencies experienced difficulties filling temporary nursing positions in the summer of 2017. Many of these positions were ordinarily filled with nurses from Denmark and Sweden, but an economic upturn in these countries has diminished interest in summer jobs in Norway.

According to Jon Paulsen, recruitment director at Manpower in Oslo, the number of Swedish nurses in Norway had diminished by 60 per cent between 2014 and 2017. He attributes the decline to better working conditions and salaries in the country of origin (Andreassen & Nguyen, 2017). To fill these shortages, registered nurses from the Baltic region were recruited for summer positions in Norwegian nursing homes in 2017.

Agency of Internationally Trained Nurses in the Recognition of Professional Qualifications

The examples above have largely represented internationally trained nurses as passive actors who await authorization decisions and fill nursing shortages. It is important to note, however, that these nurses also have agency in the recognition of professional qualifications. This section will provide several examples of how internationally trained nurses have navigated the authorization process.

The "Sweden trick" discussed above is a creative way to overturn an authorization denial. Ms. Elsbak and other Norwegian citizens had their qualifications approved in Norway after they met Swedish authorization requirements. This solution, however, is only beneficial for Nordic applicants for whom a Swedish language test is easy. Non-Nordic nurses who completed their training outside the EU/EEA would have to take several Swedish language courses to pass the test.

Nurses from the Philippines have shown a strong determination to obtain authorization in Norway. I interviewed a woman in Oslo who was trained as a nurse in the Philippines, but she did not have the financial means to move to Norway and fulfil the requirements for authorization. In order to save money and learn Norwegian, she decided to work as an au pair in Oslo. She took Norwegian language courses paid for by her host family and read nursing literature in her free time. She planned to take bridging courses after completing her two-year contract as an au pair and take the Bergenstest language exam (personal communication with author, June 2016). This trajectory has been followed by other Filipino nurses, as confirmed by Norwegian language instructors and an attaché of the Philippine Embassy in Oslo (personal communications with author, June 2016).

The upskilling trajectory with Norwegian language courses and bridging courses can take a long time. The Norwegian Directorate of Health does not set a time limit for the completion of these courses to be eligible for authorization. If an applicant has been away from the nursing profession for a long time, and the applicant's professional knowledge can be considered in need of updating, the application can be denied (Directorate of Health, email communication, 12 June 2018). This rule is more lenient than the College of Nurses of Ontario, Canada, which requires "evidence of practice" that states that applicants must have practised as a nurse in the past three years to be considered for authorization (College of Nurses of Ontario, n.d.).

The determination of foreign-trained nurses to obtain authorization is also reflected among the students who take bridging courses. In a class of twenty-nine nurses in a bridging course at Oslo and Akershus University College in 2015, approximately half of the participants used their vacation days to attend class. Others received paid time off from their employer to take the course (H. Larsen, 2015). Most of these students were employed as health care assistants. In H. Larsens's (2015) article, she mentions that Bærum, an affluent municipality that borders Oslo to the west, employs approximately sixty non-EU/EEA nurses as health care assistants. The municipality funds bridging courses for full-time employees to help them gain employment as registered nurses.

Once nurses have obtained authorization, they can improve their position in the labour market through membership in a trade union. I have discussed elsewhere that Polish nurses who became union representatives for the Norwegian Nursing Association empowered themselves and others. They learned about Norwegian labour and health regulations and became aware of rules about overtime, salary, and pensions. They used this knowledge to strengthen their own position and informed co-workers and friends about their employment rights (van Riemsdijk, 2010). This self-empowerment is also evident when foreign-trained nurses educate themselves about Norway's authorization procedures and ways to navigate its bureaucracy.

These examples show that internationally trained nurses can improve their situation while taking institutional constraints into account. These bottom-up initiatives, while sometimes unsuccessful, empower nurses to navigate the devaluation of their skills and competences.

Conclusion: Gender, International Skilled Migration, and Geographic Valuations of Professional Qualifications

The authorization requirements described in this chapter exhibit a clear spatiality in the valuation of professional qualifications. Applicants who completed their training in another Nordic country are automatically recognized in Norway. Applicants who completed their education in an EU/EEA member state have to submit more documentation but are also automatically recognized to work as nurses in Norway. The Bologna Process standards have harmonized institutions of higher education, making it easier to compare qualifications across borders. In the case of non-EU/EEA applicants, however, individual caseworkers assess whether foreign education is considered valid. As shown in this chapter, those interpretations can provide variable results. In the case of EU/EEA applicants, the decision falls on the employers, who are responsible for the assessment of language proficiency and professional qualifications. This involves a devolution of responsibility away from the authorization agency, trusting employers to ensure the health and safety of patients.

Applicants who completed their training outside the EU/EEA, however, face formidable institutional and personal obstacles before they are authorized to work as nurses in Norway. They have to pass a high level of Norwegian language proficiency, which can take years to acquire. They also have to take – and often pay for – bridging courses and pass exams to prove their nursing skills. While these courses promote the health and safety of patients, they pose considerable hurdles to entry in the profession.

The stricter authorization requirements in Norway disproportionately affect immigrant women. These women are doubly affected by the gendered valuations of professional qualifications and their immigration status. Nurses from non-EU/EEA countries are particularly vulnerable to exploitation by language schools and recruitment agencies. European nurses have a marked advantage in terms of free movement of labour and mutual recognition of professional qualifications, but they may also experience discrimination in the workplace based on their gender, race/ethnicity, nationality, and class. Thus, it is important to study gender in relation to other categories of identity and difference to gain a deeper understanding of the intersections of gender, international migration, and the valuation of skills.

As the internationalization of education intensifies, it is very likely that more nursing students will complete their degrees abroad (OECD, 2016). Thus, the assessment and recognition of nursing qualifications obtained abroad will remain a salient issue.

NOTES

1 Internationally trained nurses are defined in this chapter as nurses who completed their professional degree in a country outside of Norway. This includes Norwegian citizens who took their entire nursing degree in a country other than Norway.

2 An appeals committee (*klagenemnd*) has been established to allow nurses who have been denied authorization to appeal the decision. The decision of the complaint board is binding and creates precedents for future authorization decisions.

REFERENCES

Allan, H., Tschudin, V., & Horton, K. (2008). The devaluation of nursing: A position statement. *Nursing Ethics, 15*(4), 549–56. https://doi.org/10.1177%2F0969733008090526.

Andreassen, I., & Nguyen, L. (2017, June 21). Frykter for pasienters sikkerhet i sommer [Fears for patient safety in the summer]. NRK. https://www.nrk.no/buskerud/frykter-for-pasienters-sikkerhet-i-sommer-1.13568452.

Bach, S. (2007). Going global? The regulation of nurse migration in the UK. *British Journal of Industrial Relations, 45*(2), 383–403. http://doi.org/10.1111/j.1467-8543.2007.00619.x.

Baeten, R., & Jorens, Y. (2006). The impact of EU law and policy. In M. McKee, L. MacLehose, & E. Nolte (Eds.), *Health policy and European Union enlargement* (pp. 214–34). Open University Press.

Ball, R.E. (2004). Divergent development, racialised rights: Globalised labour markets and the trade of nurses – The case of the Philippines. *Women's Studies International Forum, 27*(2), 119–33. http://doi.org/10.1016/j.wsif.2004.06.003.

Batnitzky, A., & McDowell, L. (2011). Migration, nursing, institutional discrimination and emotional/affective labour: Ethnicity and labour stratification in the UK National Health Service. *Social and Cultural Geography, 12*(2), 181–201. http://doi.org/10.1080/14649365.2011.545142.

Buchan, J., Kingma, M., & Lorenzo, F.M. (2005). *International migration of nurses: Trends and policy implications.* The Global Nursing Review Initiative.

Choy, C. (2003). *Empire of care: Nursing and migration in Filipino American history.* Duke University Press.

College of Nurses of Ontario. (n.d.). Evidence of practice. http://www.cno.org/en/become-a-nurse/registration-requirements/evidence-of-practice/.

Dahl, K., Nortvedt, L., & Ligan, C.B. (2018). *Kompletterende sykepleierutdanning gir verdig vei mot autorisasjon* [Bridging course in nursing provides a dignified pathway to authorization]. https://sykepleien.no/forskning/2018/03/kompletterende-sykepleierutdanning-gir-verdig-vei-mot-autorisasjon.

Dahlen, K.J., & Dahl, K. (2015). Flerkulturelle sykepleiere [Multicultural nurses]. *Sykepleien Oslo, 4*, 52–4.

Davies, R. (2008). The Bologna Process: The quiet revolution in nursing higher education. *Nurse Education Today, 28*, 935–42. http://doi.org/10.1016/j.nedt.2008.05.008.

Derwing, T.M., & Waugh, E. (2012). Language skills and the social integration of Canada's adult immigrants. IRPP Study No. 31. Institute for Research on Public Policy.

Embassy of the Philippines. (2013). *Advisory: Filipino nurses & healthcare workers should be wary of advertisements & consultancy firms that peddle misleading information about nursing jobs in Norway.* http://dianaslittlehelper.blogspot.com/2013/01/advisory-to-all-filipino-nurses-hoping.html.

European Higher Education Area. (n.d.). European higher education area and Bologna process. https://www.ehea.info.

Evans, J. (1997). Men in nursing: Issues of gender segregation and hidden advantage. *Journal of Advanced Nursing, 26*(2), 226–31. http://doi.org/10.1046/j.1365-2648.1997.1997026226.x.

Girard, E.R., & Bauder, H. (2007). Assimilation and exclusion of foreign trained engineers in Canada: Inside a professional regulatory organization. *Antipode, 39*(1), 35–53. http://doi.org/10.1111/j.1467-8330.2007.00505.x.

Grignon, M., Owusu, Y., & Sweetman, A. (2013). The international migration of health professionals. IZA Discussion Papers, No. 6517, Institute for the Study of Labor (IZA), http://nbn-resolving.de/urn:nbn:de:101:1-201208096350.

Guo, S. (2009). Difference, deficiency, and devaluation: Tracing the roots of non recognition of foreign credentials for immigrant professionals in Canada. *The Canadian Journal for the Study of Adult Education, 22*(1), 37–52. https://cjsae.library.dal.ca/index.php/cjsae/article/view/1002.

Hasselhorn, H.M., Muller, B.H., & Tackenberg, P. (2005). *NEXT: Nurses' early exit study scientific report*. University of Wuppertal.

Inoue, J. (2001). Migration of nurses in the EU, the UK, and Japan. Tokyo, Hitotsubashi University: Discussion Paper Series A 526. https://ideas.repec.org/p/hit/hituec/a526.html.

Iredale, R. (2005). Gender, immigration policies and accreditation: Valuing the skills of professional women migrants. *Geoforum, 36*(2), 155–66. http://doi.org/10.1016/j.geoforum.2004.04.002.

Kofman, E. (2000). The invisibility of skilled female migrants and gender relations in studies of skilled migration in Europe. *International Journal of Population Geography, 6*(1), 45–59. http://doi.org/10.1002/(SICI)1099-1220(200001/02)6:13.0.CO;2-B.

Kofman, E., & Raghuram, P. (2006). Gender and global labour migrations: Incorporating skilled workers. *Antipode, 38*(2), 282–303. http://doi.org/10.1111/j.1467-8330.2006.00580.x.

Kofman, E., & Raghuram, P. (2015). *Gendered migrations and global social reproduction*. Palgrave Macmillan.

Larsen, H. (2015). Bruker av ferien sin for å bli godkjente sykepleiere i Norge [Use their vacation to become licensed nurses in Norway]. *Khrono*. https://khrono.no/sykepleiere/bruker-av-ferien-sin-for-a-bli-godkjente-sykepleiere-i-norge/166915.

Larsen, J.A. (2007). Embodiment of discrimination and overseas nurses' career progression. *Journal of Clinical Nursing, 16*(12), 2187–95. http://doi.org/10.1111/j.1365-2702.2007.02017.x.

Li, H., Nie, W., & Li, J. (2014). The benefits and caveats of international nurse migration. *International Journal of Nursing Sciences, 1*(3), 314–17. http://doi.org/10.1016/j.ijnss.2014.07.006.

Man, G. (2004). Gender, work and migration: Deskilling Chinese immigrant women in Canada. *Women's Studies International Forum*, *27*(2), 135–48. http://doi.org/10.1016/j.wsif.2004.06.004.

Meares, C. (2010). A fine balance: Women, work and skilled migration. *Women's Studies International Forum*, *33*(5), 473–81. http://doi.org/10.1016/j.wsif.2010.06.001.

Norwegian Broadcasting Corporation (NRK). (2015, November 10). Måtte til Sverige for å jobbe som sykepleier [Had to go to Sweden to work as a nurse]. https://www.nrk.no/dokumentar/xl/matte-til-sverige-for-a-fa-jobbe-som-sykepleier-1.12644167.

Norwegian Directorate of Health. (2016). Authorisation and license for health personnel. https://helsedirektoratet.no/english/authorisation-and-license-for-health-personnel.

Ogilvie, L., Leung, B., Gushuliak, T., McGuire, M., & Burgess-Pinto, E. (2007). Licensure of internationally educated nurses seeking professional careers in the province of Alberta in Canada. *Journal of International Migration and Integration*, *8*(2), 223–41. http://doi.org/10.1007/s12134-007-0015-y.

Organisation for Economic Co-operation and Development. (2014). *Recruiting immigrant workers: Norway 2014*. OECD Publishing.

Organisation for Economic Co-operation and Development. (2015). *International migration outlook 2015*. OECD Publishing.

Organisation for Economic Co-operation and Development. (2016). *Health workforce policies in OECD countries: Right jobs, right skills, right places*. OECD Publishing.

Seeberg, M.L. (2012). Sykepleiere fra utlandet-fra statlig til privat ansvar? [Nurses from abroad: The responsibility of the state or private actors?]. https://sykepleien.no/forskning/2012/06/sykepleiere-fra-utlandet-fra-statlig-til-privat-ansvar.

Stølen, N.M., Bråthen, R., Hjemås, G., Otnes, B., Texmon, I., & Vigran, A. (2016). Helse-og sosialpersonell 2000–2014 – Faktisk utvikling mot tidligere framskrivinger [Health-and social personnel 2000–2014 – Actual developments against earlier predictions]. Statistics Norway.

Strømman, T.W. (2001). Vi er lurt [We have been fooled]. *Dagbladet. Oslo*. https://www.dagbladet.no/nyheter/vi-er-lurt/65741895.

Svarstad, J. (2015). Sykepleiere med utenlandsk utdanning får ny sjanse [Nurses with a foreign education get another chance]. *Aftenposten*. https://www.aftenposten.no/norge/i/0xV2/sykepleiere-med-utenlandsk-utdanning-faar-ny-sjanse.

Sweetman, A., McDonald, J.T., & Hawthorne, L. (2015). Occupational regulation and foreign qualification recognition: An overview. *Canadian Public Policy*, *41*(suppl), S1–S13. http://doi.org/10.3138/cpp.41.s1.s1.

Sykepleien. (2014, September 8). Kan få autorisasjon i Sverige [Can be authorized to work in Sweden]. https://sykepleien.no/2014/09/kan-fa-autorisasjon-i-sverige.

van Riemsdijk, M. (2010). Variegated privileges of whiteness: Lived experiences of Polish nurses in Norway. *Social and Cultural Geography*, *11*(2), 117–37. http://doi.org/10.1080/14649360903514376.

van Riemsdijk, M. (2013). Obstacles to the free movement of professionals: Mutual recognition of professional qualifications in the European Union. *European Journal of Migration and Law, 15*(1), 47–68. http://doi.org/10.1163/15718166-12342023.
van Riemsdijk, M., & Wang, Q. (Eds.). (2017). *Rethinking international skilled migration.* Routledge.
Whittock, M., & Leonard, L. (2003). Stepping outside the stereotype: A pilot study of the motivations and experiences of males in the nursing profession. *Journal of Nursing Management, 11*(4), 242–9. http://doi.org/10.1046/j.1365-2834.2003.00379.x.
Windsor, C., Douglas, C., & Harvey, T. (2012). Nursing and competencies – a natural fit: The politics of skill/competency formation in nursing. *Nursing Inquiry, 19*(3), 213–22. http://doi.org/10.1111/j.1440-1800.2011.00549.x.
World Health Organization. (2010). The WHO Global Code of Practice on the international recruitment of health personnel. World Health Organization.
World Health Organization. (2014). *A universal truth: No health without a workforce.* World Health Organization.
Xu, L. (2012). *Who drives a taxi in Canada?* Citizenship and Immigration Canada.
Zhang, Z.X., Luk, W., Arthur, D., & Wong, T. (2001). Nursing competencies: Personal characteristics contributing to effective nursing performance. *Journal of Advanced Nursing, 33*(4), 467–74. http://doi.org/10.1046/j.1365-2648.2001.01688.x.

11 Ten Years of Ontario's Fair-Access Law: Has Access to Regulated Professions Improved for Internationally Educated Individuals?

NUZHAT JAFRI

It has been more than ten years since the Office of the Fairness Commissioner (OFC) opened its doors to implement Ontario's groundbreaking Fair Access to Regulated Professions and Compulsory Trades Act, 2006 (FARPACTA). The first such legislation in Canada, FARPACTA was enacted to help ensure that regulated professions in Ontario have registration practices that are transparent, objective, impartial, and fair. If implemented effectively, it would create conditions for greater mobility of internationally and domestically trained professionals across provincial and national borders. In its early days, the OFC recognized that achieving transparent, objective, impartial, and fair registration practices takes time and requires continuous adjustments to respond to the ever-changing regulatory landscape.

The last ten years have seen many changes aimed at improving access to the regulated professions in Ontario. The mindset of the province's regulatory community has evolved, and many regulators have now embedded the principles of FARPACTA and the OFC's fair-access lens into their decision-making processes. This chapter aims to share data and examples of positive changes resulting from Ontario's fair-access legislation and show how the OFC and regulators have worked together to achieve fairer registration practices. As we examine the transformation that has taken place since the office opened, we will also explore persistent challenges and discuss how we can continue our work to advance fair access.

A Pan-Canadian Framework for the Assessment and Recognition of Foreign Qualifications: The Federal Government's Approach

In 2009, the Forum of Labour Market Ministers (FLMM), at the behest of the Canadian federal government, launched the Pan-Canadian Framework for the Assessment and Recognition of Foreign Qualifications. The FLMM is comprised of federal, provincial, and territorial ministers; deputy ministers; and officials with labour market responsibilities. The Pan-Canadian Framework serves as a public commitment to ensure that regulatory authorities have foreign qualifications assessment and recognition processes based on shared principles of fairness, transparency, timeliness, and consistency.

A key commitment in the Pan-Canadian Framework is timely service. It stipulates that within one year individuals will know whether their qualifications will be recognized, or they will be informed of the additional requirements necessary for registration or be directed towards related occupations commensurate with their skills and experience.

The Pan-Canadian Framework sets out to achieve the following outcomes:

- The provision of preparation and pre-arrival supports so that immigrants may have access to reliable and accurate information and assessment services as early in the immigration process as possible.
- Reasonable and objective assessment methods with regulatory authorities sharing their approaches with their counterparts in other jurisdictions.
- Clear and comprehensive communication in a timely manner.
- Immigrants' awareness of bridge-to-licensure opportunities so that they may upgrade their qualifications for entry to practise.

The Framework includes accountability to measure achievement of the outcomes and continues to be part of the federal government's efforts to improve the assessment and recognition of international qualifications (Employment and Social Development Canada, n.d.).

Ontario's Fair-Access Legislation

Newcomers who wish to practise a regulated profession or trade in Ontario have long struggled to have their international credentials and qualifications recognized by provincial regulatory authorities. Removing barriers to access to professions and trades has been the preoccupation of governments of all stripes in Ontario since the 1980s. However, none of the efforts to tackle the problem bore fruit in the late twentieth century. No voluntary or goodwill measures on the part of the regulatory community brought about changes necessary to allow internationally educated professionals to practise their professions expeditiously. Well into this century, Ontario continued to attract large numbers of highly qualified internationally educated professionals who would land in the province only to find out that they could not work in their profession or trade without jumping through myriad hoops including costly assessments and exams, and sometimes the requirement to complete all or part of their professional education again.

In 2006, Ontario became the first jurisdiction in Canada to address the issue through legislative means by passing FARPACTA. The purpose of the legislation is to help ensure that regulated professions and individuals applying for registration by regulated professions are governed by registration practices that are transparent, objective, impartial, and fair (FARPACTA, 2006). The legislation also amended Schedule 2 of the Regulated Health Professions Act (RHPA). The essence of the legislation is to ensure that individuals who are qualified, regardless of where they were educated, have an opportunity to practise their profession or trade in Ontario.

When the legislation passed in 2006 it covered twenty-one regulated health professions and fourteen non-health professions. By the end of its tenth year, there were twenty-six health regulatory bodies and sixteen non-health bodies, including the Ontario College of Trades, covered by the legislation. The College of Trades, however, was eliminated by the newly elected government in Ontario in 2018.

Office of the Fairness Commissioner

Ontario's fair-access legislation required the appointment of a fairness commissioner and establishment of the Office of the Fairness Commissioner (OFC) to oversee the legislation's implementation and its compliance with FARPACTA and RHPA amendments.

The functions of the fairness commissioner are to:

- Set out guidelines for regulatory bodies' reports on their registration practices;
- Assess registration practices and make recommendations to changes and improvements;
- Require reviews and/or audits of registration practices if deemed necessary;
- Advise provincial government ministries about issues related to professions under their ambit;
- Monitor third parties relied upon by regulatory bodies to assess qualifications, such as credential assessment services and examining bodies;
- Provide advice and information to regulatory bodies to help them comply with the legislation;
- Advise regulatory bodies, government agencies, community agencies, colleges and universities, third parties, and others with respect to matters under the legislation;
- Report to the minister of health and long-term care about a health profession's non-compliance; and
- Report to the public and to the minister responsible for FARPACTA about OFC work.

The regulatory bodies are required to have registration, certification, and licensure practices (heretofore referred to as registration practices) that are transparent, objective, impartial, and fair, as well as entry-to-practice requirements that are necessary and relevant to practise the profession or trade. The FARPACTA spells out specific and general duties that regulatory bodies must fulfil and also includes enforcement mechanisms for non-compliance.

To comply, regulatory bodies must:

- Meet the *specific duties* to provide the following:
  - information to applicants
  - internal review or appeal processes
  - information on appeal rights
  - transparent, objective, impartial, and fair qualifications assessments
  - training to their staff
  - access for applicants to their own records
  - timely decisions, responses, and justifications to applicants
- Meet the *general duty* to have registration practices that are transparent, objective, impartial, and fair.
- Provide reports, such as Fair Registration Practices Reports, Entry-to-Practice-Review Reports, and Audit Reports.

The legislation aims to achieve systemic and incremental change to address long-standing barriers. The OFC assesses practices and asks for justification for entry-to-practice requirements. If issues are identified, the OFC makes recommendations for improvement and monitors progress. It does not intervene on behalf of individuals or directly represent applicants for registration (OFC, n.d.).

Strategy for Continuous Improvement

The OFC chose a continuous improvement approach to achieve systemic and incremental change in how regulatory bodies assess qualifications and register individuals to practise a regulated profession or trade in Ontario. It launched its Strategy for Continuous Improvement (OFC, 2015a) to enable the OFC to facilitate the improvement of registration practices, foster a culture of continuous improvement, and monitor the implementation of the fair-access legislation (see figure 11.1 for a continuous improvement strategy flowchart).

Every year on 1 March, regulatory bodies are required to submit their Fair Registration Practices Report for the preceding calendar year. These reports are the main way for the regulatory body to inform the OFC and the public about any changes and improvements it has made during the past year.

Every three years, the OFC assesses regulatory bodies' registration practices to determine whether the regulatory body is fulfilling its duties under the fair-access legislation, to identify any gaps, and to make recommendations and engage the regulatory body in action, if necessary. Upon completion of the

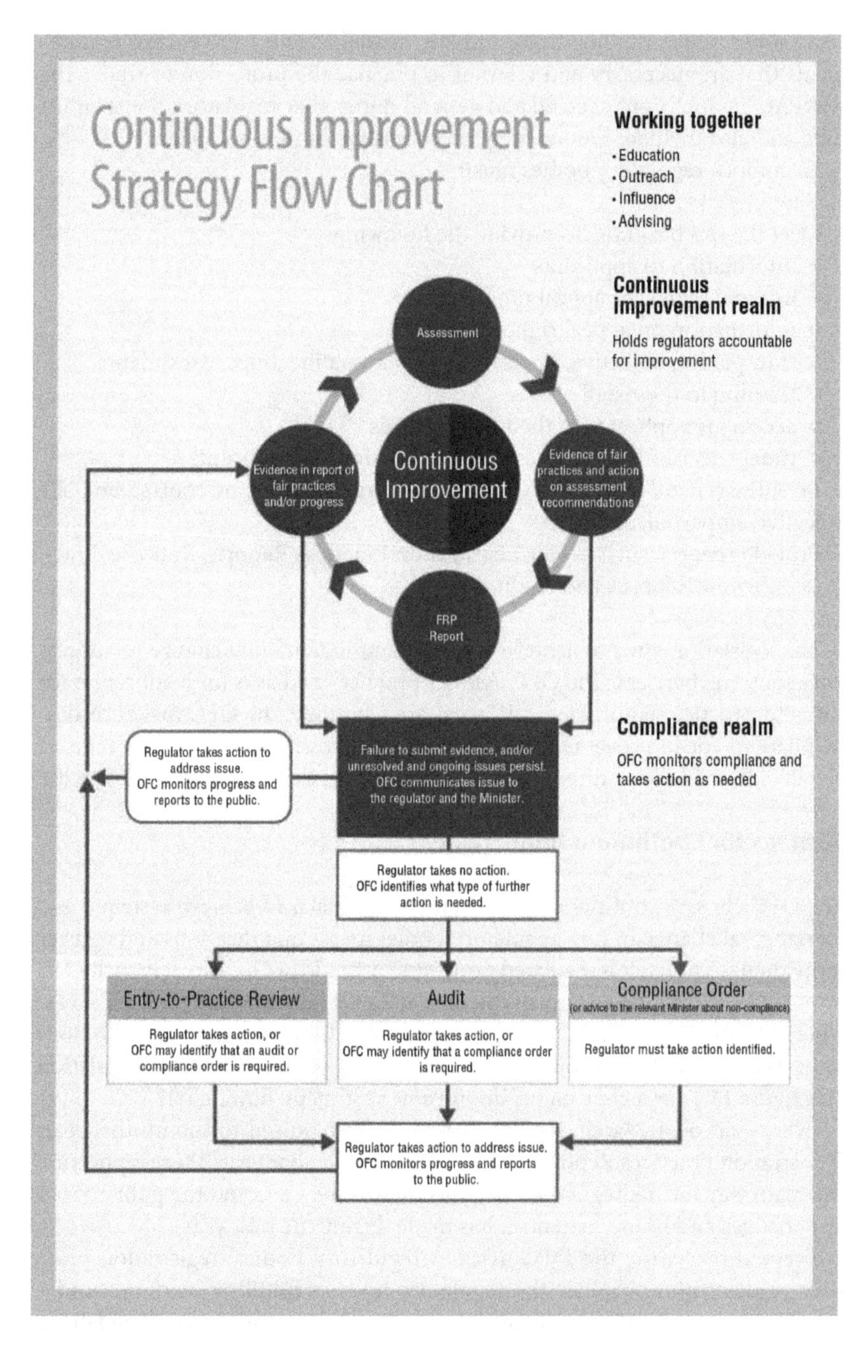

Figure 11.1. The OFC's continuous improvement strategy flow chart
Source: Adapted from OFC's 2016–2017 Annual Report and created by N. Jafri (March 2020).

assessment, regulatory bodies submit action plans to address gaps and recommendations that the OFC has identified. The regulatory body informs the OFC as actions are completed, and the OFC continues to monitor results.

If the OFC decides that it needs more information to determine whether a regulatory body is meeting its obligations, it may require an entry-to-practice review. This review allows regulatory bodies to study their own registration practices and analyse their entry-to-practice requirements to determine whether they are still necessary and relevant to practise the profession or trade.

If issues persist and the OFC requires still further information to determine whether a regulatory body is meeting its legal obligations, it may require the regulatory body to undergo an independent audit of its registration practices. An independent audit offers a wider range of techniques and methodologies to gather information and determine compliance.

While these compliance measures are critical to the implementation of the legislation, the OFC recognizes that it must work with regulatory bodies to achieve the kind of positive change that will improve access to the professions and trades for internationally educated professionals. In this spirit, the OFC works with regulatory bodies to monitor progress, identify emerging issues, share best practices, and solve problems as they arise. It meets with regulators on a regular basis and provides education and information on the fair-access law, as well as how best to implement it for long-lasting systemic change.

Through the assessments and action plans, the OFC holds the regulatory bodies accountable for identifying and removing barriers and addressing issues of fair access. Should the regulatory body fail to comply or fail to provide information to the OFC and issues remain unresolved, the OFC has recourse to further enforcement measures. It may advise the regulatory body and the appropriate provincial government minister about the issues and the situation. The OFC may issue compliance orders to regulatory bodies governed by FARPACTA. For health regulatory colleges governed by the RHPA, the OFC may recommend to the minister of health and long-term care that a health college take action.

Qualitative and Quantitative Data

To establish a baseline and measure progress over time, the OFC began collecting qualitative and quantitative data in 2007. Since 2008, regulators have been submitting data and information in their annual Fair Registration Practices Reports.

Qualitative elements of the reports focus on:

- Information for applicants
- Timely decisions, responses, and reasons

- Internal review or appeal
- Information on appeal rights
- Required documentation and acceptable alternatives
- Assessment of qualifications
- Training of assessors and decision-makers
- Access to records
- Reasonableness of fees
- Relevant and necessary entry-to-practice requirements

Quantitative data are collected for members and applicants to Ontario's regulatory bodies, including where members and applicants were originally trained in their profession – whether in Ontario or another Canadian province or territory, or internationally (US or other).

For "other international" applicants, the top source countries for original training are also identified; beginning in 2015, similar data for members has also been collected. In 2007, the OFC collected data from thirty-five regulatory bodies, and, by its tenth year, it covered forty-two regulatory bodies. As a result, the OFC has accumulated a unique body of data on registration trends on internationally educated professionals.

Qualitative Results

The data are useful in measuring results and effectiveness of the legislation. Positive trends in addressing and removing barriers include greater transparency and better information, streamlined processes, alternatives to academic requirements, removal of Canadian experience requirements, review and appeal rights in all regulatory bodies, better training for assessors and decision-makers, better accountability for third-party qualifications assessment agencies, national competency frameworks and assessments, and improved labour mobility within Canada. Each of these is reviewed below.

Greater Transparency and Better Information

Regulatory bodies provide detailed information about all of the steps and requirements for registration on their websites and in other resources. Many of them have easy to follow graphics and videos to help applicants. There is greater transparency in how decisions are made and communicated. For example, the College of Physiotherapists of Ontario established a consistent and transparent decision-making process by ensuring that staff and registration committee members are provided with training in all areas that are relevant to making fair and effective registration decisions. It maintains a staff policy and procedure manual to ensure fair, defensible, and consistent decision-making. In addition, the College uses registration criteria to assess applications in a way that is clear to both applicants and decision-makers, and finally, it has adopted the guiding principles proposed by the Ontario Regulators for Access Consortium for fairness, objectivity, transparency, accountability, and collaboration.

Streamlined Processes

Simpler, more streamlined processes have shortened the amount of time it takes to register in a regulated profession in Ontario. Some steps have been eliminated altogether, while others have been automated for greater efficiency. Most regulatory bodies have online application systems that allow applicants to trace the status of their application. For example, the Ontario College of Pharmacists requires internationally educated pharmacy graduates to complete a bridging program. However, applicants may attempt the Pharmacy Examining Board of Canada (PEBC) exams before starting the bridging program. If applicants successfully pass the PEBC qualifying exams (which are a requirement for all applicants) on the first attempt, they may ask to be exempted from the bridging program. The applicants may thereby reduce the steps in the registration process, significantly reducing its overall time and cost.

Alternatives to Academic Requirements

Completion of all academic requirements is a demonstration of foundational knowledge of a profession. Most regulatory bodies require an academic credential that is equivalent to an Ontario or Canadian academic degree, diploma, or certificate. The credential is assessed for equivalence by either a third-party assessment agency like World Education Services (WES), or by the regulatory body itself. Because of the diversity of education systems and content of professional education programs across the world, it is difficult for internationally educated applicants to meet the equivalence standard. Given this difficult situation, many regulatory bodies and their third-party assessment agencies have devised alternatives such as prior learning assessment and recognition (PLAR), evaluating exams, competency-based assessments, and bridge training programs.

For example, the PLAR of the College of Dieticians of Ontario is used for the assessment of internationally educated applicants and also for eligibility for the Internationally Educated Professionals in Nutrition bridging program offered by Ryerson University. The PLAR includes a Knowledge and Competency Assessment Tool (KCAT) and a Performance-Based Assessment (PBA). To be deemed eligible to initiate the PLAR process, an applicant must demonstrate completion of a degree reasonably related to dietetics and appropriate language proficiency.

Another example is the Internationally Educated Nurses Competency Assessment Program (IENCAP), which is administered by the Touchstone Institute and is required by the College of Nurses of Ontario (CNO) if an applicant's educational credential does not meet CNO's nursing degree requirement. IENCAP is a standardized evaluation of the knowledge, skills, and judgment of Internationally Educated Nurses (IENs) seeking to become registered nurses in Ontario. It consists of two parts: a written, multiple-choice exam with up to ninety-seven questions (two and a half hours), and an Objective Structured Clinical Examination (OSCE) consisting of twelve stations (thirteen minutes each).

Methods of competency assessment vary greatly. Some use evaluating exams, others use paper-based portfolios that are completed by the applicant and verified by a supervisor, while still others use interviews in combination with portfolios. All of these assessments can be costly both to administer and to undergo. Applicants may be required to pay anywhere between CAD$500 to as much as CAD$8,000, depending on the complexity of the assessments. Most require an academic credential assessment and a language evaluation before administering the competency assessment (Reitz et al., 2014).

Removal of Canadian Experience Requirement

Newcomers often find themselves in a circular quandary. Certain professions require applicants for registration to possess Canadian experience. This includes completion of practical training or internships and mentorships in Canadian workplaces, such as residency training for physicians. In 2013, the Ontario Human Rights Commission issued a policy on removing the "Canadian experience" barrier (Ontario Human Rights Commission, 2013). The policy was developed in consultation with the OFC and states that a strict requirement for "Canadian experience" is discriminatory on its face and can only be used in limited circumstances. The policy applies to employers and regulatory bodies. Following the release of the policy, the OFC worked with regulatory bodies to review the necessity and relevance of their Canadian experience requirement and asked them to identify specific aspects of the experience that could only be acquired in Canada. As a result of this work, some regulatory bodies have either eliminated the requirement or have devised an alternative means to meet the requirement.

For example, most internationally trained immigrant physicians are required to complete a residency, depending on whether the College of Physicians and Surgeons (CPSO) deemed this necessary. The number of residency placements in Ontario is limited, however, and such placements are very difficult to access, particularly for certain specialties. The OFC's work showed that it is usually immigrant physicians who fail to secure a residency, and who therefore are unable to become licensed. The OFC long advocated for supervised practice and practice-ready assessments to support a streamlined route to licensure for qualified international medical practitioners. A Practice-Ready Assessment (PRA) program as an alternative to a full residency requirement for family medicine was developed, involving the Ministry of Health and Long-Term Care, Touchstone Institute, CPSO, and other stakeholders. At the writing of this chapter, the program was awaiting approval for launch by the Ontario Ministry of Health and Long-Term Care.

Review and Appeal Rights in All Regulatory Bodies

When the OFC began its work, all the health regulatory colleges already included provisions under the RHPA for review and appeal rights for applicants, but some non-health regulatory bodies had no such provisions. At the end of the OFC's tenth year all regulatory bodies now allow applicants to ask for a review or appeal to registration decisions.

Better Training for Assessors and Decision-Makers

Those who assess applications and qualifications and make registration decisions are trained in the application of the fair-access law. OFC's online learning modules are used for this purpose. Assessors are also trained on the Accessibility for Ontarians with Disabilities Act and the Ontario Human Rights Code. Many have completed cultural sensitivity training.

Better Accountability for Third-Party Qualifications Assessment Agencies

Regulatory bodies are holding third parties that they rely upon for qualifications assessments to account by asking them to submit reports, and enter in memorandums of understanding and service level agreements to ensure transparent, objective, impartial, and fair assessments and exams.

National Competency Frameworks and Assessments

Many professions now rely on national competency frameworks and assessments, which are developed in collaboration with national organizations and provincial and territorial regulatory bodies. These competency frameworks and assessments eliminate duplication and ensure streamlining of requirements and assessments across the country. Moreover, there are economies of scale and cost savings for individual regulatory bodies by working together. In the end, these national competency frameworks and assessments make for fairer, faster practices. For example, the College of Occupational Therapists of

Ontario played a lead role in the development of the national Substantial Equivalency Assessment System (SEAS). The SEAS determines whether applicants' educational qualifications and competencies are substantially equivalent to an Ontario-educated occupational therapist. It is a sophisticated system in which all internationally trained applicants planning to work anywhere in Canada (with the exception of Quebec) go through a standardized assessment process that includes a competency assessment. Jurisprudence learning and assessment components were also introduced as part of the process in Ontario.

Improved Labour Mobility within Canada

Under the 1995 Agreement on Internal Trade, all regulatory bodies are subject to labour mobility legislation that allows for mobility of members of regulated professions in Ontario to practise elsewhere in Canada. Despite all of this progress on internal labour mobility, some challenges remain. For instance, internationally educated physicians still struggle to acquire residency placements as there are far fewer residency spaces available relative to the number of applicants each year. Alternatives to this requirement are under consideration for practice-ready internationally educated candidates but have not been implemented as of yet across all provinces. Likewise, internationally educated nurses still find it difficult to meet the academic requirement and must undergo multiple assessments, including Objective Structured Clinical Evaluations, before being eligible to challenge the registration exam. The amount of time and the costs involved in completing the registration process remain challenging in some professions.

Quantitative Highlights

In 2007, the Province of Ontario had 679,569 registered professionals and tradespeople. Ten years later there were 1,153,273, which included individuals in twenty-two of the twenty-three compulsory trades under the Ontario College of Trades (see table 11.1). That is an increase of 59 per cent of individuals registered in regulatory bodies overseen by the OFC since its inception.

Over this same ten-year period there were some notable trends in internationally educated individuals who were registered in the province (see table 11.2). The College of Pharmacists of Ontario had the largest proportion of internationally educated members at 45 per cent. Some professions such as audiologists, and speech language pathologists, and optometrists showed a high proportion of internationally educated members, the majority of whom were trained in the US, primarily because of a limited number of post-secondary programs available in Canada.

Table 11.1. Members and applicants, 2016

	2015	2016	Per cent change
Registered Members	1,138,996	1,153,273	1.3
Registered Internationally Trained Professionals	121,685	128,431	5.5
Applications Received for Registration	77,318	81,857	5.9
International Applications Received	11,317	12,973	14.6
Educated Individuals			
Gender of Applicants			
Male (%)	43	48	
Female (%)	57	52	
Neither (%)	0.01	0.1	
Gender of Members			
Male (%)	47	46	
Female (%)	53	54	
Neither (%)	0.001	0.001	

Source: Office of the Fairness Commissioner (2015b).

Table 11.2. Professions/trades with highest percentage (%) of internationally trained members

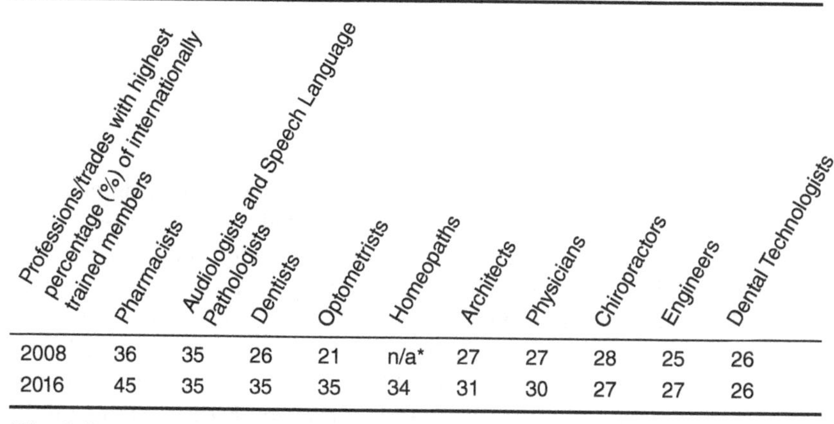

Professions/trades with highest percentage (%) of internationally trained members	Pharmacists	Audiologists and Speech Language Pathologists	Dentists	Optometrists	Homeopaths	Architects	Physicians	Chiropractors	Engineers	Dental Technologists
2008	36	35	26	21	n/a*	27	27	28	25	26
2016	45	35	35	35	34	31	30	27	27	26

*The College of Homeopaths of Ontario was established by governing law effective 1 July 1 2015.
Source: Office of the Fairness Commissioner (2015b).

Professions with the highest number of internationally educated members such as teachers, engineers, and nurses actually have lower proportions of internationally trained members (see figure 11.2). Moreover, there is a trend towards increasing numbers of Canadians obtaining their professional training abroad and returning home to register in their respective professions. These Canadians must complete all of the same steps required of newcomer professionals.

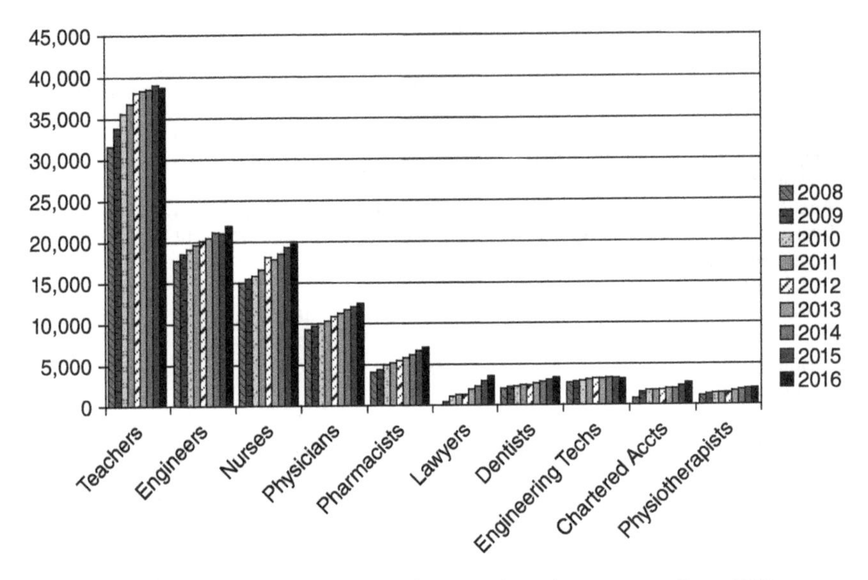

Figure 11.2. Top ten professions with highest number of internationally and US trained members
Source: Office of the Fairness Commissioner (2015b).

Profession-Specific Source Country Trends

Information gathered by the OFC has allowed for the monitoring of trends related to source countries of internationally educated applicants to the various regulated professions (see figure 11.3). For instance, Saudi Arabia was the largest source of physician applicants – these are foreign visa trainees (residents) who will return to Saudi Arabia after completing their residency, and whose residencies were paid for by the Saudi government. Iran was the second largest source of engineering applicants, behind India and ahead of China. Egypt was consistently the top source of internationally trained pharmacists in Ontario. Historically, the Philippines has been one of the main sources of internationally educated nurses, but India has also become a top source country for internationally educated nurses in Ontario. The College of Traditional Chinese Medicine began registering members in 2013 and, not surprisingly, has a significant number of new applicants and members trained in China.

Fluctuations in Numbers of Internationally Educated Applicants

There are numerous factors related to fluctuations in applications from internationally trained professionals and tradespeople. In some cases, it may take

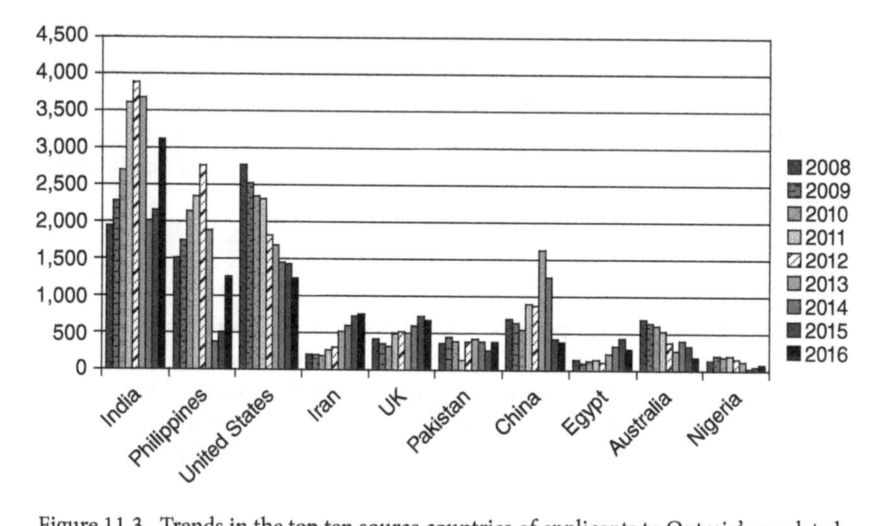

Figure 11.3. Trends in the top ten source countries of applicants to Ontario's regulated professions
Source: Office of the Fairness Commissioner (2015b).

one to three years (or longer) before newly arrived immigrants initiate their applications for licensure. Internationally educated applicants may reside in the province or country for a long period before initiating their application for registration, as income generation and establishing family stability is the focus during the initial settlement period. There has been a recent decrease in applicants initially educated in China (and other non-English speaking jurisdictions), which may indicate a response to changes in Canadian immigration policy, which now demand higher language benchmarks and pre-arrival credential assessment (Immigration, Refugees and Citizenship Canada, 2018). However, there are now increased opportunities for applicants to start the application process outside Canada, and regulators have shown increased flexibility with regard to residency requirements. There is also an increased availability of alternatives to satisfy entry-to-practice requirements, many of which have been prompted by recommendations from the OFC.

National and International Influence of the Ontario Model

Three other jurisdictions in Canada – Manitoba, Nova Scotia, and Quebec – followed Ontario's lead and now have similar legislation to FARPACTA, with similar principles of transparency, objectivity, impartiality, and fairness. Other provinces have similar functions within their immigrant integration programs.

The OFC welcomed many international visitors who wanted to learn about the Ontario legislation and how it was implemented. Representatives of Austria, France, the UK, and Germany were particularly interested in the Ontario model. Germany subsequently passed its own legislation, the Recognition Act, in 2012, which focuses on the procedure to assess professional qualifications (Eberhardt & Annen, 2014).

Conclusion

The trends presented here point to notable systemic and incremental improvements in regulatory bodies' registration practices. This has happened through a deliberate collaborative and continuous process that required the OFC to work with regulatory bodies, third-party assessment agencies, and other stakeholders to make the legislation come alive. Ontario was a trailblazer. No other Canadian jurisdiction had done what the OFC set out to do. Other Canadian provinces followed and transparent, objective, impartial, and fair registration practices have become table stakes for regulatory bodies. At a global level, regulatory bodies may look to working with their counterparts in other jurisdictions to move towards global standards that will allow for greater mobility of internationally educated professionals. Discussions are underway for the profession of dentistry, and architecture is already recognized as an international profession in which teams of architects qualified in multiple jurisdictions practice on projects across the world.

REFERENCES

Eberhardt, C., & Annen, S. (2014). What is worth a qualification? Approaches towards the recognition of vocational qualifications and competences acquired abroad. *International Journal for Cross-Disciplinary Subjects in Education*, Special Issue, *4*(2), 1991–9. doi: 10.20533/IJCDSE.2042.6364.2014.0276.

Employment and Social Development Canada. (n.d.). *A Pan-Canadian framework for the assessment and recognition of foreign qualifications.* Government of Canada. Retrieved 15 September 2021, from https://www.canada.ca/en/employment-social-development/programs/foreign-credential-recognition/funding-framework.html.

Fair Access to Regulated Professions and Compulsory Trades Act [FARPACTA]. (2006). S.O, 2006, c. 31. Government of Ontario. https://www.ontario.ca/laws/statute/06f31.

Immigration, Refugees and Citizenship Canada. (2018). *Evaluation of pre-arrival settlement services.* Government of Canada. https://www.canada.ca/en/immigration-refugees-citizenship/corporate/reports-statistics/evaluations/pre-arrival-settlement-services.html.

Office of the Fairness Commissioner. (n.d.). *About.* Retrieved 15 September 2021, from https://www.fairnesscommissioner.ca/en/About/Pages/Mandate.aspx.

Office of the Fairness Commissioner. (2015a). *Strategy for continuous improvement*. Government of Ontario. https://www.fairnesscommissioner.com/en/Resources/PDF/Strategy-for-Continuous-Improvement-Eng.pdf.

Office of the Fairness Commissioner. (2015b). *Fair registration practices reports: Guidelines for Ontario's regulatory bodies*. Government of Ontario. https://www.fairnesscommissioner.ca/en/Publications/PDF/Guidelines/guidelines_for_fair_registration_practices_reports_pdf_english_2015.pdf.

Ontario Human Rights Commission. (2013). *Policy on removing the "Canadian experience" barrier.* http://www.ohrc.on.ca/en/policy-removing-"canadian-experience"-barrier.

Reitz, J.G., Curtis, J., & Elrick, J. (2014). Immigrant skill utilization: Trends and policy issues. *Journal of International Migration and Integration, 15*(1), 1–26. https://doi.org/10.1007/s12134-012-0265-1.

12 The Changing Face of Australian Aged Care: A Precarious Dependence?

JOHN CONNELL AND JOEL NEGIN

Introduction

Globally, the world is experiencing an unprecedented demographic shift with the numbers and proportions of older people increasing significantly. This shift is acutely apparent in Australia; in 2016, 15 per cent of Australians were aged sixty-five and over; however, this proportion is projected to increase to 23 per cent by 2055. In addition, rates of disability are high among older people: 40 per cent of those aged sixty-five to sixty-nine are affected by disability, and the rate is 88 per cent for those aged ninety years and older. The number of those with dementia is projected to increase from 298,000 in 2011 to 900,000 by 2050 (Australian Institute of Health and Welfare, 2012). As a result of aging and differential longevity, the likelihood that older people could enter permanent residential aged care in their lifetime has been steadily increasing (Department of Health and Ageing, 2011). By 2050, it is expected that over 3.5 million older Australians will access aged care services, both at home and in residential facilities, and that the aged care workforce will need to triple in size (Productivity Commission, 2011). The need for carers is accelerating.

Almost 2.7 million Australians identified as carers in 2016, with about 236,000 of these working in residential care, the subject of this chapter. Residential care is increasingly shifting from community and charity organizations to for-profit companies. A widespread perception exists that the Australian aged care system is headed towards a "crisis" or a "care time bomb" (Baldassar 2017; Fine & Mitchell, 2007; Goel & Penman 2015; Hugo 2007), usually perceived as an absolute numerical shortage of care workers and of their too-rapid turnover (Howe et al., 2012; Russell, 2017). Such a shortage and crisis is not easily contextualized or defined, so that it is not necessarily obvious, as present, imminent, or impending (Adamson et al., 2017: Gao et al., 2015). The basic perception is both numerical and quantitative, but distanced from a more immeasurable quality of care that necessarily has some relationship to the capability and "quality" of carers. The shortage is more acute in rural and regional areas, in some part because recent migrants to Australia avoid these areas. In 2016 migrants (those born overseas) made up about 42 per cent of the aged care workforce, a substantial increase from 2001, when the proportion was just 35 per cent (Negin et al., 2016). While that proportion is similar to that for the total population, this is a high proportion of one particular workforce and indicative of the growing sectoral dependence on migrant workers.

In Australia as elsewhere, divorce rates are high, hence the proportion of people with a partner to look after them in old age is declining, while greater mobility means that children are less likely to live close to aging parents (Hugo, 2009). Greater longevity, a decline of traditional family structures, social restructuring and rearrangement of family units, a redefinition of family attitudes and values, and an increase in chronic illnesses have resulted in heavy pressures on both families and national welfare state regimes, alongside a shift in care to the public realm. At the same time, many Australian families employ migrants to care for their relatives, which in itself has become an issue of public concern in terms of supply and sustainability. While there is growing demand for elder care, at least by virtue of there being greater numbers of older people, household support is dwindling and is putting greater pressure on institutional contexts and non-family workers. That is particularly so in regional areas that face similar employment problems (low wages, "graveyard" shifts, difficult conditions, etc.) and where there are fewer migrant workers.

The aged and disability care workforce in Australia consists of various occupations in a continuum from the highly skilled professional workforce, including doctors and nurses, through to lower-skilled workers, which include nursing support workers and aged care workers (Hugo, 2009). The term "aged care worker" usually applies to employed carers in aged care facilities, which is the largest single occupational group employed in such facilities. In Australia, aged and disability care workers usually possess – at a minimum – an Australian Qualifications Framework (AQF) Certificate II or III in aged or disability care to be qualified to work as a personal carer. As of 2016 there were more than 160,000 non-nurse aged care workers in Australia, some of whom have international nursing credentials (Department of Health and Ageing, 2017). That represents about half of all aged care workers in residential care.

Some of the most socially important roles in human society that involve both physical and emotional elements – nurturing and caregiving – are gradually being transferred outside the household to the market, community, and state; these roles are slowly becoming the tasks of migrant workers caring for people's "most intimate needs: their bodies, their homes and their families" as "servants of globalization" (Brush & Vasupuram, 2006, pp. 181–2). The combination of emotional labour and body work, which is poorly paid and has low status, can be seen as both reason for and consequence of the recruitment of migrant workers (Dyer et al., 2008; Gray & Heinsch, 2009). For many migrants, undertaking these vital tasks means a struggle with low wages; minimum job security; no health benefits of their own; limited, even downwards, occupational mobility; and discrimination at work and beyond. As migrants dominate such sectors, their attractiveness for some local workers may decline further. With an aging population, aged care workers are in high demand, yet, in Australia, they are difficult to recruit. In Australia as elsewhere, the work is regarded as low status, poorly paid, and challenging physically and emotionally (Badkar et al., 2009; Cangiano, 2014). In Australia low pay and low status are linked to the employment of migrants, hence inevitably class, gender, and ethnic power relations are also interlinked, with poor women from vulnerable ethnic minorities being employed in the care industries (Fine & Mitchell 2007; Kaine & Ravenswood, 2014). Caring remains a "gendered profession" with 90 per cent of the direct care workforce being female (Adamson et al., 2017; Department of Health and Ageing, 2017). Skills that are essential to the provision of care are regarded as intrinsic female capabilities, and hence do not command high wages (Dhakal et al., 2017), while aged care services do not have the capacity to reshape and redefine care work as a "good job" with tangible rewards for younger skilled workers (Savy et al., 2017).

Employment of Migrant Workers, Often on a Casual Basis, Reduces Costs

In Australia, it is impossible to directly recruit aged and disability care workers from overseas under the current immigration system: carers are not classified as skilled workers and thus do not come under any current visa program. Migrants may, however, enter care work through other visa pathways, including student visas, as partners of those holding skilled work visas (457 visas), and as workers during holiday seasons (Adamson et al., 2017; Howe, 2009). The relationship between migration experience and status and either the intention to remain in the workforce or the quality of care provided remains uncertain and poorly understood. Underpinning uncertain migration and residential status are problems attached to recruiting, training and retaining care workers, alongside cultural issues such as language ability, acculturation, and decisions over residential location.

The providers within the care workforce are themselves aging, with a median age of 52 in 2016 (Dhakal et al., 2017). Care work is regarded, accurately, as poorly paid, of low status, challenging both physically and emotionally, with poor wages, and offering no career pathways. Older workers are more likely to be committed to staying in the industry, while younger workers, including overseas migrants, are more likely to leave (Dhakal et al., 2017; Gao et al., 2015), hence that aging is likely to continue. However, data and adequate understanding of the migration structure of the aging care workforce remain limited. This chapter seeks to address some of this knowledge of the care workforce profile by first analysing recent Australian census data, followed by an analysis of the limited qualitative data.

Quantitative analysis focuses on those who self-report their occupation as "aged and disabled carer" or "nursing support and personal care workers." While care workers should have a level of skill that is commensurate with the qualifications and experience equivalent to the AQF Certificate II or III, respondents defined themselves, hence some of those defining themselves as care workers may have had no formal qualification. According to census definitions, aged and disability carers provide paid general household assistance, emotional support, care and companionship for aged persons, and those living with disability in their own homes. Nursing support and personal care workers provide assistance and direct care to patients in a "variety of health, welfare and community settings." Taken together, these two groups comprise those who provide care both in home and institutional settings. The terms "care worker" or "carer" are used for both groups.

Who Cares? New Geographies of Migration

A growing proportion of Australian carers are migrants. Country of birth is an imperfect measure of an individual's country of origin, but most migrant care workers probably migrated directly from their country of birth. Both census data and alternative labour force survey data (Isherwood & King, 2017) point to the same trends. Dramatic growth in the number of care workers occurred from 2006 to 2016. In 2016, there were 215,896 care workers in Australia, up from 131,283 in 2006: a 64 per cent increase in ten years. While the majority of carers across all three census periods were Australia-born, in 2006 the next most common region of birth for carers was the UK followed by Europe. By 2016 this had changed with South Asia and Southeast Asia becoming the second and third most common regions (table 12.1). While the number of carers from all regions has grown, the increase from 2006 to 2016 has been highest for carers from South Asia (775 per cent increase) and Sub-Saharan Africa (297 per cent), with relatively modest increases in carers from the UK and Europe. Consequently the proportions from the UK, Europe, and Australia are all declining. While that parallels some of the changing structure of overall migration to Australia, especially the rise of Asia, the overall population structure is more obviously shaped by migration from the "traditional" sources of the UK, Ireland, and New Zealand.

Between 2006 and 2016, the number of carers who immigrated from some of the sending countries dramatically increased. The largest number of individuals came from Asia and Africa: Nepal with a 4,703 per cent increase (from 117

Table 12.1. Aged care workers and personal care workers in Australia from 2006, 2011, and 2016 Australian Census Data

	2006	2011	2016	% change
Australia	87,840	111,629	124,190	41
New Zealand	4,113	5,815	6,855	67
Pacific Islands	2,274	3,105	3,842	69
United Kingdom	10,675	12,252	11,905	12
Europe	7,135	7,830	7,869	10
North Africa and Middle East	974	1,650	2,416	148
South-East Asia	5,566	10,566	16,652	199
Chinese Asia	1,932	3,838	5,520	186
Southern Asia	2,164	9373	18,933	775
Central Asia	65	125	233	258
North America	512	687	856	67
South and Central America	1,636	2,159	2,473	51
Sub-Saharan Africa	2,434	5,962	9,671	297
Other	3,963	3,349	4,481	13
Total	131,283	1783,40	215,896	64

Source: Australian Bureau of Statistics (2007, 2012, 2017).

to 5,620), Nigeria with 1,614 per cent, as well as India (784 per cent increase), Kenya, and the Philippines (see table 12.2). What has emerged are two distinct and relatively new groups of carers: one group has larger numbers from South Asia (India, Nepal, and Sri Lanka), and a second group has smaller numbers from Sub-Saharan Africa. Compared to the total number of people born in each country now living in Australia, as of 2011, almost 8 per cent of all Nepalese-born people in Australia were working as carers; more than 15 per cent of all Liberia-born and 12 per cent of all Sierra Leone-born people in Australia are carers. By contrast, between 2006 and 2016, the percentage increase in carers from traditional source countries such as the UK was 7 per cent, Ireland was 33 per cent, and New Zealand was 62 per cent – much lower than for Asian and African countries (see table 12.1).

This migration has created new geographies of care employment. Within Asia, it represents a sub-regional shift from more long-standing sources, including China and Vietnam. In terms of absolute numbers, the Philippines, India, China, Nepal, and Fiji were the low- and middle-income countries that sent the greater number of carers, which also suggests a disproportionate flow from small island states such as Fiji and Tonga. That represents a migration that is shaped by economic goals and that is a function of globally uneven development, and it is often made up of health workers who come into poor working conditions, including lack of access to technology, poor promotion prospects,

Table 12.2. Percentage of increase in number of carers by country of birth (minimum 300 carers as of 2016)

Country	2006	2011	2016	% change
Nepal	117	1,809	5,620	4,703
Nigeria	64	150	1,097	1,614
Pakistan	32	91	326	919
India	1,176	5,751	10,397	784
Ghana	82	142	474	478
Liberia	93	370	522	461
Kenya	245	539	1,299	430
Bangladesh	102	152	520	410
Philippines	2,278	4,198	11,187	391
Sierra Leone	103	192	493	379
South Korea South	204	530	956	369
Hong Kong	156	212	703	351
Zimbabwe	348	1,224	1,539	342
Thailand	244	537	972	298
Japan	190	385	655	245

Source: Australian Bureau of Statistics (2007, 2012, 2017).

inadequate management, and heavy workloads (Adhikari, 2012, 2013; Connell, 2010; Walton-Roberts, 2015). What remains unclear is what migration channels these new care workers followed and whether they came as refugees, partners, or primary wage earners.

The emerging diversity of geographical origins of migrant care workers is replicated at the local level. A lone study of a single residential age care facility in Queensland recorded staff and residents from about fifty-five different linguistic backgrounds, although that degree of diversity was atypical (Gao et al., 2015). The growth in culturally diverse aged care workers is potentially of great value to the quality of aged care in Australia. Given the diversity of the Australian population, having carers from a range of backgrounds theoretically provides linguistically and culturally appropriate support to older people from those same cultures (Shanley et al., 2012; Xiao et al., 2013). Conversely, linguistic challenges lead to tensions between carers and older people in some situations and inhibit relevant care and participation, evident in Queensland and elsewhere (Cangiano et al., 2009; Gao et al., 2015; Walsh & O'Shea, 2010). Moreover, the numbers of aging Nepalese or Sierra Leonians in Australia are low, while the numbers of aging Europeans, including southern Europeans (with incomplete facility in English), are higher. An effective mismatch exists between the languages spoken by aging Australians and those of the migrant carers, an issue of concern within "the ethnic age care sector" (Adamson et al., 2017).

International migration to Australia is primarily a metropolitan phenomenon with recent migrants concentrated in metropolitan centres, despite attempts at decentralization. Consequently, migrant carers are much more prevalent in cities than in rural and regional areas. In 2011, some 47 per cent of carers in urban areas were overseas-born compared with 16 per cent in other parts of Australia; these distinctions remain and are at least suggestive of specific care deficits in rural and regional Australia.

Gendered Care

The care workforce is characterized as being predominantly female. However, while the vast majority of care workers are females, the percentage increase of male care workers, from every source (including Australia itself), is actually higher compared to that of females. While females still predominate in the workforce with 81 per cent of the workforce being made up of females in 2011, the overall increase in male workers for the five-year period was 44 per cent compared with a 29 per cent increase in females. That increase was greatest for migrants from Sub-Saharan African and South Asian countries. While this increase may be due to the physically demanding nature of some care work and a shift away from traditional gender roles, it is probably primarily due to the difficulty of obtaining alternative employment following migration to Australia. Although this trend represents a slight "defeminization" of the care workforce, men are otherwise largely concentrated elsewhere in the national workforce, which has both contributed to a probable care deficit and to ensuring that wages in this sector are substantially below national averages.

Social Contexts

Relatively little qualitative data exist on the social and economic contexts, and the working lives of migrant (or any) carers in Australia. While most information is fragmentary, it does at least suggest that the social conditions and context of carers' lives are broadly similar to those of migrant carers in other "developed" countries, which loosely indicates that most are underpaid and overworked but providing a valuable, crucial, and undervalued service.

Care workers in a single-aged care facility in Queensland reported excessive physical demands (needing to move residents and machines), a wide range of responsibilities, and psychological pressures associated with grief at the death of residents. Incomes and work schedules were also of concern; as one put it: "The wages are terrible ... If I was just packing shelves in a supermarket, you're just putting items in the right place. Whereas here you're dealing with people's lives, their emotions, their health, their wellbeing, the whole lot" (as cited in Gao et al., 2015, p. 116).

For most care workers, shift work was a problem, involving missing partners and children, for others a benefit (that enabled avoiding rush hours). However overseas-born care workers especially enjoyed the multicultural environment, feeling more relaxed in a context where they were less obviously a "minority other," although language differences often raised problems (Gao et al., 2015). Linguistic problems were widespread and difficulties in intercultural understanding were accentuated by hearing difficulties and dementia (Martin & King, 2008). Indicative of potential difficulties in providing effective care was the situation where care work was one of the scarce opportunities for overseas migrants to enter the employment market with limited language skills. Care work thus provides an opportunity for care workers to improve their English language skills and move on to "things they want to do" (as cited in Gao et al., 2015, p. 117), again indicating both that many carers do not have qualifications in care employment and do not wish to remain there. This is not a healthy context in which to develop a workforce.

Amongst a small group of overseas-born carers in rural South Australia, a particular reason for being in the care workforce was because it was easy to find work in the industry (Goel & Penman, 2015) rather than any implicit satisfaction with the job. More broadly, the increase of the male workforce reflects the situation where alternative (and superior) employment is hard to obtain. For all those frustrations, overseas workers were perceived as being valuable because of their work ethic, respect for elders, and a willingness to learn new skills (Isherwood & King, 2017).

In the single Queensland case, at least, low wages were not perceived as a major problem as long as other working conditions were adequate (which included a flexible work schedule and the convenient location of the workplace), there was organizational support, opportunities for training, and carers felt valued. Nonetheless, deciding to leave a position mainly centred on limited resources and inadequate support; those who intended to stay remained because the work was "meaningful" where it led to establishing longer-term relationships (Gao et al., 2015). By contrast, carers in South Australia (the majority from the Philippines), found it difficult to deal with frail individuals, with dementia and chronic illnesses, and with emotional problems and terminal conditions; most felt they were given the worst jobs and the most difficult shifts (Goel & Penman, 2015). However, it was felt that perseverance might be rewarded – Asian migrant carers believed that helping old people would accumulate blessings for them in the future (Gao et al., 2015). In both contexts, cultural differences, especially language problems both with residents and with co-workers, were seen by the care workers as barriers to care but of lesser significance than access to adequate resources, both physical and human. Most care workers felt more valued at their workplace than they did within wider society, which accounted for their ultimate frustrations.

Positive perceptions of the working environment were less evident in metropolitan centres such as Sydney where journeys to work could be long and reasonable housing was difficult to find. The cost of housing in cities like Sydney has substantially increased relative to wages over the past decade, so that many carers are commuting a much longer distance to get to work than ever before, and by public transport. Thus many Nepalese carers make journeys of well over ninety minutes twice a day from western Sydney to the northern beaches. However, care workers are far from a homogeneous group: they have different objectives and attitudes, to their work, to their place at work and in wider society, and in what they seek to do.

Skill Drain: Care Workers as "Drop-Outs"

Proportionately, the largest increase in foreign workers in the care workforce has been from Nepal followed by Nigeria, Pakistan, and India. These countries are also significant sources of other cadres of health workers such as nurses (Negin et al., 2013) alongside other countries, especially in Sub-Saharan Africa, where care workers are needed domestically. That loss to other countries is accentuated and further complicated by "deskilling." Many care workers arrive as either partners or skilled migrants and, since carers are, by immigration classifications, unskilled, those who have entered Australia in skilled categories must have skills in "superior" activities, mainly in nursing.

While data on deskilling are scarce in Australia, this is evident in Canada where 44 per cent of the foreign-born long-term care workers are registered nurses in their country of origin (Bourgeault et al., 2009). Numerous examples of the same deskilling process exist elsewhere, where nurses and other professional health workers find work as carers because their qualifications are not recognized in their new country (Jonson & Giertz 2013; Pratt, 1999; Salami et al., 2014; Schultz & Rijks, 2014). All of a small group of eight nurses from Portuguese-speaking Africa who immigrated to Lisbon had initially been unable to find a nursing job, citing racism and the non-recognition of qualifications as the main reasons, consequently obtaining unskilled work as nursing aides or cleaners, without either social security or long-term job security (Luck et al., 2000). Filipina nurses in Vancouver routinely became nannies and caregivers to secure migration opportunities, and had limited subsequent chances of regaining their old status (Pratt, 1999).

Such brain and skill waste is not unusual. Deskilling is also occurring in Australia, although the share of foreign-born care workers with a nursing qualification is unknown (Colombo et al., 2011). The largest increase in aged and disability carers in the Australian workforce who are foreign-born is among workers from Nepal, and at least some of these workers were nurses in their home country, as is true of Nepalese migration to the UK (Adhikari 2012,

2013). The most significant skill loss comes where health workers (especially nurses) cannot gain registration overseas because their qualifications are deemed inadequate or they cannot afford registration or the necessary local courses. Losses are substantial but immeasurable (Connell, 2010). A secondary skill loss occurs where nurses are deliberately employed as caregivers in nursing homes rather than working in hospitals, a widely reported situation, sometimes where unscrupulous employers have taken advantage of nurses as a source of cheap labour. Deskilling produces dissatisfaction, which has resulted in loss of morale and high turnover rates.

Nurses particularly have sometimes migrated as partners rather than workers, and taken up "temporary" unskilled work to support families, before eventually seeking to get their qualifications recognized and take up nursing positions. A substantial number of Tongan nurses in Australia gave up nursing partly because their qualifications were not recognized (and they needed jobs immediately to pay off debts, pay rent, and so on), thus "finding work was often a matter of subsistence." Health work was sometimes distant from homes, new family structures posed different demands and jobs such as taxi-driving allowed necessary flexibility (Fusitu'a, 2000, p. 60). The same situation occurred in New Zealand, where many migrant Samoan or Tongan nurses were employed in rest homes and residential care, having failed to have their qualifications recognized or registered, and so earned lower incomes. Compared with nurses as a whole, Pacific islanders were disproportionately likely to be employed in public hospitals, rest homes, or residential care (Koloto, 2003). Such circumstance are widespread and well documented for various migrants in Australian, New Zealand, and American nursing homes in the UK and elsewhere (Connell, 2010). Many senior Nigerian nurses worked in UK nursing homes, though none had sought to do so, found the work routine, tedious, and restricted, while also experiencing racist, discourteous, and domineering treatment by white carers and management who challenged and ignored their professional and social status (Aboderin, 2007, p. 2243). Deskilling and disempowerment are thus widespread, ironically especially where overseas workers are recruited because of their skills (Allan, 2007; Dyer et al., 2008; O'Brien, 2007), and also where their migrant status is insecure (McGregor, 2007).

Once deskilling occurs, migrants rarely regain positions relevant to their skills and are more likely to leave the workforce altogether. Indeed, there is widespread despondency amongst care workers that they have "descended" into positions that they would not have deliberately chosen other than out of necessity and that they would like to relinquish if circumstances were to change (Cuban, 2013; International Organization for Migration, 2012). In turn, that raises questions about the quality and value of the work undertaken by disenfranchised and disempowered workers. Indeed, employment as a carer can be seen as "characterised by uncertainty, low income, limited social benefits, and statutory entitlements, and which is further shaped by social context and social location" (Vosko, 2010, p. 2): circumstances typical of a precariat. Deskilling is effectively reinforced by gendered flows into this somewhat precarious and unregulated part of the labour market. Expensive training is almost completely lost, incomes are lower than anticipated, and significant gains are not made by the health systems, migrant workers, or their kin at home who wait for remittances. Migration is no great success story in the short term, other than in keeping low-cost health care structures of developed countries relatively functional.

Deskilling emphasizes what has been seen as a "race to the bottom" (Cuban, 2008, p. 83), where skilled health workers are in the contradictory position of being well-educated but hired as low-cost labour: a form of "contradictory class mobility" (Parreñas, 2001). Migrant status and ethnicity reinforce nursing and labour market hierarchies, so that race, class, and gender remain divisive, keeping migrants long-time strangers in a strange land. Carers travel thousands of kilometres to look after the relatives of others: the quintessential emotional labour, that is "the induction or suppression of feeling in order to sustain an outward appearance that produces in others a sense of being cared for in a convivial, safe place" (Hochschild, 1983, p. 7), at the expense of the suspension of direct emotion for their own relatives. Thus carers in Cumbria, UK, achieved little social mobility but had become a "hidden, silenced group" despite their crucial role in providing care and reducing costs within an increasingly liberalized and privatized "audit culture" (Cuban, 2008). Essential work is ultimately undervalued, and not only for migrants. The increased reliance on migrant women carers in Australia raises similar important questions regarding the impact on the source nations that provide the migrant care workers, especially as those countries also experience aging (Connell & Walton-Roberts, 2016; Harper et al., 2008, pp. 166–8). Within Australia that issue has been invariably ignored.

Conclusions

The number of foreign-born care workers in Australia both in absolute numbers and as a percentage of the total number of carers has increased substantially between 2006 and 2016. The largest numbers come from the UK, South Asia, and Southeast Asia, with a recent shift towards Nepal and Sub-Saharan Africa. Beyond these basic data few studies of migrant care workers in Australia exist, hence conclusions are necessarily tentative and temporary. However, clear parallels exist with circumstances in other "developed" countries, accentuated by neglected welfare policies and migration policies that offer few opportunities for unskilled workers or workers with unrecognized skills (Misra et al., 2006; Pajnik, 2016). Complex questions arise over whether recipients of care are customers or citizens, and whether efficiency and competition are alien concepts to good social labour and whether they corrode the foundations of trust, continuity, and judgment that are critical to care. The steady increase in the proportion of the carer workforce coming from overseas parallels that across the OECD, as in Canada, the UK, Ireland, the US, and New Zealand, where similar structures and patterns occur (Badkar et al., 2009; Bourgeault et al., 2009; Cangiano et al., 2009; Doyle & Timonen 2009; Paraprofessional Healthcare Institute 2013). It is highly likely that in Australia, as elsewhere in the OECD, this will be an ongoing trend; as Australia ages, more aged and disability care workers will be required, and the locally born disdain such employment.

Despite their substantial numbers, care givers are both ubiquitous and invisible, without champions or effective advocacy, marginalized, non-unionized, and socially undervalued. Any improvements in work conditions, and even simple respect and kindness, are therefore hard won, especially when the workforce changes rapidly. Most are unwillingly embedded in precarious work that offers poor employment conditions, which may extend to their being often casual and short term; tenuous links to their education and skills; and limited potential as stepping stones to the jobs they want, where opportunities for decent work (and decent wages) have been eroded: a set of overlapping insecurities. Carers perfectly fit the definition of precarious workers as those who "provide hyperflexible labour, available when required, undemanding when not" (Anderson, 2010, p. 300). Minimal social mobility emphasizes inequality. A marginalized workforce cares for a vulnerable segment of the national population. That both raises questions about who cares for the carers, the need for an ethics of care to be extended to them, and the failure to recognize that care requires specialist knowledge, expertise, and training. Meanwhile, elder care in Australia is increasingly provisioned through an international migration process that represents a systemic loss of value in terms of credentials for the migrant and the sending nation.

While it is impossible to determine from census data how long carers have been in Australia, the rise in numbers from south Asia and Sub-Saharan Africa parallels a general recent increase in migration from those countries, hence a high proportion of carers are new arrivals to Australia, finding their way to the "bottom" of the employment market as an emerging underclass or precariat (Isherwood & King, 2017). Indeed, the albeit-limited information on work choices of care workers indicates that few actually chose care work but were effectively forced into it by being unable to gain alternative employment, a situation that replicates the experience of Filipinas in New Zealand and Zimbabweans in the UK (Lovelock & Martin, 2016; McGregor, 2007). While there is no necessary reason why such workers should not be effective, this situation points to significant subsequent churn and turnover.

These are realities that must be managed in circumstances where opposition to migration is considerable. Employers in care facilities have used a wide variety of visa types to employ workers, including occupational training visas and student visas. Appropriate care qualifications do not exist in many countries, especially those of significance for recent migration to Australia. As care workers cannot obtain a visa based on skills as care workers, they must either be migrating with a recognized skill or are migrating through other pathways such as family migration. In the former case, the migration of care workers has been further complicated by the process of deskilling, a situation that benefits no one. Meanwhile such mixed-migration trajectories and ownerships of skills point to the substantial problem of having an adequate level of competence and adequate skill mix to sustainably meet the needs of a growing and culturally diverse aging population: complex and disconcerting questions, attached to an Australian crisis with repercussions in other continents.

REFERENCES

Aboderin, I. (2007). Contexts, motives and experiences of Nigerian overseas nurses: Understanding links to globalization. *Journal of Clinical Nursing*, 16, 2237–45. https://doi.org/10.1111/j.1365-2702.2007.01999.x.

Adamson, E., Cortis, N., Brennan, D., & Charlesworth, S. (2017). Social care and migration policy in Australia: Emerging intersections? *Australian Journal of Social Issues*, *52*(1), 78–94. https://doi.org/10.1002/ajs4.1.

Adhikari, R. (2012). *Perils and prospects of international nurse migration from Nepal*. Centre for the Study of Labour and Mobility.

Adhikari, R. (2013). Empowered wives and frustrated husbands: Nursing, gender and migrant Nepali in the UK. *International Migration*, *51*(6), 168–79. https://doi.org/10.1111/imig.12107.

Anderson, B. (2010). Migration, immigration controls and the fashioning of precarious workers. *Work, Employment and Society*, *24*(2), 300–17. https://doi.org/10.1177/0950017010362141.

Allan, H. (2007). The rhetoric of caring and the recruitment of overseas nurses: The social production of a care gap. *Journal of Clinical Nursing*, *16*(12), 2204–12. https://doi.org/10.1111/j.1365-2702.2007.02095.x.

Australian Bureau of Statistics. (2007). *2006 census of population and housing*. Australian Bureau of Statistics.

Australian Bureau of Statistics. (2012). *2011 census of population and housing*. Australian Bureau of Statistics.

Australian Bureau of Statistics. (2017). *2016 census of population and housing*. Australian Bureau of Statistics.

Australian Institute of Health and Welfare. (2012). *Dementia in Australia*. Australian Institute of Health and Welfare.

Badkar, J., Callister, P., & Didham, R. (2009). *Ageing New Zealand: The growing reliance on migrant caregivers*. Institute of Policy Studies.

Baldassar, L. (2017). Who cares? The unintended consequence of policy for migrant families. In D. Tittensor & F. Mansouri (Eds.), *The politics of women and migration in the Global South* (pp. 105–23). Palgrave Macmillan.

Bourgeault, I., Atanackovic, J., LeBrun, J., Parpia, R., Rashid, A., & Winkup, J. (2009). *The role of immigrant care workers in an aging society: The Canadian context and experience.* University of Ottawa Press.

Brush, B., & Vasupuram, R. (2006). Nurses, nannies and caring work: Importation, visibility and marketability. *Nursing Inquiry, 13*(3), 181–5. https://doi.org/10.1111/j.1440-1800.2006.00320.x.

Cangiano, A. (2014). Elder care and migrant labor in Europe: A demographic outlook. *Population and Development Review, 40*(1), 131–54. https://doi.org/10.1111/j.1728-4457.2014.00653.x.

Cangiano, A., Shutes, I., Spencer, S., & Leeson, G. (2009). *Migrant care workers in ageing societies: Research findings in the United Kingdom.* COMPAS. https://www.compas.ox.ac.uk/wp-content/uploads/PR-2009-Care_Workers_Ageing_UK.pdf.

Colombo, F., Llena-Nozal, A., Mercier, J., & Tjadens, F. (2011). *Help wanted? Providing and paying for long-term care.* OECD Publishing.

Connell, J. (2010). *Migration and the globalisation of health care.* Edward Elgar.

Connell, J., & Walton-Roberts, M. (2016). What about the workers? The missing geographies of health care. *Progress in Human Geography, 40*(2), 158–76. https://doi.org/10.1177/0309132515570513.

Cuban, S. (2008). Home/work: The roles of education, literacy and learning in the networks and mobility of professional women migrant carers in Cumbria. *Ethnography and Education, 3*(1), 81–96. https://doi.org/10.1080/17457820801899132.

Cuban, S. (2013). *Deskilling migrant women in the global care industry.* Palgrave Macmillan.

Department of Health and Ageing. (2011). Technical paper on the changing dynamics of residential aged care: Prepared to assist the productivity commission inquiry caring for older Australians. Department of Health and Ageing.

Department of Health and Ageing. (2017). The aged care workforce, 2016. Department of Health and Ageing.

Dhakal, S., Nankervis, A., Connell, J., Fitzgerald, S., & Burgess, J. (2017). Attracting and retaining personal care assistants in to the Western Australia (WA) residential aged care sector. *Labour and Industry, 27*(4), 333–49. https://doi.org/10.1080/10301763.2017.1418236.

Doyle, M., & Timonen, V. (2009). The different faces of care work: Understanding the experiences of the multi-cultural care workforce. *Social Politics, 17*(1), 29–52. https://doi.org/10.1017/S0144686X08007708.

Dyer, S., McDowell, L., & Batnitzky, A. (2008). Emotional labour/body work: The caring labours of migrants in the UK's National Health Service. *Geoforum, 39*(6), 2030–8. https://doi.org/10.1016/j.geoforum.2008.08.005.

Fine, M., & Mitchell, A. (2007). Review article: Immigration and the aged care workforce in Australia: Meeting the deficit. *Australasian Journal of Ageing, 26*(4), 157–61. https://doi.org/10.1111/j.1741-6612.2007.00259.x.

Fusitu'a, P. (2000). *My island homes: Tonga, migration and identity* [Unpublished BSc honours thesis]. University of Sydney.

Gao, F., Tilse, C., Wilson, J., Tuckett, A., & Newcombe, P. (2015). Perceptions and employment intentions among aged care nurses and nursing assistants from diverse cultural backgrounds: A qualitative interview study. *Journal of Aging Studies*, 35, 111–22. https://doi.org/10.1016/j.jaging.2015.08.006.

Goel, K., & Penman, J. (2015). Employment experiences of immigrant workers in aged care in regional South Australia. *Rural and Remote Health, 15*(1), 2693. https://doi .org/10.22605/RRH2693.

Gray, M., & Heinsch, M. (2009). Ageing in Australia and the increased need for care. *Ageing International, 34*(3), 102–18. https://doi.org/10.1007/s12126-009-9046-3.

Harper, S., Aboderin, I., & Ruchieva, I. (2008). The impact of the outmigration of female care workers on informal family care in Nigeria and Bulgaria. In J. Connell (Ed.), *The international migration of health workers* (pp. 163–71). Routledge.

Hochschild, A. (1983). *The managed heart: Commercialization of human feeling.* University of California Press.

Howe, A. (2009). Migrant care workers or migrants working in long-term care? A review of Australian experience. *Journal of Aging and Social Policy, 21*(4), 374–92. https://doi.org/10.1080/08959420903167140.

Howe, A., King, D., Ellis, J., Wells, Y., Wei, Z., & Teshuva, K. (2012). Stabilising the aged care workforce: An analysis of worker retention and intention. *Australian Health Review, 36*(1), 83–91. https://doi.org/10.1071/AH11009.

Hugo, G. (2007). Contextualising the "crisis in aged care" in Australia: A demographic perspective. *Australian Journal of Social Issues, 42*(2), 169–82. https://doi.org /10.1002/j.1839-4655.2007.tb00047.x.

Hugo, G. (2009). Care worker migration, Australia and development. *Population Space and Place, 15*(2), 189–203. https://doi.org/10.1002/psp.534.

International Organization for Migration. (2012). *Crushed hopes: Underemployment and deskilling among skilled migrant women.* International Organization for Migration.

Isherwood, L., & King, D. (2017). Targeting workforce strategies: Understanding intra-group differences between Asian migrants in the Australian aged care workforce. *International Journal of Care and Caring, 1*(2), 191–207. https://doi.org/10.1332 /239788217X14937990731721.

Jonson, H., & Giertz, A. (2013). Migrant care workers in Swedish elderly and disability care: Are they disadvantaged? *Journal of Ethnic and Migration Studies, 39*(5), 809–25. https://doi.org/10.1080/1369183X.2013.756686.

Kaine, S., & Ravenswood, K. (2014). Working in residential aged care: A Trans-Tasman comparison. *New Zealand Journal of Employment Relations, 38*(2), 33–46. https://

www.researchgate.net/publication/273440938_'Working_in_Residential_Aged_Care_A_Trans-Tasman_Comparison.
Koloto, A. (2003). *National survey of Pacific nurses and nursing students.* Koloto and Associates.
Lovelock, K., & Martin, G. (2016). Eldercare work, migrant care workers, affective care and subjective proximity. *Ethnicity and Health, 21*(4), 379–96. https://doi.org/10.1080/13557858.2015.1045407.
Luck, M., Fernandes, M., & Ferrinho, P. (2004). At the other end of the brain-drain: African nurses living in Lisbon. *Studies in Health Service Organisation and Policy, 16*, 157–69.
Martin, B., & King, D. (2008). *Who cares for older Australians? A picture of the residential and community based aged care force, 2007.* Commonwealth of Australia.
McGregor, J. (2007). Joining the BBC (British Bottom Cleaners): Zimbabwean migrants and the UK care industry. *Journal of Ethnic and Migration Studies, 33*(5), 801–24. https://doi.org/10.1080/13691830701359249.
Misra, J., Woodring, J., & Merz, S. (2006). The globalization of care work: Neoliberal economic restructuring and migration policy. *Globalizations, 3*(3), 317–32. https://doi.org/10.1080/14747730600870035.
Negin, J., Coffmann, J., Connell, J., & Short, S. (2016). Foreign-born aged care workers in Australia: A growing trend. *Australasian Journal on Ageing, 35*(4), E13–E17. https://doi.org/10.1111/ajag.12321.
Negin, J., Rozea, A., Cloyd, B., & Martiniuk, A. (2013). Foreign-born health workers in Australia: An analysis of census data. *Human Resources for Health, 11*(69). https://doi.org/10.1186/1478-4491-11-69.
O'Brien, T. (2007). Overseas nurses in the National Health Service: A process of deskilling. *Journal of Clinical Nursing, 16*(2), 2229–36. https://doi.org/10.1111/j.1365-2702.2007.02096.x.
Pajnik, M. (2016). "Wasted precariat": Migrant work in European societies. *Progress in Development Studies, 16*(2), 159–72. http://dx.doi.org/10.1177/1464993415623130.
Paraprofessional Healthcare Institute. (2013). America's direct-care workforce. PHI.
Parreñas, R.S. (2001). *Servants of globalization: Women, migration and domestic work.* Stanford University Press.
Pratt, G. (1999). From registered nurse to registered nanny: Discursive geographies of Filipina domestic workers in Vancouver, BC. *Economic Geography, 75*(3), 215–36. https://doi.org/10.1002/9780470755716.ch24.
Productivity Commission. (2011). *Caring for older Australians: Final inquiry report.* Productivity Commission.
Russell, S. (2017). *Living well in an aged care home.* Research Matters.
Salami, B., Nelson, S., Hawthorne, L., Muntaner, C., & McGillis Hall, L. (2014). Motivations of nurses who migrate to Canada as domestic workers. *International Nursing Review, 61*(4), 479–86. https://doi.org/10.1111/inr.12125.

Savy, P., Warburton, J., & Hodgkin, S. (2017). Challenges to the provision of community aged care services across rural Australia: Perceptions of service managers. *Rural and Remote Health, 17*(2), 4059. https://doi.org/10.22605/RRH4059.

Schultz, C., & Rijks, B. (2014). *Mobility of health professionals to, from and within the European Union.* International Organization for Migration.

Shanley, C., Boughtwood, D., Adams, J., Santalucia, Y., Kyriazopoulos, H., Pond, D., & Rowland, J. (2012). A qualitative study into the use of formal services for dementia by carers from culturally and linguistically diverse (CALD) communities. *BMC Health Services Research*, 12, 354. https://doi.org/10.1186/1472-6963-12-354.

Vosko, L. (2010). *Managing the margins: Gender, citizenship, and the international regulation of precarious employment.* Oxford University Press.

Walsh, K., & O'Shea, E. (2010). Marginalised care: Migrant workers caring for older people in Ireland. *Journal of Population Ageing, 3*(1-2), 17–37. https://doi.org/10.1007/S12062-010-9030-4.

Walton-Roberts, M. (2015). International migration of health professionals and the marketization and privatization of health education in India: From push-pull to global political economy. *Social Science and Medicine*, 124, 374–82. https://doi.org/10.1016/j.socscimed.2014.10.004.

Xiao, L., De Bellis, A., Habel, L., & Kyriazopoulos, H. (2013). The experiences of culturally and linguistically diverse family caregivers in utilising dementia services in Australia. *BMC Health Services Research,* 13, 427. https://doi.org/10.1186/1472-6963-13-427.

13 Care Worker Migration and Robotics in Japan's Aged Care Sector

HÉCTOR GOLDAR PERROTE AND MARGARET WALTON-ROBERTS

Introduction: Why Look at Japan?

Japan has been dealing with the demographic challenges associated with an aging population for decades – challenges so significant and ubiquitous that "no social domain, institution or individual has remained unaffected" (Coulmas, 2007, p. 2). As we shall learn in this chapter, the country's Long-Term Care Insurance (LTCI) system is currently under considerable stress and will likely continue to be for the years to come, particularly due to questions of financial sustainability and understaffing. To address the very pressing labour shortage in the sector, the Japanese government has adopted various measures that include, among others, raising wages for caregivers, admitting a larger number of foreign health care workers, and relying more markedly on robotics and artificial intelligence. Having said this, it is debatable that such measures will successfully alleviate the pressures of an aging population, and that the 340,000 workers needed in this sector by 2025 will be secured.

In this chapter, we delve into these questions in more detail, hoping to understand how Japan is managing its care needs. We believe that doing so can be instructive with regard to assessing approaches towards transferring the value of skills between different distinct national systems of training and health care. From our perspective, the case of Japan is particularly interesting: not only is the country the front-runner of super-aged societies (Arai et al., 2015) but it is also the third largest world economy and a dominant player in global affairs – so the decisions that are made in the East Asian nation have a considerable impact on global value chains. We begin the chapter with a short overview of the Japanese context in terms of demography and long-term care. Then, we analyse recent policy developments in the area of immigration law, focusing more specifically on health care workers. The third section of this chapter looks at the role that robotics and artificial intelligence can play in the field of care. To conclude, we briefly examine some of the various changes that we are likely to witness in this area as robots are introduced in facilities and households.

Modern-Day Japan and the Challenging Provision of Elderly Care in the Context of Changing Demographics

Modern-day Japan is characterized by its political stability under the Liberal Democratic Party (*Jimintō*) and by a somewhat stagnant economy that has been on a moderate growth track since 2011. This stability needs to be understood within the difficult geopolitical context of the region and the global turn to neo-nationalism that we have come to witness since the mid-2010s. Indeed, there is a case to be made that nationalism is becoming increasingly stronger in Japan. That being said, the East Asian country has taken the global leadership in promoting free trade, fully engaging with economic globalization. Culturally speaking, and despite the existing diversity, Japanese society predominantly adheres to a "conservative/essentialist" understanding of culture and national identity to this day (Befu, 2009, p. 35). This prevailing view of "Japanese-ness" is based on the conception of a single, homogenous culture shared by an egalitarian, harmonious society with minimal regional differences and small minorities (Sugimoto, 2009, p. 3). These dominant ideas have clear policy implications in all sorts of arenas.

An additional point, which deserves special attention in this particular case, is that of demography. Japan's population structure has changed quite profoundly during the post-1945 period. Currently, the country is aging fast and its overall population, as well as its working age cohort, continue to shrink (National Institute of Population and Social Security Research [IPSS], 2012, pp. 2–3). As all projections indicate, these phenomena will only accelerate in the future (IPSS, 2012, pp. 2–3). On top of this, fertility rates remain well below replacement levels (World Bank, 2017), and the phenomenon of depopulation in secondary regions becomes more and more evident every year (Matanle & Rausch, 2011). Although Japan is by no means an outlier in this regard, no other nation has arguably experienced such trends so acutely and for such a prolonged period.[1] Among all of these, population aging is perhaps the most salient question. Japan is the world's "front-runner of super-aged societies" (Arai et al., 2015); 28.4 per cent of the population is currently over sixty-five years of age ("Elderly citizens," 2019).

Even though most of the previously discussed demographic trends are not necessarily *negative* in and of themselves, there is an undeniable set of economic and social policy challenges that are generally associated with the demographic stage in which Japan finds itself today. The population changes that the country is experiencing represent significant modifications to the demographic variable embedded in many of our widely accepted socio-economic models and theories. As Robbins and Smith (2016) point out, "our core theories of political economy were forged both within a context of assumed growing capitalist expansion and with a hidden bias of constant and absolute (demographic) growth" (p. 6). And so it follows that any consequential changes to this variable are likely to create substantial disruptions in various domains and challenge the status quo. For instance, it is evident that "an older and smaller population has implications for productivity and long-term growth" (Hong & Schneider, 2020, p. 21). Also, we can observe that high old-age dependency ratios are making it more complicated to ensure the sustainability of the country's welfare state (Organisation for Economic Co-operation and Development [OECD], 2018). Indeed, social security costs, which already accounted for 33.3 per cent of Japan's total expenditure as of 2017, are on an upward trend as more funds need to be allocated for pensions and health care, among other things (Ministry of Finance, Japan, 2017). The country's universal health care system is under considerable stress, and the provision of a number of other basic services, like those covered under the LTCI system, is becoming relatively difficult. Some may argue that these pressures can act as an incentive to foster greater efficiency and innovation and that they actually represent an opportunity. However, it is undeniable at this point that they also pose formidable dilemmas for policymakers. Nowhere is this clearer than in the case of the publicly funded LTCI system.

The current LTCI – or *Kaigo Hoken* – is a public insurance program that facilitates the provision of care for the elderly in Japan (Mitchell et al., 2004).[2] The LTCI was introduced in 2000 as traditional family arrangements were being renegotiated and the proportion of elderly citizens was on the rise (Ministry of Health, Labour and Welfare, Japan [MHLW], 2016, p. 4). All people over the age of 65 are eligible, as well as those 30–64 with aged-related disabilities (Mitchell et al., 2004, p. 8). Under the LTCI system, two main types of services are offered by various different providers: (1) *in-home services* administered by both public and private entities; and (2) *services at facilities* in not-for-profit and government-run nursing homes and hospitals (Mitchell et al., 2004, pp. 9–10). Essentially, the LTCI program is a pay-as-you-go system funded primarily by means-tested premiums levied on insured persons (which account for 50 per cent of the program's revenue), and by general tax revenue (which accounts for the other half) (Mitchell et al., 2004, p. 27). With such funds, municipalities pay individual service providers 90 per cent of the cost of any service they offer. The remaining 10 per cent – as well as housing costs and meals – is generally covered by users (MHLW, 2016, p. 8). Fees for all services are set by the central government. One of the main features of long-term care in Japan is that community care is encouraged over institutionalization (Iwagami & Tamiya, 2019). This is justified on fiscal grounds, but also as the most supportive of seniors' well-being (Kavedzija, 2018).

For two decades, the LTCI system has helped Japanese seniors to have their care needs met fairly successfully. With that in mind, we should also indicate that a number of flaws and limitations have been identified both by researchers and practitioners; for example, by 2004 rationing was a concern (Mitchell et al., 2004). Nowadays, the system is struggling to provide for all, which is apparent in the long waiting lists that exist to access institutionalized care (Ogawa et al., 2018). The LTCI program has also been criticized for exacerbating already existing regional inequalities (Mitchell et al., 2004, p. 13), and for failing to provide family caregivers with enough support (Iwagami & Tamiya, 2019, p. 68). Elder abuse in nursing homes is also on the rise ("Elderly abuse," 2019). Arguably, all these issues are either connected to or directly stem from the two main problems that this scheme faces today: fiscal sustainability and the severe and chronic problem of understaffing (Kato, 2015).

Leaving fiscal sustainability questions aside, we will concentrate on the problem of understaffing within the care sector. At the time of writing, Japan had been experiencing a general and very severe labour shortage since 2015 (Martin, 2015). The care industry was one of the most affected sectors. In 2017, there were 3.5 caregiving positions for every job seeker – more than double the overall industry figure of 1.54 ("Fill the gap," 2018), with a projected shortage of 340,000 caregivers by 2025 ("Foreigners who flunked," 2019). This phenomenon, although certainly more marked in the case of Japan, is still in line with the trends that we are witnessing in other developed nations; the World Health Organization estimated that by 2030 health care worker shortages will rise to 80 million. For some perspective, this is double the 2013 stock of health care workers (Liu et al., 2017). In an economy with full employment, prospective job seekers have been turning to better opportunities in other fields, particularly because of the very demanding character of the job, relatively low wages,[3] long hours, workplace harassment, and difficult workplace atmosphere (Yashuhiro, 2020). To make matters worse, staff turnover has traditionally been quite high, with a current 60 per cent of workers quitting after three years on the job (Yashuhiro, 2020). This applies to foreign caregivers too, with 19 to 34 per cent of workers quitting their jobs and returning home despite having passed their certification exams (Siripala, 2018).

Given this relatively bleak scenario, it is unclear how Japan will manage to secure enough health care workers to keep the LTCI system afloat, especially when the demand will continue to grow in the future, and considering projections suggest that 7 million citizens will suffer some form of dementia (Siripala, 2018). Most analysts seem to agree that the program will most certainly be under tremendous pressure in the future, regardless of potential policy measures. Kato (2015, p. 2), for instance, is of the opinion that it will be "almost impossible" to increase care service supply in order to fully meet rising demand for services. Still, various proposals have been made with a view to improving the situation. Kato (2015) suggests striving to improve the overall working conditions of care work professionals by providing them with better contracts, career path assistance, and training to further enhance their skills. In his opinion, it is also fundamental to consider the very important role that information and communication technologies are likely to have in the future, making the system more efficient (and possibly less labour intensive). Yashuhiro (2020) indicates that nursing care facility managers would greatly benefit from better training on how to make the workplace more appealing for their employees. Iwagami and Tamiya (2019) place the focus on the very important role of accident/injury prevention and of community networking with a view to curtailing demand. While these measures are interesting and could potentially be considered, there are two main proposals that have become recurrent and that are almost always alluded to within this conversation. The first one is to expand the very modest current pool of foreign caregivers. The second entails investing more heavily and relying more markedly on robotics. We will examine these ideas in more detail in the next two sections.

Japan's Immigration Policy at a Crossroads and the Increasing Admission of Foreign Health Care Professionals

A salient trend that we can observe when studying global migration patterns is that a substantial number of high-income countries have become more and more reliant on foreign health care professionals, especially caregivers (OECD, 2010). This is a reality in many Asian nations and territories such as Hong Kong, Singapore, and increasingly Taiwan and Korea. Japan, however, had not quite utilized foreign labour in this sector until fairly recently. Actually, the first foreign nurses and caregivers who moved to the country for work purposes arrived only after 2008, once bilateral economic partnership agreements (EPAs) were signed with the Philippines (2006) and Indonesia (2007) (Ohno, 2012). Governmental attitudes towards the acceptance of foreign health care workers have changed since then, primarily motivated by the severity of labour gaps in care facilities (where the vast majority of these professionals are employed). More admission pathways have been facilitated since 2008, and in fact, Japan's immigration policy as a whole can be said to be in the midst of a paradigm shift.

There is no denying that there is a certain difficulty to understanding Japan's immigration policy. On the one hand, albeit modest if compared to other Organisation for Economic Co-operation and Development (OECD) members, the share of non-Japanese residents has increased substantially since the late 1980s (Ministry of Justice, Japan [MOJ], 2015a, p. 4).[4] The number of foreign workers surpassed the 1 million threshold for the first time in 2016 ("Japan foreign workers top record 1 million," 2017) – and sat at 1.66 million in 2020 ("No. of foreign workers," 2020). However, for decades the political establishment in Japan has been very invested in preserving the façade that there is "no national immigration policy." In fact, on numerous occasions, Shinzo Abe was reported to have said that Japan "will not adopt an immigration policy" (see

Kodama, 2015, p. 9). Arguably, this stance is an attempted institutionalization of the "homogenous people" discourse (Burgess, 2014). But regardless of political statements, it is fair to say that Japan *does* have a "de facto" immigration policy, even if it is not a fully-fledged, comprehensive legislation (Goldar Perrote, 2017). The amendments that have been introduced as of late do not consider immigration explicitly to be a replacement for the loss in population. They do, however, facilitate the mobility of foreign workers with a view to alleviating labour shortages in specific sectors and contributing to the economy as a whole (Prime Minister's Office of Japan, 2014), all while respecting the country's "nationhood regulatory principle." As Peng argues (2016), the idea of migration and multiculturalism "remains disconnected from that of a shared national identity premised on ethnic and cultural homogeneity" (p. 278).

We do not discount the fact that these conservative forces have played a central role in the evolution of Japan's immigration legislation over the decades. To our mind, the essentialist conception of national identity that Japanese citizens and political elites predominantly adhere to has a lot to do with the fact that, for instance, the path to permanent residency is blocked for most newcomers. Having said this, we are of the opinion that institutional politics, power struggles among influential players, economic considerations, and other variables (such as Japan being a latecomer in this regard or the comparative disadvantage of the Japanese language) must be factored in when discussing policy change. In fact, we believe that a number of analyses and opinion pieces on these matters excessively (and at times even, exclusively) concentrate on the "public opposition against immigration" narrative; they take antagonism towards immigration as a given when actually, the reality of attitudes towards newcomers in Japan, especially in the context of an aging population, is more complicated and often times quite different from what we tend to assume (see Kobayashi et al., 2015; Stokes & Devlin, 2018). For instance, specific polls showed that 54 per cent of Japanese citizens were in favour of accepting more foreign workers in 2018 (Ebuchi & Takeuchi, 2018).

Within the current immigration framework, the highly skilled are targeted – although retention remains quite low for many reasons (see Oishi, 2012). Skilled professionals are also allowed to work in Japan. That being said, visa restrictions limit their intake and prospects for settlement. Up until 2018, unskilled and semi-skilled workers were invited on a temporary basis, mostly through loopholes and in small numbers (Goldar Perrote, 2017). However, the nature of the labour shortage in Japan has been so severe since the mid-2010s that policymakers had no choice but to revise the way they had been dealing with the semi-skilled up until that point, and to introduce something akin to a guest worker program with an amendment to the immigration law in 2018 (the "Specified Skilled Worker" visa program; see Ministry of Foreign Affairs, Japan [MOFA], n.d.). Active since 2019, this system has been designed to adjust flexibly to economic conditions (Milly, 2020) and now allows a greater number of blue-collar workers to earn a living in Japan in one of the fourteen targeted industries (including caregiving) (MOFA, n.d.). The government is hoping to attract 345,000 such workers by 2025 (Menju, 2019a), which, for Japanese standards, is a fairly ambitious target. Although the program does not necessarily challenge Japan's overall stance regarding immigration (Menju, 2019b), its inception arguably symbolizes the beginning of a new era within Japan's immigration control management.

The Specified Skilled Worker visa program does mark a departure from the previous regime, from our perspective at least, for (1) it acknowledges the fact that blue-collar workers are being invited into the country; (2) it accepts these workers through the front door, not resorting to loopholes such as occurred through the Technical Intern Trainee Program (see Goldar Perrote, 2017); and (3) it allows a number of these workers to apply for permanent residency so long as they meet certain requirements (although by now, the vast majority will only stay in the country for up to five years). For practical purposes, we can describe this new visa system as a temporary foreign worker program. In the future, this scheme could perhaps become a pathway for permanent immigration into Japan – although it is certainly too early to say. The cautious way in which Japan is moving forward with this program, tinkering with different options and easing restrictions step by step, is in line with the state's "highly selective, cautious, progressive incrementalism" immigration management strategy (Goldar Perrote, 2017). Along with this new status of residence, two additional sets of reforms were undertaken in 2019: (1) the Ministry of Justice's Immigration Bureau was reorganized and expanded into the Immigration Services Agency; and (2) a set of measures to assist foreign residents was passed (Yamawaki, 2019). Such measures represent a step forward within the arena of integration.

Foreign health care professionals have been working in Japan since 2008. Since that time, a number of different residence statuses have been introduced to accommodate the growing demand for workers in the sector (there was one for nurses and four for caregivers at the time of writing). That said, there are specific forces that discourage the admission of health care workers, including the economic, the political, and the practical. Economic reasons include the expense of the LTCI program and the severe cost of addressing labour shortages that fall to the state (Yanagida, 2012). Some have argued that if a large number of unskilled workers were granted permanent settlement and family reunification rights, a net loss could be generated in terms of social welfare redistribution (for more details, see Ganelli & Miake, 2015; Goldar Perrote, 2017; Yoshida, 2014). It has also been suggested that a tight labour supply could incentivize investment in labour-saving technologies and automation, however, the situation in a substantial number of nursing homes is so dire that their operation is compromised. In terms of political reasons, for decades, institutions such as the Ministry of Health, Labour and Welfare (MHLW) and particularly the Japanese Nursing Association (JNA) lobbied successfully against any changes to the immigration law on the grounds that an influx of cheap labour could undermine the conditions of domestic workers and negatively affect the quality of the service provided (Ogawa, 2014). Finally, practical issues include concerns about caregivers being able to learn Japanese and culturally adapt. Despite these constraints, the pool of foreign workers within the sector was eventually expanded through various schemes. Next, we take a brief look at the different types of work that foreign health care professionals do and the visa systems that enable them to earn a living in the country.

Nurses

Nurses differ from caregivers because of the very nature of their codified expertise and their ability to perform complex medical tasks. Nursing licences obtained overseas are not accepted in Japan, so those foreign residents who wish to work in the sector must meet specific educational requirements and pass the national nursing examination (Japanese Nursing Association, n.d.). Those living overseas who wish to move to Japan to work as nurses have had the opportunity to do so since 2008 under a series of bilateral agreements (EPAs) that were signed with a number of Southeast Asian countries.[5] While the demographic argument and the need for these professionals may have played a role when deciding to sign these agreements, many argue that the main rationale was economic: Japanese bureaucrats may have accepted these workers in exchange for lower tariffs on certain goods (Otake, 2015). White papers from the MHLW are even more specific: "Approving potential nurses ... is not a response to the labour shortages in the health services; this training program has been agreed under the EPAs on the basis of strong requests from [these] countries" (MHLW, 2011). Through EPA programs, foreign nurses – who must have two years of prior experience in their home countries – receive on-the-job training and language lessons. Candidates are then required to pass the national qualification exam and acquire licences to be able to work in Japan (Yanagida, 2012). Successful applicants are granted a three-year renewable visa, but few pass the exams at the end of their four-year period of stay. Only 413 out of the 1,300 applicants had passed the national exams as of 2018 ("Japan's foreign nurses," 2019). The low passing rates and the linguistic challenges these EPA nurses face results in them mostly being relegated to nursing assistants during their training years (Tarumoto, 2012). Health facilities have also shown their dissatisfaction with the scheme because of their inability to retain these workers and the elevated training costs (Lopez, 2012).

Caregivers

The first foreign caregivers to move to Japan for the purpose of work arrived under the same EPA programs discussed above (Ohno, 2012). These candidates faced difficulties similar to those of nurses (language hurdles, discrimination in certain cases, difficulty in passing the exams and obtaining their licences, etc.), but the passing rate has improved since the program was first instituted and currently stands at about 60 per cent. By 2018, 4,300 people had enrolled, 1,724 had taken the national exam, and 985 had passed ("Foreigners who flunked," 2019).

Since the EPA agreements were signed, three other programs have been adopted that are aimed at increasing the number of foreign caregivers in the country. The first scheme is linked to the Nursing Care Status of Residence, which was introduced in 2016, and which has enabled foreign nationals who study nursing care in Japanese colleges and vocational schools to work in the country as long as they graduate and pass the national qualification exam (MOJ, 2015c, p. 6). In that event, successful candidates are offered renewable visas if they secure a labour contract with a nursing care facility. By September 2019, there were 2,037 non-Japanese students of nursing care in the country who accounted for about 30 per cent of the total number of new enrolees ("Foreigners make 30% of new students," 2019).

The second scheme is the Technical Intern Trainee Program (TITP), which has effectively functioned as Japan's labour rotation system to accept temporary unskilled and semi-skilled workers since it was started in 1993 (Shindo, 2019). In 2015, it was announced that care work would be incorporated into the TITP (Japan International Trainee & Skilled Worker Cooperation Organization, 2015). By the end of 2018, only 247 caregivers had participated in the TITP, disappointing government expectations that had been set at 5,000.

The third scheme is the Specified Skilled Worker visa program (previously discussed), which started operating in 2019, and which includes nursing care as one of its fourteen targeted industries. As of 2020, foreign caregivers under this program were only allowed to remain in the country for up to five years. This, we suggest, could potentially change in the near future given the clear need for these workers, the demands of employers to let them stay for longer, and the fact that other foreign workers on this same visa are eligible to stay in Japan indefinitely under specific conditions. Within the first year, roughly 1,700 applicants had passed the standardized exam for obtaining the worker visa for the nursing care industry, falling short of government expectations (Kyozuka, 2020). The numbers of foreign caregivers on this specific visa are therefore expected to grow steadily over the next few years as the system becomes more settled.

All in all, it is evident that Japan is trying to secure a larger number of caregivers through various means and primarily on a rotation basis. However, policymakers are keenly aware of the fact that there are undeniable and clear cultural and linguistic barriers to smoothly integrating a large number of non-Japanese caregivers within the LTCI system. At the time of writing, the Abe administration had indicated their wish of filling about 60,000 of the projected 340,000 vacant positions in Japan by 2025 with foreign caregivers ("Fewer foreigners," 2018) – and particularly with those coming in through the new Specified Skilled Worker visa scheme. Economic and practical reasons mainly explain why the state will only rely *partially* on foreign labour to fill the shortage (those 60,000 out of 340,000). Steps have been taken to attain this goal but, to our mind, the 60,000 target may prove overly optimistic for various reasons, including the potential for greater international competition for care workers in the future.

Housekeepers

A relatively new housekeeping visa has enabled a number of foreign housekeepers to work in Japan under quite a restrictive framework since 2017 (MOJ, 2015b; Reynolds & Aquino, 2017). Arguably, the main goal of this initiative is to get more Japanese women into the labour market by outsourcing their domestic tasks. Housekeepers are expected to take care of cooking, shopping, and child-minding duties – although no nursing or health care related responsibilities are permitted. Practically speaking, however, housekeepers may be asked to look after elderly parents and the line between domestic tasks and elder care will be difficult to enforce. As of 2018, only six companies were licensed to provide housekeepers and only 270 had been hired by July of that year. Their plan was to employ 3,000 by 2022 (Takao & Shimono, 2018).

In summary, Japan's immigration policy changes have resulted in the admission of various types of foreign health care professionals since 2008. The country has clearly prioritized the acceptance of foreign caregivers by introducing various programs over the last few years. Most of these only allow such workers to stay in Japan temporarily, have failed to meet expected targets, and have been criticized by scholars and visa holders alike for various reasons. The framework for admission is somewhat of a haphazard patchwork of various schemes that feature different conditions and requirements. As the country adopts more comprehensive immigration legislation, the way in which unskilled and semi-skilled foreign workers are dealt with has been changing. Japan now recognizes the need to rely on this type of labour and recently introduced something akin to a guest worker program. Some 60,000 foreign caregivers are expected to help partially alleviate the projected shortage of 340,000 in the care sector by 2025, according to official plans. Past experience and the current trajectory of these programs make us sceptical that this ambitious number will be achieved. In any case, the government's willingness to utilize foreign labour in this regard is an indicator that labour gaps in long-term care are extremely pressing, and that foreign resources will be a (but not *the*) fundamental tool when attempting to solve the understaffing problem in the LTCI program. Most likely, efforts will also be made to make care labour more attractive for local workers. But if recruiting a large number of foreign caregivers is not *the* answer, then what is?

Robotics and AI: Revolutionary in the Context of Long-Term Care?

The recent administration highlighted the crucial role that robots and artificial intelligence can play in addressing caregiving challenges. Soon after he took office in 2012, Shinzo Abe earmarked the equivalent of USD$21 million with a view to developing nursing care robots and promoting their use (Lewis, 2017). That very same year, the Ministry of Economy, Trade and Industry (METI) and the MHLW put forward a list of "priority areas of using robot technology" in this field, signalling the growing eagerness to invest in this sector (Ministry of Economy, Trade and Industry [METI], 2017). The care-related initiatives exist within the larger government's "New Robot Strategy," which includes three pillars: (1) becoming the robot innovation hub of the world; (2) becoming the world's leading robot utilization society (particularly in small and medium-sized enterprises as well as in nursing/medical care and infrastructure); and (3) leading the world with robotics in the Internet of Things Era – especially robots with IT utilizing big-data, networks, and AI (METI, 2015, p. 32). As it has been suggested, the use of robotics and artificial intelligence has the potential to allow for a more efficient way of organizing the factors of production, to optimize the operation of the country's health care and LTCI system, and to stabilize costs in the long term. Robotics and artificial intelligence may take over some of the difficult tasks currently performed by caregivers, increase overall staff efficiency, and perhaps reduce the need for human labour.

These are bold proposals, but Japan is at a privileged position when it comes to developing this infrastructure for a number of reasons. First, the country has relied on this sort of technology during the post 1945 era, and particularly since the 1970s, which saw the implementation of industrial robots at a large scale (Rathmann, 2016). While robots in the service sector remain fairly underdeveloped when compared to their industrial counterparts, Japan nevertheless ranks fourth in the global Automation Readiness Index (The Economist Intelligence Unit, 2018). Some of the main research and development clusters in the field of robotics at a global scale are located in Japan (World Intellectual Property Organization, 2015), and the country has the second most robot-intensive economy in the world in terms of industrial robots measured against manufacturing value added. In 2012, Japan accounted for about 23 per cent of the number of robot units in operation globally (METI, 2015, p. 2). Japan is also the world's main industrial robot manufacturer, producing 55 per cent of the global supply (West, 2019). The METI was hoping to see an expansion in the domestic market of nursing care robots to 50 billion yen by 2020 (METI, 2015).

The Japanese are positive about robots in general, stemming perhaps from positive representation in popular culture (Bartneck et al., 2005). Others, such as Robertson (2014, p. 576), hypothesize that it may have to do with the animist character of the Shinto substratum that underlies Japanese culture. This animism allows robots to be considered as part of a network of beings and to "be experienced as 'living' things"; Shintoism does not make ontological distinctions between organic/inorganic, animate/inanimate, and human/non-human forms. In any case, what we do observe is a general readiness and a consensus to develop and utilize robots in many areas of everyday life. Moreover, while the current labour shortage is seen as a negative factor, a number of voices have suggested that it can also be thought of as a "blessing in disguise," as it represents a major incentive to devise more efficient labour-saving strategies and technologies (Min Lan, 2017).

Within the domain of care, robots and other digital technologies are being developed both by public and private entities to support and service not just nursing homes and hospitals but also those seniors who live in their own homes and who have specific care needs. Rathmann (2016, p. 13) identifies three main areas of care robot technology currently under development. The first one is that of *care assistance*, which aims to reduce the burden for employees and home caregivers and to improve the overall quality of care. This includes lifting aids of different sorts (exoskeletons or various systems like the lifting polar bear robot RIBA[6] or the RoboBed[7]); indoor and outdoor mobility aids to assist those with reduced mobility; robotic shower systems; toileting aids; and monitoring systems consisting of devices with sensors, which may also feature external communication functions (Japan Agency for Medical Research and Development, 2020, pp. 13–14). The second area is that of *interaction*. Here, we can talk about socially assistive robots (SARs) like PARO the seal,[8] those designed for communication purposes such as NAO[9] or KOBIAN-R,[10] and the ones whose goal is to entertain seniors. The third and final is that of *therapy*, where robots such as Pepper[11] have been envisioned with the goal of keeping elders active or to assist care workers with various rehabilitation tasks.

The development of this infrastructure is a promising endeavour that represents an opportunity to enhance the quality of life of the elderly and to potentially reduce the burden for those who are employed in the care sector. We can already appreciate some of the positive impacts of these new technologies: Caregivers and elders alike have indicated their satisfaction with some of their applications (Wright, 2019). For example, many dementia patients find it easier to communicate with the robots than they do with other humans (Sung-sun, 2018). We should, however, be cautious of overly optimistic scenarios that paint a not-so-distant future of independent robots replacing human workers in care facilities and private households. The reality is that, as of today, robots in the service sector in general – including those in care – are very limited in what they can do. Robots tend to be very accurate at performing one specific task, but they will not necessarily replicate or be a substitute for the "human touch" in many situations (e.g., emotional support). Generally, they cannot perform a variety of tasks. On top of that, a number of robots are somewhat difficult to operate, and a substantial number of care managers have reported feeling uneasy with robots; they are concerned not only about their price and limitations but also about their own lack of knowledge to operate them correctly and about potentially receiving complaints from patients (Niemela, 2018). Also, assuming that Japanese elders across the board are comfortable with robots taking care of them is a mistake (Peckitt, 2018). A final point to consider is that robots are not autonomous; they need to be directed, maintained, and sometimes repaired (Wright, 2019). This means that robots will not be replacing human caregivers anytime soon.

Although promising, the robots employed in the field of care are at an early stage of development and the "robotization" process has proved relatively slow due to technological, design, and integration reasons (Niemela, 2018, p. 8). As of 2018, only 8 per cent of Japanese nursing homes had robots of some sort in their facilities (Hurst, 2018). Perhaps, as Tohoku University robotics professor Kazuhiro Kosuge has suggested, our expectations may have been too high all along, "not only because of their technical limitations or the economics of their use, but also because of the high levels of human expertise required to put them to work" (Lewis, 2019). Robots alone will certainly not be the solution to Japan's understaffing crisis in the field of caregiving, especially in the short term. In this future, human caregivers will be crucial, necessary, and possibly, as Wright (2019) argues, not replaced but progressively *dis*placed, particularly those with a foreign background. As robots are introduced in the workplace, we will also be likely observers to a change in the nature of care work (that is, care duties will be transformed, different sets of skills will probably be demanded, and so on). An interesting and emerging area of research is that of the interaction between caregivers and care robots, which we briefly examine below.

Interactions between Robots Employed in the Field of Care and Health Professionals

Here we reflect on the many other implications that progressive robotization may have for those employed in the care sector, particularly working conditions. Many of these questions are informed by some of the perspectives offered within the emerging multidisciplinary field of Human-Robot interaction (HRI), which focuses on "the study of the functionality and the usability of robots when performing tasks that involve human beings" (Olaronke et al., 2017, p. 46).

With regards to the experiences of patients, researchers seem to agree that HRI does "have the capability to improve the quality and accessibility to health care services – which in turn increases patient's health outcome" (Olaronke et al., 2017, p. 48). Going beyond technical capabilities and potential usage, it is becoming more and more evident that innovations are also raising concerns over questions of privacy, trust, deception, emotions, and safety. Research in this field has explored, among other things, questions of dignity (Arkin et al., 2014); the connection between privacy and monitoring (Caine et al., 2012); and whether or not robot carers are able to provide authentic and/or ethical care, which is perhaps one of the most crucial debates (see Tronto, 2005). For example, Meacham and Studley (2017) argue that robot carers will be able to provide authentic care, as they satisfy the first three of the four elements of care: attentiveness, competence, responsiveness (and responsibility).

For those employed in the field of caregiving, the introduction of robots both in institutional and private settings inevitably leads to changes in working arrangements. Robotic innovations can protect workers physically (e.g., exoskeletons) and assist them in key moments (e.g., companion robots). It has also been suggested that they can help caregivers provide better quality of care. Wright (2019) has shown that robots also have the potential to foster creativity in the workplace, making it more engaging for seniors and staff alike. Further, digital and AI advances will enable on-demand interaction to improve patient experiences. Many of these innovations will allow caregivers to spend more time providing care and less time documenting it, improving not only the quality of the services provided but also overall efficiency. AI can also arguably enhance the professional development and learning of caregivers (Deloitte, 2018).

Automation in health care is likely to reduce the number of health care professionals needed in the future, and the ratio of workers per patient – generally 1:3 in Japan (Yanagida, 2012) – will lessen, hopefully without compromising the overall quality of care. However, that does not mean that the need for labour will be slashed overnight. Indeed, recent studies at Japanese nursing homes show that robots do not currently reduce the need for human labour, nor do they allow caregivers to take on more meaningful care tasks by fully automating certain duties (Wright, 2019). Robots in the service sector are at an early stage of development, so there are not many tasks that can be partially automated – let alone fully automated. Another factor that we should consider is that, as operating costs are cut, automation may create new demands and therefore change the nature of task distribution across occupational hierarchies, which is quite a complex area in health care. With the introduction of these innovations, we are likely to see a change in the tasks demanded from human workers and a skills reconfiguration.

Many in the private sphere[12] as well as in government have emphasized that the goal of employing robots in the sector is to *supplement* human workers (to assist and protect them from physical burden) and *not replace* humans (Lewis, 2019). For example, METI (2015) states:

> Work environments where nursing workers can provide services with satisfaction will be created by making the best use of robotic nursing equipment while maintaining the basic concept that care is given by human hands, and a paradigm shift to enhance work efficiency and reducing the number of workers will be aided by the use of robotic nursing equipment at care sites. (p. 63)

With this in mind, what will this idea of "supplementing caregivers" mean in practice? Likely it means that "the more robots are introduced in work environments, the more that the human interaction with those systems becomes crucial" (Moniz & Krings, 2016, p. 7). Increasingly, we can expect to see workers take on tasks that include greater complexity, more multitasking, interfacing, as well as the need to anticipate robot action (Moniz & Krings, 2016, p. 8). This can possibly lead to a more stratified workplace and to the emergence of a division between more qualified workers, who can get involved in the decision-making process, and those who will be expected to only be passively involved in the technical system (Moniz & Krings, 2016, p. 11). Wright (2019), for instance, has indicated that this phenomenon may be already occurring in Japan. He has also argued that, because robot infrastructure requires more caregivers to operate them for the time being, foreign workers are being introduced in certain facilities in more precarious ways. He suggests that, moving forward, an increasing number of migrants could take on less visible roles, which brings about all sorts of ethical implications. This new arrangement would involve more alienating tasks, less communication, and less tactile contact, and therefore entail deskilling, precarization, and overall displacement. Nevertheless, in practical terms, this assemblage of care may be favoured by policymakers because it arguably overcomes linguistic and cultural barriers – rendering the caregiving process "culturally odorless" (Wright, 2019, p. 349). As Wright also acknowledges, the net impact for workers is ambiguous and difficult to quantify. Deskilling may indeed happen, but the robotization of the workplace can also lead to some forms of upskilling.

Concluding Thoughts

Japan provides an interesting example of a super-aging society that is experimenting with multiple strategies to address its care needs. It is projected that 340,000 caregivers will be required by 2025 to ensure the optimal operation of the country's severely understaffed LTCI system. The government has taken steps to rely more heavily on foreign health care professionals and to invest in robotics and artificial intelligence. As far as foreign health care professionals are concerned, various visa programs have been introduced since 2008, and currently Japan's (im)migration policy is in the midst of a paradigm shift. The government is hoping to attract 60,000 foreign caregivers by 2025, mostly on a temporary basis, although there are several reasons to believe that this plan will prove unrealistic. Regarding robotics and AI, the development of these technologies in the area of care work is arguably one of the pillars of Japan's "New Robot Strategy" (METI, 2015). Interesting innovations have been developed in this field, and there is little doubt that these technologies can potentially benefit both users and workers in various ways. That being said, we should be wary of overly optimistic scenarios where robots manage to replace human labour, especially in a not-so-distant future. Care work cannot fully be replaced; rather, these technologies can supplement the work of human caregivers and be a beneficial addition to the workplace. In any case, the robotization process in the sector has proved fairly slow thus far and it and seems unlikely that we will experience a profound revolution in this sense, at least in the short term. This may in turn frustrate the government's ambitious plans in this regard – and indeed, given the various factors discussed in this paper, it seems reasonable to believe that the LTCI system will be under tremendous pressure in the years to come.

Japan is representative of the difficulties that many industrialized countries are going through when it comes to ensuring an effective provision of care. The Japanese case poses interesting questions regarding the various emerging ethical concerns derived from the introduction of robotics and AI in the workplace. It also invites us to reflect on the changing nature of care workplaces, the different skills demanded, and the emergence of new and distinct workforce hierarchies, which may include non-human or robotic workers.

ACKNOWLEDGMENTS

We would like to thank Dr. Konrad Kalicki of the Department of Japanese Studies, National University of Singapore for his helpful comments on an earlier version of this chapter.

NOTES

1 There are country-specific exceptions. For instance, Bulgaria is currently shrinking faster than Japan is (World Bank, 2018). This, however, is mostly due to emigration.
2 There are also private insurance options, but they tend to be complementary to the LTCI.

3 Wages for caregivers are determined within the framework of the LTCI remuneration system (Kato, 2015).
4 There were 2.23 million foreign residents in Japan out of the total 127.1 million in 2015 ("Japan foreign population climbs," 2019; Statistics Bureau of Japan, 2016). That amounts to 1.75 per cent. For some perspective, in 2015 the foreign resident stock in the US was 14.3 per cent, in the UK it was 13.2 per cent, and in France it was 11.1 per cent (United Nations, Department of Social and Economic Affairs, 2016). Japan's foreign resident population is comparatively modest, but it is growing every year.
5 Initially signed with the Philippines (in 2006) and Indonesia (in 2007), and later Vietnam (Ohno, 2012). Free trade agreements have also made it possible for Thai nurses to work in Japan under similar conditions (Save & Shingal, 2014).
6 RIKEN-TRI Collaboration Center for Human-Interactive Robot Research, "RIBA," http://rtc.nagoya.riken.jp/RIBA/index-e.html.
7 Panasonic, "Resyone Plus Robotic Care Bed," https://www.panasonic.com/global/corporate/technology-design/ud/welfare.htl.
8 PARO Robots, "PARO Therapeutic Robot," http://www.parorobots.com/.
9 SoftBank Robotics, "NAO," https://www.softbankrobotics.com/emea/en/nao.
10 Takanishi Laboratory, "KOBIAN-R," http://www.takanishi.mech.waseda.ac.jp/top/research/kobian/KOBIAN-R/index.htm.
11 SoftBank Robotics, "Pepper," https://www.softbankrobotics.com/emea/en/pepper.
12 See, for instance, Lewis (2019).

REFERENCES

Arai, H., Ouchi, Y., Toba, K., Endo, T., Shimokado, K., Tsubota, K., Matsuo, S., Mori, H., Yumura, W., Yokode, M., Rakugi, H., & Ohshima, S. (2015). Japan as the front-runner of super-aged societies: Perspectives from medicine and medical care in Japan. *Geriatrics & Gerontology International, 15*(6), 673–87. https://doi.org/10.1111/ggi.12450.

Arkin, R.C., Scheutz M., & Tickle-Degnen, L. (2014). Preserving dignity in patient caregiver relationships using moral emotions and robots. In *2014 IEEE international symposium on ethics in science, technology and engineering* (pp. 1–5). https://doi.org/10.1109/ETHICS.2014.6893414.

Bartneck, C., Nomura, T., Kanda, T., Suzuki T., & Kato, K. (2005). Cultural differences in attitudes towards robots. In *Proceedings of the AISB symposium on robot companions: Hard problems and open challenges in human-robot interaction* (pp. 1–4). https://doi.org/10.13140/RG.2.2.22507.34085.

Befu, H. (2009). Concepts of Japan, Japanese culture and the Japanese. In Y. Sugimoto (Ed.), *The Cambridge companion to modern Japanese culture* (pp. 21–37). Cambridge University Press.

Burgess, C. (2014, June 18). Japan's "no immigration principle" as solid as ever. *Japan Times*. http://www.japantimes.co.jp/community/2014/06/18/voices/japans-immigration-principle-looking-solid-ever/.

Caine, K., Šabanovic, S., & Carter, M. (2012). The effect of monitoring by cameras and robots on the privacy enhancing behaviors of older adults. In *Proceedings of the seventh annual ACM/IEEE international conference on human-robot interaction* (pp. 343–50). https://doi.org/10.1145/2157689.2157807

Coulmas, F. (2007). *Population decline and ageing in Japan – the social consequences.* Routledge.

Deloitte. (2018, January 9). *Deloitte global health care outlook: Working towards smart health care* [Press release]. https://www2.deloitte.com/global/en/pages/about-deloitte/articles/global-health-care-sector-outlook-working-towards-smart-health-care.html.

Ebuchi, T., & Takeuchi, Y. (2018, October 29). 54% of Japanese in favour of accepting more foreign workers. *Nikkei Asia.* https://asia.nikkei.com/Spotlight/Japan-immigration/54-of-Japanese-in-favor-of-accepting-more-foreign-workers2.

The Economist Intelligence Unit. (2018). The automation readiness index. *The Economist.* https://www.automationreadiness.eiu.com.

Elderly abuse a growing problem in ageing Japan. (2019, April 11). *Nippon.com.* https://www.nippon.com/en/japan-data/h00428/elderly-abuse-a-growing-problem-in-aging-japan.html.

Elderly citizens accounted for record 28.4% of Japan's population in 2018, data show. (2019, September 15). *Japan Times.* https://www.japantimes.co.jp/news/2019/09/15/national/elderly-citizens-accounted-record-28-4-japans-population-2018-data-show/#.Xo_PIOzRY1J.

Fewer foreigners than expected coming to Japan to work as caregiver trainees, data shows. (2018, December 1). *Japan Times.* https://www.japantimes.co.jp/news/2018/12/01/national/fewer-foreigners-expected-coming-japan-work-caregiver-trainees-data-shows/.

Fill the gap in nursing care workers. (2018, June 26). *Japan Times.* https://www.japantimes.co.jp/opinion/2018/06/26/editorials/fill-gap-nursing-care-workers/#.XpUUd-zRY1L.

Foreigners make 30% of new students at nursing care schools in Japan. (2019, September 4). *Yomiuri Shimbun.* https://the-japan-news.com/news/article/0005983433.

Foreigners who flunked nurse's exam may get OK as care workers. (2019, September 5). *Asahi Shimbun.* https://web.archive.org/web/20200702144503/http://www.asahi.com/ajw/articles/AJ201909050004.html.

Ganelli, G., & Miake, N. (2015). Foreign help wanted: Easing Japan's labour shortages. *IMF Working Paper.* https://www.imf.org/external/pubs/ft/wp/2015/wp15181.pdf.

Goldar Perrote, H. (2017). *Newcomers, welcome? Exploring the connection between demographic change, immigration legislation design and policy mobilities in ageing Japan* [Master's thesis, Wilfrid Laurier University]. Scholar's Commons @ Laurier. https://scholars.wlu.ca/cgi/viewcontent.cgi?article=3096&context=etd.

Hong, G.H., & Schneider, T. (2020). SHRINKONOMICS: Lessons from Japan. *Finance & Development, 57*(1), 20–3. https://www.imf.org/external/pubs/ft/fandd/2020/03/pdf/fd0320.pdf.

Hurst, D. (2018, February 6). Japan lays groundwork for boom in robot carers. *The Guardian*. https://www.theguardian.com/world/2018/feb/06/japan-robots-will-care-for-80-of-elderly-by-2020.

Iwagami, M., & Tamiya, N. (2019). The long-term care insurance system in Japan: Past, present, and future. *JMA Journal*, *2*(1), 67–9. https://doi.org/10.31662/jmaj.2018-0015.

Japan Agency for Medical Research and Development. (2020). *Project to promote the development and introduction of robotic devices for nursing care*. http://robotcare.jp/data/etc/ROBOT-CARE-pamphlet_eng.pdf.

Japan foreign workers top record 1 million. (2017, January 18). *Rappler*. https://www.rappler.com/world/asia-pacific/japan-foreign-workers-record-1-million.

Japan International Trainee & Skilled Worker Cooperation Organization. (2015). *Technical intern trainee program: Care worker*. https://www.jitco.or.jp/en/regulation/care.html.

Japan's foreign nurses: Disinformation, lack of support shows struggles for arrivals. (2019, December 12). *The Mainichi*. https://mainichi.jp/english/articles/20191211/p2a/00m/0na/023000c.

Japan's foreign population climbs to all-time high. (2019, March 29). *Nippon.com*. https://www.nippon.com/en/features/h00137/japan%E2%80%99s-foreign-population-climbs-to-all-time-high.html.

Japanese Nursing Association. (n.d.). Overview of Japanese nursing system. https://www.nurse.or.jp/jna/english/nursing/system.html.

Kato, H. (2015). Is the long-term care insurance system sustainable? – There are two issues to solve. *Discuss Japan: Foreign Policy Forum*, no. 30. https://www.japanpolicyforum.jp/pdf/2016/no30/DJweb_30_eco_03.pdf.

Kavedzija, I. (2018, June 1). What can the world learn about Japan's social care system? *World Economic Forum*. https://www.weforum.org/agenda/2018/06/social-care-japanese-style-what-we-can-learn-from-the-world-s-oldest-population.

Kobayashi, T., Collet, C., Iyengar S., & Hahn, K.S. (2015). Who deserves citizenship? An experimental study of Japanese attitudes toward immigrant workers. *Social Science Japan Journal*, *18*(1), 3–22. https://doi.org/10.1093/ssjj/jyu035.

Kodama, T. (2015). *Japan's immigration problem: Looking at immigration through the experiences of other countries*. Daiwa Institute of Research. https://www.dir.co.jp/english/research/report/others/20150529_009776.pdf.

Kyozuka, T. (2020, March 20). 40% of Japan's foreign caregivers trained by specialized agency. *Nikkei Asia*. https://asia.nikkei.com/business/companies/40-of-japan-s-foreign-caregivers-trained-by-specialized-agency.

Lewis, L. (2017, October 17). Can robots make up for Japan's care home shortfall? *Financial Times*. https://www.ft.com/content/418ffd08-9e10-11e7-8b50-0b9f565a23e1.

Lewis, L. (2019, June 27). G20 reality is that robots alone cannot solve the problems of ageing. *Financial Times*. https://www.ft.com/content/ae0bd03a-4a51-11e9-bde6-79eaea5acb64.

Liu, J.X., Goryakin, Y., Maeda, A., Bruckner, T., & Scheffler, R. (2017). Global health workforce labor market projections for 2030. *Human Resources for Health*, *15*(11). https://doi.org/10.1186/s12960-017-0187-2.

Lopez, M. (2012). Reconstituting the affective labour of Filipinos as care workers in Japan. *Global Networks*, *12*(2), 252–68. https://doi.org/10.1111/j.1471-0374.2012.00350.x.

Martin, A. (2015, November 15). Lack of workers hobbles Japan's growth. *Wall Street Journal*. https://www.wsj.com/articles/lack-of-workers-hobbles-japans-growth-1447635365.

Matanle, P., & Rausch, A.S. (2011). *Japan's shrinking regions in the 21st century: Contemporary responses to depopulation and socioeconomic decline*. Cambria.

Meacham, D., & Studley, M. (2017). Could a robot care? It's all in the movement. In P. Lin, K. Abney, & R. Jenkins (Eds.), *Robot ethics 2.0: From autonomous cars to artificial intelligence* (pp. 97–113). Oxford University Press.

Menju T. (2019a, February 6). Japan's historic immigration reform. *Nippon.com*. https://www.nippon.com/en/in-depth/a06004/japan%E2%80%99s-historic-immigration-reform-a-work-in-progress.html.

Menju, T. (2019b, November 18). Japan's immigration policy put to the test. *Nippon.com*. https://www.nippon.com/en/in-depth/d00515/japan%E2%80%99s-immigration-policies-put-to-the-test.html.

Milly, J. (2020, February 20). *Japan's labor migration reforms: Breaking with the past?* Migration Policy Institute. https://www.migrationpolicy.org/article/japan-labor-migration-reforms-breaking-past.

Ministry of Economy, Trade and Industry of Japan. (2015). *Japan's robot strategy: Vision, strategy, action plan*. The Headquarters for Japan's Economic Revitalization. http://www.meti.go.jp/english/press/2015/pdf/0123_01b.pdf.

Ministry of Economy, Trade and Industry of Japan. (2017, October 12). *Revision of the priority areas to which robot technology is to be introduced in nursing care* [Press release]. https://www.meti.go.jp/english/press/2017/1012_002.html.

Ministry of Finance of Japan. (2017). Japanese public finance factsheet. https://dl.ndl.go.jp/view/download/digidepo_11096623_po_04.pdf?contentNo=1&alternativeNo=.

Ministry of Foreign Affairs of Japan. (n.d.). *Specified skilled workers*. https://www.mofa.go.jp/mofaj/ca/fna/ssw/us/index.html.

Ministry of Health, Labour and Welfare of Japan. (2011). *Acceptance of foreign nurses / care worker candidates from Indonesia, Philippines and Vietnam*. https://www.mhlw.go.jp/stf/seisakunitsuite/bunya/koyou_roudou/koyou/gaikokujin/other22/index.html.

Ministry of Health, Labour and Welfare of Japan. (2016). *Long-term care insurance system of Japan*. https://www.mhlw.go.jp/english/policy/care-welfare/care-welfare-elderly/dl/ltcisj_e.pdf.

Ministry of Justice of Japan. (2015a). *Basic plan for immigration control* (5th ed.). http://www.moj.go.jp/isa/content/930003137.pdf.

Ministry of Justice, Japan. (2015b). *Immigration control* [Report]. Part two: Major policies related to immigration control. http://www.moj.go.jp/content/001166925.pdf [site no longer exists].

Ministry of Justice of Japan. (2015c). *Immigration control* [Report]. Part one: Immigration control in recent years. http://www.moj.go.jp/content/001166925.pdf [site no longer exists].

Min Lan, T. (2017, April 26). Japan's labor shortage could be blessing in disguise, not just for women. *Nikkei Asia*. http://asia.nikkei.com/Viewpoints/Min-Lan-Tan/Japan-s-labor-shortage-could-be-blessing-in-disguise-not-just-for-women.

Mitchell, O.S., Piggott, J., & Shimizutani, S. (2004). *Aged-care support in Japan: Perspectives and challenges*. National Bureau of Economic Research No. W10882. https://doi.org/10.3386/w10882.

Moniz, A.B., & Krings, B.J. (2016). Robots working with humans or humans working with robots? Searching for social dimensions in new human-robot interaction in industry. *Societies*, *6*(3), 1–21. https://doi.org/10.3390/soc6030023.

National Institute of Population and Social Security Research (IPSS). (2012). *Population projections for Japan (January 2012): 2011 to 2060*. http://www.ipss.go.jp/site-ad/index_english/esuikei/ppfj2012.pdf.

Niemela, M. (2018). Notes from a research visit to AITS, Tokyo. VTT. http://roseproject.aalto.fi/images/publications/Robotic-care-in-Japan-6-2018-M.Niemel-ROSE.pdf.

No. of foreign workers in Japan totals 1.66 million. (2020, January 31). *Kyodo News*. https://english.kyodonews.net/news/2020/01/c485e8e3a4ac-no-of-foreign-workers-in-japan-totals-record-166-million.html.

Ogawa, R. (2014). Globalization and transformation of care in Japan. In M.M. Merviö (Ed.), *Contemporary Social Issues in East Asian Societies: Examining the Spectrum of Public and Private Spheres* (pp. 86–105). IGI Global.

Ogawa, R., Chan, R.K., Oishi, A.S., & Wang, L.R. (Eds.). (2018). *Gender, care and migration in East Asia*. Palgrave Macmillan.

Ohno, S. (2012). Southeast Asian nurses and caregiving workers transcending the national boundaries. *Japanese Journal of Southeast Asian Studies*, *49*(4), 541–69. https://www.jstage.jst.go.jp/article/tak/49/4/49_KJ00007716039/_pdf.

Oishi, N. (2012). The limits of immigration policies: The challenges of highly skilled migration in Japan. *American Behavioural Scientist*, *56*(8), 1080–100. https://doi.org/10.1177/0002764212441787.

Olaronke, I., Oluwaseun, O., & Rhoda, I., (2017). State of the art: A study of human-robot interaction in healthcare. *International Journal of Information Engineering and Electronic Business*, *9*(3), 43–55. https://doi.org/10.5815/ijieeb.2017.03.06.

Organisation for Economic Co-operation and Development. (2010). *International migration of health workers*. OECD Policy Brief. https://www.who.int/hrh/resources/oecd-who_policy_brief_en.pdf?ua=1.

Organisation for Economic Co-operation and Development. (2018). *Working better with age: Japan*. OECD Publishing. https://www.oecd.org/els/emp/Working-better-with-age-Japan-EN.pdf.

Otake, T. (2015, April 19). For foreign caregivers, role remains ambiguous. *Japan Times*. https://www.japantimes.co.jp/news/2015/04/19/national/for-foreign-caregivers-role-remains-ambiguous/.

Peckitt, M. (2018, October 14). Do the elderly and disabled people in Japan want robots to look after them? *Japan Times*. https://www.japantimes.co.jp/community/2018/10/14/voices/elderly-disabled-people-japan-want-robots-look/#.XpbLDOzRY1J.

Peng, I. (2016). Testing the limits of welfare state changes: The slow-moving immigration policy reform in Japan. *Social Policy & Administration*, *50*(2), 278–95. https://doi.org/10.1111/spol.12215.

Prime Minister's Office of Japan. (2014). *Japan revitalization strategy*. https://www.kantei.go.jp/jp/singi/keizaisaisei/pdf/honbunEN.pdf.

Rathmann, M. (2016). Care robots for an over ageing society: A technical solution to Japan's demographic problem? http://www.kuasu.cpier.kyoto-u.ac.jp/wp-content/uploads/2015/10/Care-Robots-for-an-Over-Aging-Society.pdf.

Reynolds, I., & Aquino, N. (2017, January 10). Learning to bow: Japan reluctantly opens door to foreign housemaids. *Japan Times*. https://www.japantimes.co.jp/news/2017/01/10/national/learning-bow-japan-reluctantly-opens-door-foreign-housemaids/.

Robbins, P., & Smith, S.H. (2016). Baby bust: Towards political demography. *Progress in Human Geography*, *41*(2), 199–219. https://doi.org/10.1177/0309132516633321.

Robertson, J. (2014). Human rights vs. robot rights: Forecasts from Japan. *Critical Asian Studies*, *46*(4), 571–98. https://doi.org/10.1080/14672715.2014.960707.

Save, P., & Shingal, A. (2014). *The preferential liberalization of trade in services: Comparative regionalism*. Edward Elgar Publishing.

Shindo, R. (2019). *Belonging in translation: Solidarity and migrant activism in Japan*. Bristol University Press.

Siripala, T. (2018, June 9). Japan's robot revolution in elderly care. *Japan Times*. https://www.japantimes.co.jp/opinion/2018/06/09/commentary/japan-commentary/japans-robot-revolution-senior-care/#.XpUTSOzRY1J.

Statistics Bureau of Japan. (2016, December 27). *Basic complete tabulation on population and households of the 2015 population census of Japan was released* [Press release]. http://www.stat.go.jp/english/info/news/20161227.htm.

Stokes, B., & Devlin, K. (2018). *Perception of immigrants, immigration and emigration*. Pew Research Center. https://www.pewresearch.org/global/2018/11/12/perceptions-of-immigrants-immigration-and-emigration/.

Sugimoto, Y. (2009). Japanese culture: An overview. In Y. Sugimoto (Ed.), *The Cambridge companion to modern Japanese culture* (pp. 1–20). Cambridge University Press.

Sung-sun, K. (2018, December 18). Japanese dementia patients communicate better with robots. *Korea Biomedical Review*. http://m.koreabiomed.com/news/articleView.html?idxno=4775.

Takao, Y., & Shimono, Y. (2018, July 21). A day in the life of a foreign housekeeper in Japan. *Nikkei Asia*. https://asia.nikkei.com/Life-Arts/Life/A-day-in-the-life-of-a-foreign- housekeeper-in-Japan.

Tarumoto, H. (2012). Towards a new migration management: Care immigration policy in Japan. In M. Geiger & A. Pécoud (Eds.), *The new politics of international mobility: Migration management and its discontents* (pp. 157–72). Institut für Migrationsforschungund Interkulturelle Studien.

Tronto, J.C. (2005). Care as the work of citizens: A modest proposal. In M. Friedman (Ed.), *Women and citizenship* (pp. 130–45). Oxford University Press.

United Nations, Department of Social and Economic Affairs. (2016). *International migration report 2015: Highlights*. https://www.un.org/en/development/desa/population/migration/publications/migrationreport/docs/MigrationReport2015_Highlights.pdf.

West, J. (2019, April 16). Robotics improves the fate of demographic decline in Japan. *Brink News*. https://www.brinknews.com/has-japan-found-its-robotics-sweet-spot/.

World Bank. (2017). Fertility rate, total (births per woman) – Japan. https://data.worldbank.org/indicator/SP.DYN.TFRT.IN?locations=JP.

World Bank. (2018). Population growth (annual %). https://data.worldbank.org/indicator/sp.pop.grow.

World Intellectual Property Organization. (2015). *World intellectual property report: Breakthrough innovation and economic growth*. https://www.wipo.int/edocs/pubdocs/en/wipo_pub_944_2015.pdf.

Wright, J. (2019). Robots vs migrants? Reconfiguring the future of Japanese institutional eldercare. *Critical Asian Studies*, *51*(3), 331–54. https://doi.org/10.1080/14672715.2019.1612765.

Yamawaki, K. (2019, June 26). Is Japan becoming a country of immigration? *Japan Times*. https://www.japantimes.co.jp/opinion/2019/06/26/commentary/japan-commentary/japan-becoming-country-immigration/#.Xo6yaOzRY1J.

Yanagida, K. (2012). *Coping with an ageing population in Japan: Towards a policy framework for attracting high quality nurses and caregivers from South-East Asia under the Economic Partnership Agreements* [Unpublished manuscript].

Yashuhiro, Y. (2020, February 5). The challenges facing Japan's long-term care services. *Nippon.com*. https://www.nippon.com/en/in-depth/d00530/the-challenges-facing-japan%E2%80%99s-long-term-care-services.html?cx_recs_click=true.

Yoshida, R. (2014, May 18). Success of "Abenomics" hinges on immigration policy. *Japan Times*. https://www.japantimes.co.jp/news/2014/05/18/national/success-abenomics-hinges-immigration-policy/.

SECTION 5

Recasting Brain Drain and Global Circulation

14 Nursing the Nation: The Intellectual Labour of Early Migrant Nurses in the US, 1935–1965

CHRISTINE PERALTA

The Rise of Nursing Leadership in the Philippines

In September 1943, a Filipina graduate student at the University of Chicago submitted her master's thesis on nursing administration, which outlined her plans to create the first university-level nursing program in the Philippines. Julita Sotejo's academic milestone is significant for the history of Filipino migration and labour because it created the nurse education apparatus that led to the massive outmigration of nurses from the Philippines to multiple sites across the globe.[1] Sotejo was part of a group of understudied Filipina women scholars who migrated to the US from the Philippines in the early half of the twentieth century. Their migration patterns foreshadowed the transnational linkages that prefigured later waves of Filipina women working and living internationally after the Second World War.

Despite being an early migrant nurse, Sotejo's intention was not to help facilitate the production of nurses whose credentials could make them more desirable as employees in the Global North. Instead, her aims were to improve the training of nurses so that the profession could better address the health needs of the Philippines. Anticipating the brutal devastation of the war while she was abroad, Sotejo (1943) stated that the "post-war picture will be very dark," and therefore she thought that nurse education will be significant because "there will be a pressing need for large-scale health and social planning … It is the health needs, of which nursing needs are a part, of the people of the Philippines which led to this study" (p. 1).

Before Sotejo studied abroad, she served ten years as the head principal at the largest nursing school in the Philippines, the Philippine General Hospital (PGH). The hospital and nursing school were first established by the American colonial medical regime.[2] At the PGH, the nurse training model was based on an apprenticeship, where they studied for a year and then completed three years of practice at the hospital. The student nurses completed a few required courses, but for the most part, they learned the profession through hands-on training. For example, a head nurse would manage several student nurses who were assigned low-level tasks, such as receiving patients, dressing wounds, and emptying bed pans. Sotejo's major criticism for this training model was that it resulted in the exploitation of student nurses who were expected to work long hours at the expense of their training.

She was particularly concerned that students who underwent this form of training did not have enough time for planning and carrying out nursing procedures because their labour supplemented the hiring of permanent nursing staff. The exploitation of nurses in the Philippines was a well-known concern for a number of nursing faculty members, including Sotejo, who stated that the origin of nursing, originally started by American colonial officials, "came into existence for reasons other than to provide sound education for future nurses of the Philippines, one among which is to supply nursing service at low cost to the hospital through student labor" (Sotejo, 1949, p. 23). She argued that by relying on student nurses, there was a risk of "overloading and overworking of students which are detrimental to both their education and their health" (Sotejo, 1943, p. 40). Therefore, she was convinced that nursing students had to develop as students before they could practise as nurses. She believed that students could do this in a similar fashion to how medical students became doctors, by first completing a four-year bachelor's degree before going to medical school.

Sotejo placed value in a well-rounded liberal education and wanted to create a pathway for nurses to participate in college life and receive an education in the sciences and humanities. However, she wanted Filipina nurses to be well-equipped in helping Filipinos first. Therefore, in addition to redesigning the curriculum to enable a deeper understanding of scientific principles, she also wanted the nurses to learn about the social and political conditions of the country so that they would be able to "comprehend the nature of social problems" that faced the communities that they would be serving (Sotejo, 1943, p. 43). During Sotejo's tenure, she had witnessed economic decline, the recategorization of Filipinos' legal status and relationship with the US, and, most significantly, the Japanese occupation of the Philippines during the Second World War. Therefore, Sotejo envisioned that a university education would create a feminized professional workforce that could help rebuild the Philippines after the war. As a visionary of the profession that the Philippines is known for, Sotejo's work in the Philippines and abroad was significant, but she is a figure whose intellectual labour has largely been neglected in the literature.[3]

One of the reasons why Sotejo is absent is the common idea in the scholarship that nursing came to the Philippines due to colonialism. This idea is held by scholars both critical and ambivalent towards the idea of the US empire in the Philippines, such as Vecoli (1999) who stated: "The Filipino nurses … who migrated to the United States surely enjoyed improved life opportunities because of the colonial history of the Philippines. Is it possible that some good have come from imperialism?" (p. 120). What is interesting is that whatever the historians' positions are on the US empire, they overlap when it comes to the idea of nursing. Even Choy (2003), whose work is one of the most critical of the US imperial legacy in the development of nursing, states: "Rendered invisible are the ways that U.S. colonialism in the Philippines created an Americanized training hospital system that eventually prepared Filipino women to work as nurses in the United States as opposed to the Philippines" (p. 5). In Choy's framing of the history of nursing in the Philippines, the transition from colonized subject to exploited worker is an inevitability. Therefore, despite having different positions on the US empire, all positions gloss over the intellectual labour that Filipina women did in order to modernize nursing in the Philippines. However, when we examine the experiences of early migrant nurses in the US, we can see how American faculty obstructed the efforts of those who were working to professionalize nursing in the Philippines if it threatened the racial order that they were invested in maintaining in their own programs. Therefore, Sotejo's life work in Philippine nursing education represents a rupture in the US colonial legacy of nursing. This distinction is the focus of this chapter.

Another reason for this absence is that the predominance of historical work on Filipino migration focuses on two periods: the formal colonization of the Philippines by the US, and the large migration of Filipinos under the Hart Cellar Act of 1965.[4] By 1935, the Philippines was still a US possession that was under the authority of Congress and the president, although one that was in a transition towards independence. Therefore, Sotejo is in periodization limbo, developing her program after the Philippines became a commonwealth and studying abroad in the US two decades prior to the mass movement of Filipino nurses to the US.

The problem with neglecting the period from 1935 to 1965 is that with the withdrawal of US medical professionals in key leadership positions in the country, several Filipino health workers ascended into positions of power that were previously unattainable to natives. Therefore, this period can be characterized as a time when Filipino health professionals had unprecedented influence on the foundational development of the health infrastructure of the country. Additionally, Sotejo was not an anomaly, but instead part of a larger group of health physicians, dieticians, pharmacists, social workers, and other nurses who had come of age during the US empire and were now participating in restructuring, redefining, and rebuilding the health system in the country after the war. For example, if we focus specifically on the field of nursing administration, from 1948 to 1964, four other early migrant nurses who completed post-graduate research in North America established or led their own college-level nursing programs: Cessara Tan developed the College of Nursing at Southwestern College (1951), Socorro Salamanca became dean of nursing at Manila Central

University (1951), Purita Asperilla founded the University East's nursing program (1959; see Choy, 2003, p. 56), and Rosario Sison-Diamante established the nursing program at the Philippine Women's University in 1964. In addition, these programs created new positions for women in the form of nursing faculty and administration.

In considering Sotejo's life, this chapter argues that the Filipino nurse's intellectual labour was a critical step in the professionalization of nursing in the Philippines. In reviewing first her experience abroad and then her efforts to redesign nursing education, I demonstrate the work that Filipino nurses did in designing their own health care beyond the US empire. This will challenge the commonly held notion that nursing was a gift of the empire by showing how the colonial structure in which nursing functioned hindered nurses like Sotejo from reaching their objective of modernizing nursing in the Philippines. Additionally, I will demonstrate that the establishment of modern nursing in the Philippines was a critique of the empire, not a gift from it.

Highlighting Filipina women's intellectual labour and leadership in the profession of nursing leads me to one of the major research questions of this chapter: We know that there was a mass nursing emigration following the war to the US, but who were the medical professionals on the ground who were training nurses in the Philippines? Ostensibly, Filipino nursing faculty were training students at a sufficiently advanced level to give their education parity with nations that have been historically understood as more medically "advanced." In that case, who were the nursing professors that engineered a program extensive enough to accommodate such a large migration of specialized health care workers? How do the experiences of early migrant Filipino nurses allow us to understand emergent processes of stratification that emerged with regard to how racialized women were inserted into workplaces in North America? Most importantly, what motivated women like Julita Sotejo to educate nurses, who provided the medical talent to educate Filipino nursing cohorts from the 1950s to the 1970s?

To begin to answer these questions, my method is to construct the biography of Sotejo through a combination of different archival perspectives that show Sotejo in different facets: Rockefeller Foundation (RF) records that cover her time studying abroad in the US and Canada, her personal papers at the University of the Philippines, and a collection of her own published works housed at the National Library of Medicine. Sotejo's biography fills an important gap in the history of nursing in the Philippines, because her life and career in nursing expanded from the American colonial period to after the Second World War. By centering the biography of a nurse that challenges historical narratives, we are able to reinterpret this period in a new light. As Choy (2010) states, using biographies of nurses as a method of research can be used to "to foreground nurses who have been ignored, misunderstood, or forgotten" (p. 20).

It is also important to point out that the mark that Sotejo left in multiple archives is rather exceptional. Through my research of nurses from this time period, it is difficult to trace the life history of one nurse for the entirety of her career. Even rarer is to have extensive work available that captures Sotejo's ideas in her own words on medicine, the Philippines, and her profession.

The disappearance of Sotejo in the history of modern medicine in the Philippines becomes a layered form of erasure, a palimpsest, since she represents so many forms of knowledge that should be legible, particularly to a Western audience. For instance, she is the only Filipino nurse that has her own archival papers anywhere. It is important for her disappearance from the historical record to be approached as a multi-dimensional erasure embedded in her race, citizenship, and gender. For example, in attempts to critique empire, the role that native practitioners played has been de-emphasized in histories of modern medicine in the Philippines.

Additionally, in order to delegitimize Filipina women's attempts to professionalize nursing, male patients and health workers started a verbal and written campaign lamenting the loss of "nurses who nursed," women who performed feminized hospital labour. What this ultimately meant was that they missed nurses performing low-skilled care work in the hospital wards, without acknowledging the highly specialized labour that these nurses were now able to perform as a result of their education designed by nursing intellectuals like Sotejo (Sotejo & Beltran-Jackson, 1965).

Prestigious Scholar and Menial Labourer: The Fungible Status of Early Nurse Migrants

During the American colonial period, the Rockefeller Foundation (RF) forged a partnership with the Philippine government, creating special fellowships for gifted Filipinos to travel and study abroad in the US, Canada, and Europe. Health work abroad was of interest to the RF and, by 1921, in partnership with the Philippine government, the Rockefeller Foundation's Nurse Fellowship Program designated two fellowships to be awarded specifically to Filipina women who would be sent abroad to learn specialized techniques that had yet to be developed in the Philippines. The goal of these fellowships was not only to produce brilliant nurses but also to create nursing leaders. The ideal candidates were individuals who would use the education that they learned abroad as a foundation to expand structural opportunities for other Filipina women. From their inception, the scholarships had two purposes: to be a prestigious scholarship for students, and to be an opportunity for greater professionalization. For example, a nurse who specialized in public health work would be assigned to work at a settlement house, or a nurse specializing in nutrition would be assigned to work in collaboration with a dietician. Essentially, the professional component of the nursing fellowship was a way for nurses to be exposed to a more established and better-funded programs in order to gain a better familiarity with health organizations, create life-long partnerships, and reproduce similar systems in the Philippines.

I argue that the fungibility of the status of nurses abroad prefigured a number of pitfalls of international nursing recruitment, particularly the Exchange Visitor Program (EVP), a US program that recruited professionals, including nurses, to work and gain new skills under the Education Act of 1948. As Reddy (2015) argues, the EVP was modelled after the colonial Rockefeller fellowship program. It is therefore necessary to consider how these programs used the prestige and language of advancement to attract international professionals to perform work abroad as vulnerable temporary or guest workers in the hopes of gaining new skills not available in their home country. Simultaneously, the program was attractive to employers, because they could fill labour shortages with a population base that was highly vulnerable due to their short contracts, lack of citizenship, and the muddled characterization of their program as being for both work and study. Nurses who wanted to advance their careers professionally using these programs often represented a class of professionals who had the desire to do well, and this demanded they follow authority figures. This reflects the hierarchization of nursing as well as nursing's subservience to medicine (Gamarnikow, 1978). Additionally, this created the perfect conditions for the exploitation of a highly skilled class of people. Hence, when I use the term "fungible,"[5] which typically refers to commodities that are exchangeable, when referring to Filipino nurse migrants, I am highlighting how a condition of their labour in North American hospitals required them to shift between the different categories of menial labourer, pupil, and expert. In a similar vein, the hospital only had to think about Filipino nurses in terms of their labour, providing them the bare minimum to carve out a temporary life because there was no expectation that the nurses would stay.

Choy (2003) examined the EVP by looking at several nurses' experiences on the ground; she concluded that nurses' experiences abroad varied and had a lot to do with the employees who hired them. In contrast, I argue that EVP and the Rockefeller nursing fellowship program that inspired it were inherently exploitative because a wide variety of migrant nurses' experiences were left to the whims of these employees, and issues raised between these parties were easier to ignore because of the "fungible" nature of the nurses (Choy, 2003). This condition of variability was also an aspect of other forms of temporary labour policy developed at this time, including Mexican farm migrants who worked under bracero contracts, and nannies who obtained visas to work within family households (Chang, 2000; Loza, 2000). Although these forms of labour may seem distinct from women with medical expertise seeking education or professional development, the lack of universal protections in all three labour programs highlight the precarity that temporary foreign workers of all skill backgrounds experienced.

Using the biography of Sotejo, this chapter looks at the conditions of Filipina women who worked and studied abroad and were asked to perform duties that were below their level of training or outside of the objectives of their fellowships' plan of study. To understand why Sotejo essentially experienced an early form of professional deskilling, which is a common experience for internationally trained health workers currently, it is important to examine the racial and class shifts that were occurring in the field of nursing in the US during her time abroad. During this time period, in order to address decade-long nursing shortages, hospitals began hiring both licensed practical nurses and nursing aides. These were shorter programs designed for nursing work that was supposed to be performed under the supervision of a registered nurse. Due to the shorter training period, there was less prestige associated with these forms of nursing, and therefore these programs oftentimes recruited women of colour from working-class backgrounds. Sotejo herself was part of this nursing shortage solution, and therefore had unwittingly entered into a labour force that had forged a social hierarchy that reinforced racial bias. This racial bias was also structurally reinforced. Since the program was both for intellectual and professional advancement, it loosely defined what a nurse should be required to do abroad. Therefore, the categorical ambiguity of both work and study created a state of fungibility, where Filipino nursing scholars could be treated as prestigious international scholars and as menial (student) labourers simultaneously.

Unequal to the Responsibility

Starting in the 1921, the Rockefeller Foundation's Nurse Fellowship Program was designed to mentor leaders in the field of nursing, who were in turn expected to help build educational programming in their home country. The type of program that a prospective nurse proposed to study had to be a nursing field that no one else in the country had expertise in or for which there was dire need. For example, one of the first RF nursing fellows was Socorro Salamanca, who used the fellowship to study public health nursing (Salamanca, 1922, Fellowship Recorder Cards). Public health nursing was a field of study that the US colonial government deemed there was a need for. Additionally, nurses who underwent this fellowship had already received training as professional nurses and had been active in the field for many years. Therefore, they were considered senior nurses in the Philippines and it was a great honour to receive this award.

In the summer of 1941, Julita Sotejo and the other RF nurse fellow that year, Leonor Malay, arrived at Yale University to study nursing. Immediately, the relations between these nurses and the faculty became fractured because of the visiting nurses' unwillingness to do ward duty. Ward duty involved low-skill menial labour such as changing diapers, sanitizing test tubes, emptying bed pans, and making hospital beds. In the PGH where Sotejo was the principal and Malay was an instructor, nursing students performed ward duty. In response to the fellows' recalcitrance, Yale Professor Effie Taylor wrote a letter to the RF criticizing Sotejo's and Malay's work (these complaints were recorded in the RF's Fellowship Recorder Cards that were used to track a nurse's performance and behaviour).[6] She stated that the two women never did ward duty because they did not feel "equal to the responsibility" (Sotejo, 1941, Fellowship Recorder Card, p. 2). Therefore, the faculty presented a narrative to the RF that the nurses refused to do menial labour because they lacked the knowledge to do such work.

In response to the nurses' refusals, the Yale faculty members required Sotejo and Malay to do six hours of ward duty per day as well as refresher courses in nursing (Sotejo, 1941, Fellowship Recorder Card; Malay, 1941, Fellowship Recorder Card). However, both Sotejo and Malay had experience with ward duty when they were nursing students nearly a decade prior, and both were instructors at the Philippine General Hospital nursing school, meaning they had likely once taught the courses they were being required to take. Hence, if Sotejo and Malay had read the critique that they were unequal to the task, the nurses would have agreed with Taylor not because of any intellectual shortcomings but because they were overqualified as nursing instructors to perform student duties. In defiance of these requirements, Sotejo and Malay did not report to ward duty or show up to the remedial courses.

Ward duty fulfilled neither the RF's educational nor professional goals. During a fellow's tenure at a university, the dean of the school of nursing was responsible for designing the program of study, which was supposed to be related to the specialization that the nurse was pursuing. Sotejo's file states that she had applied for the fellowship in order to study nursing administration and education, but instead the dean had assigned her to ward duty, where she would gain no new insights on how to run a university-level nursing program (Sotejo, 1941, Fellowship Recorder Card).

As the first university-level program in the US, Yale had several opportunities that would have been beneficial for Sotejo. For example, Taylor was mentored by Annie Goodrich, the woman who had replaced the Connecticut Training School for Nurses with the Yale School of Nursing in 1926. Sotejo was trying to accomplish a similar goal in Manila. Taylor had also written a series of publications arguing for more universities to establish departments of nursing and had mentored several women who went on to become deans of new nursing programs (Barrett, 1949). Indeed, Taylor's article "The Right of the School of Nursing to the Resources of the University," which was published in 1934, was one of the first academic nursing pieces to articulate the type of training that Sotejo was striving to establish in the Philippines. The article was a foundational work for Sotejo, who continued to use it in her own writing long after she had finished at Yale (Sotejo, 1943, 1949).

Therefore, Taylor should have been the ideal mentor to help Sotejo reach her professional goals. There were several opportunities Taylor, the dean of nursing, could have offered a woman with ambitions of becoming a nursing dean. For example, Sotejo could have shadowed Taylor, or she could have taught a course as a visiting faculty member in the department. As a nurse who worked in a tropical region, Sotejo had experience treating patients on a range of diseases that would be less familiar for nursing students at Yale. Indeed, twenty years later, Sotejo was asked to do a series of lectures in the American South at the University of Florida (Sotejo, 1960). In previous years, other nurse fellowships required special partnerships with non-profits and government organizations, and the Yale campus was filled with opportunities that would have been helpful for Sotejo to fulfill her objectives abroad. However, Sotejo's access to these opportunities were hindered by the Yale nursing faculty who were unwilling to mentor her.

While the intentions of the Yale nursing faculty are not known with certainty, it appears that by making ward duty a daily requirement for Sotejo and Malay, Yale nursing faculty were attempting to use the fellows as an extra source of labour during a national nursing shortage that had started during the Great Depression and reached increasingly dire levels during the Second World War (Whelan, n.d.). The faculty interpreted the visiting nurses' refusal to do menial work as a sign that they were deficient in knowledge; the subject position they were seen to hold prior to arrival.

Effectively, the deficiencies of the health system in the Philippines were embodied in the nurses regardless of their own extensive training. In fact, Sotejo's record is filled with glowing reviews about her exceptional intelligence. In the RF's own internal documentation, it reads: "[Sotejo] is highly intelligent and will return to a key position when she is able to get back to Manila" (Sotejo, 1941, Fellowship Recorder Card, p. 3). Sotejo scored at the top of her class in most of her courses, but was evaluated by her faculty as being too narrowly focused on her objectives. After their time at Yale, both Sotejo and Malay moved on to the University of Toronto, where they had a more productive time and were not asked to do ward duty. Nettie Douglas Fidler, nurse faculty at the University of Toronto, had this to say about Sotejo: "It is difficult to know what Miss S[otejo] thinks. She is the best public speaker in the whole group of students ... in teaching and administration. Miss S[otejo] has written 2 tomes on a nursing curriculum for the P.I. She has an unusually good mind, knows what she wants to do and does not let her instructors interfere in the program she wants carried" (Sotejo, 1941, Fellowship Recorder Card, p. 2).

Fidler's tone is a mixture of respect and trepidation. Fidler says Sotejo was an excellent communicator, scoring high in her course on nurse pedagogies, but she opens her assessment by saying that it is hard to really penetrate what Sotejo is thinking. Additionally, throughout her tenure abroad, Sotejo's instructors often said that Sotejo was extremely worried about her friends and family who were under attack by the Japanese in the Philippines.

During their time abroad, Sotejo's and Malay's positions were ripe with vulnerabilities that were similar to what temporary contract workers as well as international graduate students face today. For example, they were expected to be brilliant scholars and reliable workers despite being emotionally exhausted by crises that their families experienced back home. For Sotejo and Malay, with the Japanese occupying the Philippines, they could often feel stranded, or guilty that they were relatively safe while their families were in harm's way. When they were struggling in their programs, what choices did they have? They could not go home.

Therefore, they were subjected to multiple systems of order situated in different temporalities and geographies as well as a different set of rules. In the case of Sotejo and Malay, there was the present and immediate order of things that the Yale administration was establishing, which positioned the nurses as both scholars and menial labourers, who they were interested in extracting labour or prestige from at the current moment. The RF positioned the nurses as scholars, but mostly appraised the nurses on their promise and what they would be able to accomplish once they left the US and returned to the Philippines (Brush, 1995; Reddy, 2015). Similarly, there was the order of the Philippine government, which saw the nurses as ambassadors of Filipino humanity, expecting the women to demonstrate the modernity and equality of medical training in the Philippines. Therefore, Filipino nurses attempted to conform to a dizzying number of roles that oftentimes clashed with one another, which meant that Sotejo and Malay were in a difficult position. The space that they did manage to carve out for themselves abroad was a tense negotiation. This can explain why they refused to do ward duty despite the increasingly forceful demands of their hosts.

During their time at Yale, nursing faculty attempted to discipline Sotejo by cornering her and showing her how to bathe a baby. This was obviously an attempt to humiliate a woman who was an international guest. The gesture was confrontational and was meant to force Sotejo into a specific position of subservience. Racial and gendered hierarchies at Yale allowed for the faculty to reframe the fellows' resistance as merely ignorance. This reflects a rigid power structure in the nursing school: while Sotejo and Malay may have thought they were equals within the nursing faculty, they were not. Despite the high status these nurses had achieved in the Philippines, a US colony, the faculty objected to the agency these "third world" subjects were displaying.

The hostility and confusion that the nurses faced can best be understood as what Kramer (2006) calls "inclusionary racial formation" (p. 5). This is the belief that colonized and racialized subjects would one day prove they were ready for self-rule but required American tutelage and supervision. Therefore, while this tutelage was in effect, it rationalized inferior treatment and the idea that even the most prestigious nursing scholars should perform menial care work. In a similar case study, Ines Cayaban, a Filipino nurse, received a fellowship from the Daughters of the American Revolution to pursue graduate training at the University of Chicago. However, when a minister associated with the organization became a widowed father, the association changed the terms of Cayaban's funding from that of a fellow to that of a domestic servant; she would now have to earn her room and board for school by being a full-time nanny to a widower's son (Choy, 2010).

Both cases reveal an assumption that Filipino nurses should be honoured to do the low-paid or unpaid care work of North American women. This also meant that professionally trained Filipina women were valued at the same level as untrained North American nursing students or care workers. The fungibility of Sotejo's, Malay's, and Cagayan's positionality represents an early example of the process of racial stratification, which emanates from a legacy of colonialism and uneven development. This legacy of imposing and projecting subservient status onto even professional Filipinos persisted long after the country's independence.

It is important to emphasize that the opportunity to study abroad was critical because the Philippines lacked any sort of post-graduate training aside from a public health nursing program established in the 1920s (Fitzgerald, 1922). Any offer of financial support would be a potentially life-changing proposal, yet the nursing scholarships were designed during the American colonial period in the Philippines, and premised on the colonial racial hierarchies that endured long after Philippine independence.

Just as Filipino nurses learned at an early stage the importance to their careers of being good visitors when they studied abroad, they also learned the importance of being good hosts. Sotejo, Malay, and Cayaban were with the PGH in 1929 when the hospital formed a partnership with the RF and the Philippines and when the Siam (Thailand) government began sending Siamese women to the Philippines to study nursing. In fact, Ines Cayaban, the nurse whose fellowship from the Daughters of the American Revolution was altered on a whim, knew intimately the importance of diplomacy and generosity that a nursing faculty member should use with visiting nurses. Cayaban had been assigned to be a Siam monarch's daughter's personal chaperone while the daughter studied nursing in the Philippines (Cayaban, 1981). In comparison to the Filipino nurses' experiences abroad, as hosts, the nurses took great care to integrate the Siamese students into the activities of the domestic nursing students, making sure that all of them took classes with the Filipino nursing students and resided in dormitories. They also made sure that the Siamese students had time to experience the Philippines culturally, by arranging trips for the students to go sightsee on the weekends.

Therefore, as hosts, the faculty at the PGH created a study abroad program that reflected the broader humanistic goals of enriching the lives of the students and not merely capturing their labour.

A Filipino Nursing Order

As stated previously, Filipino nurses abroad had to negotiate foundational, institutional, and governmental hierarchies, but there was also a network of Filipino nurses who established their own order, creating forms of support. For example, it appeared that Yale University instructors felt especially aggrieved when *both* Sotejo and Malay refused to do ward duty. As a result, they could not isolate just one offender, although it appeared that Sotejo experienced the brunt of the criticism. The unequal scrutiny placed on Sotejo quite possibly had to do with her promise and her temperament, which was considered by some to be quite disagreeable, but also that Malay was known to be more easygoing than Sotejo. In a separate incident in which Malay complained that the instruction she was receiving at the University of Toronto was merely a repeat of the work she had already done in the Philippines, the RF had a meeting with Malay to "point out the limitations of her background and what Toronto had to offer" (Malay, 1941, Fellowship Recorder Card, p. 2). Malay never again complained about the program at the University of Toronto, but when Sotejo was critical of a course, the faculty managed her criticism differently by creating a new course for her. Sotejo was Malay's boss back at the PGH, so Malay had to concur with Sotejo's acts of resistance.

The different orders of authority that nurses had to negotiate had their own distinct temporalities and spatialities. At Yale it was uncomfortable for Malay to disobey Yale's request, but Malay's professional career was tied to Sotejo. For example, seven years later, they would work together as dean and assistant dean of the College of Nursing at the University of the Philippines. This meant that as fellows abroad they inhabited a double hierarchy.

In chapter 6 of this volume, Caitlin Henry's theorization of the "global intimate" clarifies the positionality of early migrant nurses like Sotejo and Malay by offering a means to think of different networks of power that circulated around and between them. In demonstrating the ways multiple scales collide within nursing, Henry argues, "the global intimate entails the entanglement and interconnections rooted in everyday intimacies, but also made across time and place." This further highlights the complicated global formations that impacted the decisions that Filipino nurse migrants made during the Second World War. First, they were transimperial colonial subjects who were in between the Japanese and the US empires. Second, at the local and national scale they were working within the racialized hierarchy of the American nursing workforce, which devalued the labour of women of colour. Furthermore, since most educational institutions prepare students for the future, Sotejo and Malay were also operating within the space of the aspirational, which was a future Filipino independent nation that did not yet exist.

Developing a New Model of Filipino Nursing

On 23 June 1942, when Sotejo's fellowship was supposed to end, the RF granted her an extension. The RF cited her exceptional work at the University of Toronto, and additionally deemed it "impossible" to send her back to the Philippines while it was still at war with Japan. During this year long extension of her fellowship, Sotejo completed a Master of Science degree in nursing at the University of Chicago under the supervision of nursing faculty member Nellie Hawkinson.

In her dissertation, Sotejo stated that a poor evaluation of Filipino nurses in a national nursing survey in 1939 inspired her to study abroad. The survey stated that nurses "f[e]ll short in the application of … technical knowledge in their relation with sick people due to the lack of human understanding and sympathy" (Committee on the Training School for Nurses, 1939, p. 1; Sotejo, 1943). Sotejo was motivated to phase out the hospital model because nursing schools in the Philippines "were not meeting the needs of the time" (Sotejo, 1943, p. 18).

Therefore, Sotejo's entire mission revolved around a motivation to train nurses who could effectively serve the people of the Philippines, not to reproduce colonial models of US nursing. Beyond merely implementing an American model of college nursing in the Philippines, Sotejo's dissertation reveals the level of work needed in order to establish a nursing program at the University of the Philippines. First, there was not one model of training in North America: there were three types of university training models, ranging from a loose affiliation to an official department within the university. During this time period, it was not self-evident in North America and the Philippines that nurses needed college degrees. Since Sotejo wanted nurses with a broad education who could better serve Filipinos, she believed that a four-year bachelors' degree would be the best. In a similar line of logic, Sotejo also argued that a curriculum that included scientific training, social sciences, and humanities would help foster better connections with people. This meant that nurses would be able to better understand the motivations and political conditions that produced poverty. A nursing colleague of Sotejo, Marcela Gabatin, also explained why training in the humanities would be useful for nurses. Since the majority of nurses would be working within communities that were not their own, a broad training in the cultural practices of different Filipino ethnicities would mean that nurses would be able understand and form bonds with the people not from their own cultural group (Gabatin, 1950). Therefore, Sotejo's work to establish a nursing model in the Philippines was not merely the implementation of an American model but also involved creating a nursing curriculum and programs that would address the challenges uniquely faced by nurses in the Philippines and improve their efficacy at helping a uniquely Filipino patient care population.

Upon returning to the Philippines in 1945 after completing her master's degree in nurse education, Sotejo established the first college of nursing at the University of the Philippines, which would phase out the exploitative hospital training system.

This training model was replicated in universities throughout the country. Although this model was intended to train a population of nurses for the benefit of the Filipino nation, it inadvertently created a permanent migratory labour force. Sotejo's intellectual work had lasting consequences for the Philippines, as it created a groundbreaking program that successfully integrated a large number of women into the Philippines and the global care workforce. Sotejo's experiences motivated her to establish the university-level nursing program in the Philippines. This program would allow nursing students to receive a well-rounded college education and to pursue advanced training in the profession without having to leave the country. Ironically, her work also created the intellectual possibilities that facilitated one of the biggest emigration waves from the Philippines to the Western world. Sotejo's experience abroad presaged the complications that were to arise with the US recruitment of foreign nurses.

Conclusion: Filipino Medical Knowledge in the World

Although Julita Sotejo did not forge a long-lasting professional relationship with Effie Taylor and the other nursing faculty at Yale Universty, Sotejo's ability to establish a university-level nursing program in the Philippines was dependent on her ability to create productive collaborations with other Filipino nurses, including nursing fellow and assistant dean Lenor Malay. Beyond the Philippines, Sotejo became a globally recognized authority figure in the field of nursing. For example, the American University of Beirut consulted Sotejo's research on nursing education, while *Kango*, the Japanese Nurse Association's publication, translated her article "Nursing Education in the Philippines" into Japanese. She gave lectures to the Associazioni Infermiere Professionali in Rome, Italy, and to the nursing program at the University of Florida Gainesville (Sotejo, 1954, 1958, 1963). Sotejo's understanding of nursing administration and education was sought after and replicated in countries and regions that in the 1950s had yet to develop formal nursing education. Some of the countries that once sought Sotejo's advice on implementing nursing education represent some of the largest receiving nations for Filipino nurses in the present day.

This chapter tells an untold story of Filipino nurses who established university level nursing programs in the Philippines, which gave thousands of nurses the credentials to obtain US citizenship under the Immigration and Nationality Act of 1965. By examining Sotejo's biography, we see that nursing education was not merely a colonial model grafted into a Filipino context, but that it took the intellectual labour of Filipina women, like Sotejo, who were hindered by racial discrimination from American faculty, to improve nursing education in the Philippines. In chapter 17 of this volume, the authors argue on behalf of going beyond narrating migration through the language of drain and instead focusing on "the places and spaces through which medical knowledge is acquired, adapted, and performed."

Therefore, in focusing on the intellectual labour of early migrant nurses like Sotejo and her motivation to create a nursing force for an independent Filipino nation, this chapter highlights the ways that Filipino nurses' objectives and aspirations often worked in contradiction to the US colonial project that produced them.

While Sotejo's complete vision of nursing in the Philippines did not end up nursing the nation in the way that she and her fellow colleagues had originally envisioned, her biography provides insights that are still relevant to the present-day discussions of overseas nursing migration from the Philippines. For example, in her dissertation she wrote that "a society has the responsibility of providing the type of preparation needed for professional nursing" (Sotejo, 1943, p. 2). Although she was talking specifically about the responsibility that the Philippines had to educate its own people to become nurses, her words offer North American countries something to consider regarding their responsibility to incentivize and expand their own nurse education programs, as well as the responsibility they have to nurses from sending countries like the Philippines.

NOTES

1 To give a broader sense of these migrations after 1965: the first nursing exodus to the US lasted from 1965 to 1969, which was facilitated by the passing of the Hart Cellar Act of 1965; subsequent waves of emigration took place from 1975 to 1980 to the US as well as other Western countries, including Canada, the UK, and the Arab States of the Persian Gulf.

2 For more on the history of the Philippine General Hospital, see Catherine Choy, "The Usual Subjects: The Preconditions of the Professional Migration" in Choy (2003).

3 Two nursing biographies were written about her in the 1960s and 1970s: Aragon (1978), and Samson (1965). Also, she was featured briefly in Choy (2003, pp. 56, 88) and in Duque (2009).

4 For more on Filipino Migration, see Posadas and Guyotte (1990); A. Espiritu (2005); Y.L. Espiritu (1995); Baldoz (2011); España-Maram (2006); Mabalon (2013); and Parreñas (1998).
5 The idea is influenced by Day's (2014) use of the word "fungible" to describe the status of Chinese migrant labour in North America.
6 When the Rockefeller Foundation sponsored Filipino scholars to study abroad, the Foundation collected copious records on each fellow. This included letters from instructors and supervisors, as well as notes on conversations that people affiliated with the Foundation had about and with the fellows. They also included personal details about the Fellow's physical appearance and attitude. Notes on the Fellows continued even after their fellowship ended. Rockefeller Archive Center took these records on Filipino fellows and created abridged versions of the information downsized to the size of index cards that were typed. Therefore, the direct records have been mediated by the invisible labour of the archive.

REFERENCES

Aragon, L. (1978). Life's work and philosophy of Dean Julita V. Sotejo. *ANPHI, 13*(4), 2–9. PMID: 383009.

Baldoz, R. (2011). *The third Asiatic invasion: Migration and empire in Filipino America, 1898–1946*. New York University Press.

Barrett, J. (1949). *Ward management and teaching*. Appleton-Century-Crofts.

Brush, B. (1995). The Rockefeller agenda for American/Philippines nursing relations. *Western Journal of Nursing Research, 17*(5), 540–55. https://doi.org/10.1177/019394599501700506.

Cayaban, I. V. (1981). *A goodly heritage*. Gulliver Books.

Chang, G. (2000). *Disposable domestics: Immigrant women workers in the global economy.* South End Press.

Choy, C.C. (2003). *Empire of care: Nursing and migration in Filipino American history.* Duke University Press.

Choy, C.C. (2010). Nurses across borders: Foregrounding international migration in the nursing history. *Nursing History Review, 18*, 12–28. https://doi.org/10.1891/1062-8061.18.12.

Committee on the Training School for Nurses. (1939, September 11). The Philippine General Hospital Committee on the Training School of Nurses Report to the Director, Philippine General Hospital, Manila.

Day, I. (2014). *Alien capital: Asian racialization and the logic of settler colonial capitalism.* Duke University Press.

Duque, E. (2009). Modern tropical architecture: Medicalisation of space in early twentieth-century Philippines. *Architectural Research Quarterly, 13*(3/4), 261–71. https://doi.org/10.1017/S1359135510000114.

España-Maram, L. (2006). *Creating masculinity in Los Angeles's Little Manila: Working-class Filipinos and popular culture in the United States*. Columbia University Press.

Espiritu, A. (2005). *Five faces of exile: The nation and Filipino American intellectuals*. Stanford University Press.

Espiritu, Y.L. (1995). *Filipino American lives*. Temple University Press.

Fitzgerald, A. (1922). Report on the nursing situation in the Philippine Islands, April to December. Folder 6, Box 5, Series 242C, RG 1.1, RAC. Rockefeller Archive Center.

Gabatin, M. (1950). The need for cultural development in professional growth. *The Filipino Nurse*, 190.

Gamarnikow, E. (1978). Sexual division of labor: The case of nursing. In A. Kuhn & A. Wolpe (Eds.), *Feminism and materialism: Women and modes of production* (pp. 96–123). Routledge.

Kramer, P. (2006). *The blood of government: Race, empire, the United States, and the Philippines*. Ateneo de Manila University Press.

Loza, M. (2000). *Defiant braceros: How migrant workers fought racial, sexual, and political freedom*. University of North Carolina Press.

Mabalon, D. (2013). *Little Manila is in the heart: The making of the Filipina/o American community*. Duke University Press.

Parreñas, R.S. (1998). "White Trash" meets the "Little Brown Monkeys": The taxi dance hall as a site of interracial and gender alliances between white working-class women and Filipino immigrant men in the 1920s and 30s. *Amerasia Journal*, *24*(2), 115–34. https://doi.org/10.17953/amer.24.2.760h5w08630ql643.

Posadas, B.M., & Guyotte, R.L. (1990). Unintentional immigrants: Chicago's Filipino foreign students become settlers, 1900–1941. *Journal of American Ethnic History*, *9*(2), 26–48. https://www.jstor.org/stable/27500756.

Reddy, S. (2015). *Nursing and empire: Gendered labor migration from India to the United States*. University of North Carolina Press.

Salamanca, S. (1922). Public Health and Nursing Fellowship, Recorder Cards, RG 10.1, RFA, RAC. Rockefeller Archive Center.

Samson, J.A. (1965). Julita V. Sotejo, arbiter of charity. *St. Thomas Nursing Journal*, *4*, 183–5.

Sotejo, J. (1941). Nursing Education and Administration Fellowship Recorder Cards, RG 10.1, RFA, RAC. Rockefeller Archive Center.

Sotejo, J. (1943). *A university school of nursing in the University of the Philippines* (Unpublished master's thesis at the University of Chicago).

Sotejo, J. (1949). Do our schools of nursing meet the ICN standard of nursing education. *The Filipino Nurse*.

Sotejo, J. (1954, 1958, 1963). Personal papers of Julita Sotejo, correspondence with Febe Tedesco (19 October 1958); Dorothy Smith (9 July 1963); and Yasuko Otake (2 June 1954, written by Jeanette White on her behalf). University of the Philippines Library, Diliman, Quezon City, Philippines.

Sotejo, J. (1960). Personal papers of Julita Sotejo, correspondence with the University of Florida. University of the Philippines Library, Diliman, Quezon City, Philippines.

Sotejo, J., & Beltran-Jackson, M. (1965). *Learning nursing at the bedside*. University of the Philippines Press.

Taylor, E. (1934). The right of the school of nursing to the resources of the university. *The American Journal of Nursing, 34*(12), 1187–94.

Vecoli, R. (1999). Comment: We study the present to understand the past. *Journal of American Ethnic History, 18*(4), 115–25. https://www.jstor.org/stable/27502473.

Whelan, J. (n.d.). Where did all the nurses go? Penn Nursing, University of Pennsylvania. https://www.nursing.upenn.edu/nhhc/workforce-issues/where-did-all-the-nurses-go/.

15 From Brain Drain to Brain Retrain: A Case of Nigerian Nurses in Canada

SHERI ADEKOLA

Introduction

In this chapter, I examine Nigerian-trained health care professionals and their education and experience in Canada. This chapter begins with a brief overview of Nigerian migration in Canada. I then examine the negative and positive discourses surrounding skilled migrants followed by an overview of the methodology used in the course of my research. The chapter closes with the study's findings and conclusions.

Over the last two decades, the bulk of research on skilled migrants has used macro- and meso-scale analysis based on occupation, skill, or country of origin (see Couton, 2013; de Haas, 2010; de Haas & Vezzoli, 2013; Koser & Salt, 1997; Remennick, 2013). Less noticed, but equally important, is the perspective skilled migrants themselves have of the discourses that are used to explain their skilled migration and its causes and consequences for them, their families, and their home and host countries. This chapter focuses on the migrants' own perspectives regarding their migration experience and reveals how continuous educational development and training are key aspects of their migration process.

Nigeria has witnessed several waves of emigration since the early 1970s. In the past two decades, educated and highly skilled individuals have comprised a sizeable portion of outgoing international migration. Approximately 20 per cent of global migrants in 2010 originated from Africa (IOM, 2013). In 2016, 13.4 per cent of recent immigrants to Canada were born in Africa, a fourfold increase from the 1971 Census of Canada (3.2 per cent). In 2014, Nigeria ranked eleventh as a source country for permanent residents to Canada (Statistics Canada, 2016). The migration of Africans to Canada is particularly important when we account for the skills migrants bring to Canada in the health care field. According to the College of Nurses of Ontario (CNO), Nigeria was ranked in the top five between 2008 and 2013 as the country of education for registered nurses (CNO, 2014).

Nigeria is one of the top twenty source countries for migrants to Canada and ranks in the top fifteen source countries for permanent residents. Migrants from Nigeria mostly arrive in Canada as skilled workers. Yet relatively less is known about Nigerian migrants on the whole as compared to migrants from other lead countries of immigration to Canada such as India, China, or the Philippines. Both current and anticipated trends indicate that skilled African migrants will comprise an important part of Canada's projected population growth. According to Clemens and Pettersson (2008), Africa is the second-highest supplier of nurses globally, and Nigeria ranked third in Africa by graduating 7 per cent of African nurses. In Canada, the CNO reported in 2015 that internationally trained nurses accounted for 25.7 per cent of their membership the previous year, and that Nigeria ranked in the top five source countries in 2013 and 2014 (with 2.1 per cent of applicants and 2 per cent of registered practical nurses), and ranked in the top three in 2012 and 2013 (at 3.0 per cent and 3.1 percent, respectively) for registered nurses (just after the Philippines and India) (CNO, 2016). Even though a significant share of immigrant health care professionals come from Nigeria, relatively little research focuses on this group.

Research on health care labour migration from Sub-Saharan Africa yields valuable insights into the nuances of migration decisions. Authors have highlighted the motivations for Nigerian-educated health care professionals to leave Nigeria (Aboderin, 2007; Adelakun, 2013; Astor et al., 2005; Kalipeni et al., 2012; Salami et al., 2016). The social, cultural, political, professional, economic, and institutional contexts in which migrants make decisions in developing countries differ from that of developed countries (Guilmoto & Sandron, 2001; Vujicic et al., 2004). The demand for internationally trained nurses in the developed country posits a strong pull for Nigerian nurses (Brush et al., 2004). Salami et al. (2016) identify push factors, retention rate, and inability to secure a job as the major causes of nurse migration from Nigeria. Their study notes that although there is a strong push for Nigerian nurses who are motivated to emigrate from Nigeria due to work environments, the desire to advance in their careers and improve economic prospects is also part of their calculation. Oni (2000) notes that Nigerian educational infrastructure is underdeveloped, with the ratio of lecturers to students low, and the number of universities too small to meet the country's needs.

While destination countries such as Canada consider skilled immigration favourably, there are widespread worries about what has been interpreted as "brain drain" (Bhagwati & Hamada, 1974), and the negative consequences of high skilled emigration on countries of origin. For example, recent population data for Nigeria reports 18.25 skilled health personnel per 10,000, compared to Canada's 123.82 per 10,000 (World Health Organization, 2017). An outflow of skilled workers from already poorly resourced health systems feeds into the motivating factors that draw health workers from low- and middle-income countries to developing ones. However, research has also shown that internationally educated nurses can experience racism, deskilling, cultural shock, discrimination, and barriers to integration in these destination countries, suggesting brain drain may be amplified in both the sending and receiving context. This chapter examines this issue by first examining the pre-migration context, then reviewing the range of discourses that accompany this type of migration. The second half of the chapter discusses the methodology used to collect the data, followed by the empirical findings and conclusions.

Pre-migration Context in Nigeria

There are several credentials (certificate, diploma, higher national diploma/advanced diploma, and undergraduate degrees) that are possible in Nigeria. One can become a nurse in Nigeria through formal training or as an apprentice; however, nursing education at the baccalaureate level and beyond is only offered by a few universities in the country (Ezeonwu, 2013). Apprenticeships normally occur in the private sector and can last three years or more (see Fajobi et al., 2017; Teal, 2016). It must be noted that only after completing their education and passing their final qualification exams can nurses apply for a basic license with the Nursing and Midwifery Council of Nigeria – the governing body for nursing and midwifery programs. The council currently has 120,000 nursing professionals registered (Nursing and Midwifery Council of Nigeria, n.d.).

Theoretical Context: Discourses of Skilled Migration

Multiple conceptual discourses associated with international skilled worker migration reflect a spectrum of development consequences from positive to negative. De Haas (2010) has represented this general debate as a pendulum that swings from positive to negative. For example, in the 1970s, debate was dominated by the negative aspects of migration (brain drain), but, by the 2000s, there was a shift to positive debates of brain circulation. De Haas (2012) refers to theorists of the positive discourses as "migration optimists." The migration optimist approach perceives migration largely as having a positive impact on the development process. Migration is seen as benefiting societies that send migrants because the movement can generate flows of return capital, remittances, and investment that can stimulate development and modernization in the sending countries. "Brain gain" also generates return migrants who can become positive agents of economic growth by expanding technological skills, investment, and production systems (Harvey,

Table 15.1. Skilled migrant discourse spectrum negative to positive

Skilled migration discourse	Brief overview	Overall developmental consequence
Brain drain (Bhagwati & Hamada, 1974)	Emigration of highly qualified, talented professionals from one country to another, part of the broader process of uneven development.	Negative
Brain abuse (Bauder, 2003)	Improper utilization of the quality and the skills of people they brought from their country of origin.	Negative
Brain desertification (Faist, 2008)	Highly skilled migrants do not return and do not sustain any ties with those who stayed in the countries of origin.	Negative
Brain circulation (Saxenian, 2005)	Highly skilled migrants who move in and out of their host and home countries for business, work, and investment purposes.	Positive
Brain exchange (Williams & Baláž, 2003)	Balanced and effective use of human capital within systems of international movement. This is evident when multinational firms move skilled workers between their operations in different countries.	Positive
Brain gain (Ciumasu, 2010)	Accumulation of human capital in the destination country.	Positive
Brain networking (Ciumasu, 2010)	Brain networking (longtime commitment of expatriates to distance collaboration) can facilitate decision-making among undecided ones to return home, then to return (brain circulation).	Positive
Brain train (Beckles et al., 2002)	The brain train is the idea that higher education is essential to economic development. Serial mobility (including onward and return migration) in order to accumulate credentials and retraining has been identified as brain retrain.	Positive

2008; Pellegrino, 2001; Spoonley et al., 2003). Table 15.1 outlines various theories that represent this spectrum. While a vast amount of literature examines these concepts, there has been relatively little analysis of how skilled migrants themselves view such discourses, or what definition aligns most closely to how they view their own migration decisions, experiences, and outcomes.

Negative Discourses

Brain Drain

Brain drain describes the phenomenon of the emigration of highly qualified, talented professionals from one country to another in search of better opportunities. The outmigration of health professionals from Africa has perhaps been the exemplar of the brain drain debate. El-Khawas (2004) argues that receiving countries acquire talent without spending resources to educate or train the skilled workers. Multinational companies also benefit from the profit acquired through the lower wages of foreign workers, and migration contributes to the economic growth of the receiving countries through tax payments and increased productivity. As a migrant-sending continent, Africa may benefit from remittances, but the gap between initial investment in human capital development and returns through remittances are not seen as equivalent by brain drain advocates (Bhagwati & Hamada, 1974; Lucas, 1988). Papademetriou (1985) concluded that migration will cause depletion of human resources from sending countries by enticing their most productive, healthy, and dynamic members.

Brain Abuse

Bauder (2003), in his study based in Vancouver, focused on professional immigrants from South Asia and the former Yugoslavia, many of whom faced extensive challenges in having their credentials recognized. Bauder termed this "brain abuse," and claims this non-recognition of foreign credentials represents the systemic exclusion of immigrant workers from the upper segments of the labour market. Workers are segmented through immigration status and place of origin, constructing barriers against the full recognition of foreign credentials (also see Ku et al., 2019; Samers, 1998). Immigrants thus lose access to their previously held occupational status and income. This devaluation of immigrant cultural capital is viewed as wasting human capital. It has a negative effect on the migrants and their families because the loss of labour market status means diminished social status.

Brain Desertification

Another negative theme includes "brain desertification," which refers to highly skilled migrants who do not return to their countries of origin nor sustain any ties with people there. This is the worst case of brain drain. According to Faist (2008), brain desertification is characterized by the perception of the lack of sustained ties with those who stayed in the country of origin and the lack of migrants' potential for return.

Positive Discourses

Brain Circulation

"Brain circulation" is used to describe highly skilled migrants who move in and out of their host and home countries for business, work, and investment purposes (Harvey, 2008). Brain circulation allows a transfer of knowledge through expatriate nationals and occurs when migrants who have left their origin country return home to establish relationships for business and scientific purposes, or to start new companies while maintaining their social and professional ties with the destination country. Saxenian (2005) refers to these former immigrants as "the new Argonauts" who have now become transnational entrepreneurs by structuring a two-way flow of capital, skills, and technology. Vinokur (2006) explains brain circulation in terms of reversible migration including the return of expatriates, which acts as an equalizing transfer of technological and organizational capacity back to source countries. This results in a positive-sum game where former migrants become transnational entrepreneurs, structuring a two-way flow of skills, technology, and capital between destination and source countries.

Brain Exchange

"Brain exchange" represents the effective use of human capital in the relatively balanced and mostly temporary flow of migrants between core economies (Williams & Baláž, 2003).

Brain Gain

Brain gain is defined as the accumulation of further specialized human capital for highly skilled migrants in the destination country after initial migration (Ciumasu, 2010). Human capital represents a form of investment, and migration decisions represent one way to alter the potential rate of return on that investment (Liebig, 2003). When migration is engaged for ongoing specialized training, this generates an increase in human capital that is not solely provided by the sending countries. Human capital accumulation then adds another driver to migration decision-making. Migration allows for the potential of greater returns on the initial investment in education and training.

Brain Networking

"Brain circulation" and "brain networking" are used interchangeably by some scholars (e.g., Saxenian, 2005). Ciumasu (2010), however, warns against using brain circulation and brain networking as synonyms because he sees them as exclusive. Brain networking is a long-term commitment of expatriates to distant collaboration, which can facilitate decisions among the undecided to return home as they fall in and out of the network.

Brain Train

"Brain train" is used to represent the power of higher education to contribute to Caribbean development (Beckles et al., 2002). Concerning international migration, it captures the idea that individuals engage in migration to accumulate credentials and training, with the intention for further onward or return migration. Varma and Kapur (2013) refer to this as brain retrain. I use the idea of a global brain train to indicate the experience of international migration leading to the accumulation of credentials and further education.

Methodology

The study uses qualitative analysis in order to gain insights into the experiences of Nigerian-educated health care professionals in Canada. Using a qualitative research approach allowed us to delve deeper into "how people make sense of their lives, experiences, and their structures of the world" (Creswell, 1994, p. 145). The research for this study was conducted in Canada and Nigeria. Using modified surveys and semi-structured interviews, data were collected from a total of fifty-nine (n = 59) migrant workers in Canada. Email, phone, and face-to-face interviews were conducted with the eight key informants. The key informant interviewees were employers, community leaders, and social group activists interviewed in order to obtain insights from community leaders concerning the effects of migration. Table 15.2 shows the distribution of participants by occupation. For the purpose of this chapter, registered nurses and registered practical nurses are the focus, representing just under two-thirds (57.7 per cent) of the Canadian sample.

A modified survey (with open-ended questions), was used to capture detailed information on the demographic characteristics of Nigerian health care professional migrants including sex, age, education, occupation, housing status, marital status, and migration experience. In-depth, open-ended interviews were also conducted with twenty (n = 20) migrant health care professionals willing to be interviewed. These interviews elicited detailed information on their migration experiences in Canada, including their personal experiences settling down and coping mechanisms and changes they attributed to their migration experience.

Taken together, the in-depth interviews and survey responses provide important quantitative data and personal narratives about migration experiences.

Table 15.2. Sample distribution

Occupation	Count	Percentage (%)
Registered nurse	29	49.2
Personal support worker	13	22.0
Other health professional	8	13.6
Registered practical nurse	5	8.5
Student	4	6.7
Totals	59	100.0

Surveys were distributed using a paper-based format. Individuals were invited to participate, and the survey was delivered to them personally by hand, email, or post. It was also sent to several people through faith and cultural groups; to students at colleges; to members of Nigerian organizations and professional associations; and to some colleges with health worker bridging programs including Humber, Sheridan, Fanshawe and Centennial Colleges, and York University. In addition, personal acquaintances employed as health care workers distributed some questionnaires at their schools or workplaces.

This kind of distribution methodology functioned as a form of snowball sampling, ensuring access to social groups beyond the researcher's immediate social circle and extending the data collection across a diverse population. The survey was also mailed to connections found on social media sites like Facebook and LinkedIn. Using purposive and snowball sampling, the sample population for this project consisted of skilled immigrants from Nigeria who had lived in Canada for at least four years. Analysing data from the modified open-ended survey, the interviews, key informant interviews, and field notes elicited major themes that were central to understanding the migrant experience and how these skilled migrants perceived the consequences of migration.

Empirical Findings

This section starts with skilled migrant discourse and ends with the migrants' current education, documenting the skill discourse of Canadian international nurse migrants from Nigeria and the consequences of migration.

Respondent Perceptions of International Skilled Migration Discourses

During the interviews and surveys in Canada and Nigeria, the participants were presented with different skilled migrant discourses (see table 15.1 above). They were asked to choose the ones with which they most strongly identified. The findings reveal the dominance of skilled migration discourses for Nigerian trained nurses; this includes a strong alignment with the idea of brain train, that

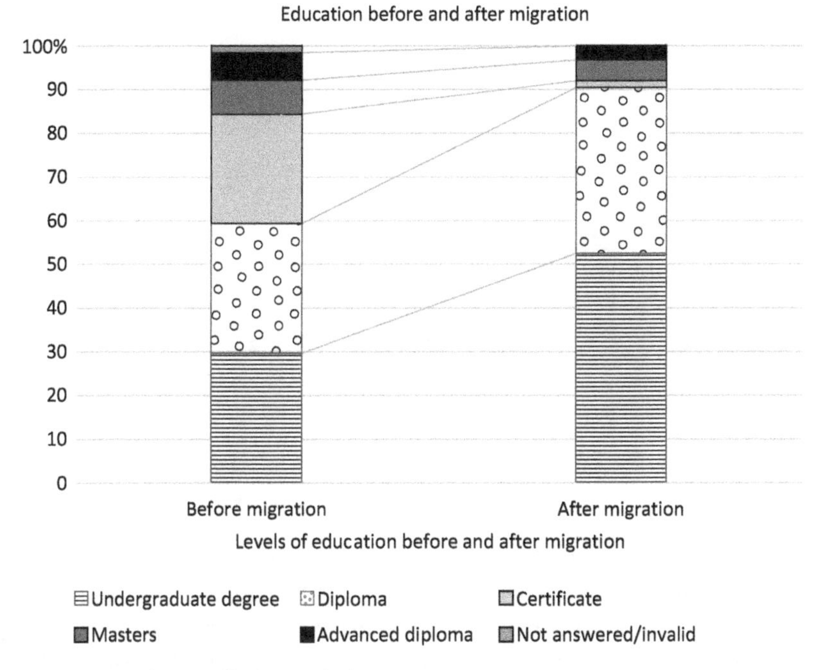

Figure 15.1. Education before and after migration

of continued educational upgrading throughout the migration journey. Only 3 per cent of respondents identified brain drain as most representative of their experiences. The remainder associated with more positive or neutral discourses of brain train, brain gain, brain exchange, or brain circulation. Furthermore, surveys also revealed that 83 per cent of participants said they acquired new skills after migration, while 76 per cent stated they gained new credentials. This is a noticeable finding, as most of the participants were already skilled workers with previous knowledge and qualifications before arriving in Canada. There was also a notable increase in levels of education after migration (see figure 15.1).

In terms of the occupations the respondents held before and after migration, figure 15.2 demonstrates that participants were engaged in nursing as well as teaching, training, and several occupations not necessarily included in the health field. But after migration, a number became personal support workers (PSWs), and twice as many people started to work in health-related occupational fields including nursing. While working as a PSW might be seen as deskilling for registered and practical nurses, interviews with PSWs suggest they were satisfied with this employment and found wider satisfaction in their general quality of life in Canada. For example, one respondent reported that

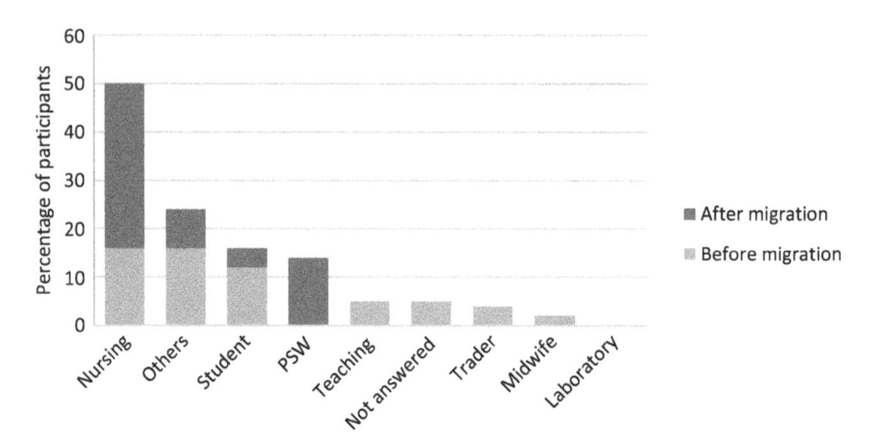

Figure 15.2. Pre- and post-migration occupation

completing a PSW qualification was useful career advice they followed: "While here, I learnt that some courses are more lucrative, so I switched and I completed the Personal Support Worker Diploma so I can work [in this area until] I get papers and I can pursue my Registered Nursing degree" (Registered Practical Nurse, under 30 years old, Interview Participant 14). When asked why she made this career change, the participant answered that she did it for the job prospects and because her parents preferred the career change; although this may seem like a process of deskilling, the participant did not see it as such.

Education/Career Change

A key theme that emerged from the research was the participants' ability to adapt to a new environment by increasing their level of education and changing their careers. The majority of participants saw education and career change as positive. Many of those that had degrees prior to their migration felt the frustration of having to start over again with their career: "I studied finance but I never really use my diploma to do anything other than my business. I had to go back to high school after applying to Humber College and no admission and then start all over." (Registered Nurse, over 60 years old, Interview Participant 7).

Another participant received advice concerning changing their occupation and moved into nursing:

R: When we reached here, [a] couple of my husband's friends came to our house to give me the lesson of my life.

I: What do you mean by that?

R: They told me health care, nursing or RPN [registered practical nurse], is the only lucrative job I can do here when I told them I am an accountant from back home. (Registered Practical Nurse, 40–50 years old, Interview Participant 13)

Key informant interviews reflected that Nigerian migrant professionals are strongly focused on using migration to advance their professional careers:

> Nigerian professional migrants are goal-oriented and focused. They have the mindset and determination to attain the level [of career they want] within a very short period of time. Most of them sacrifice whatever it takes to get them certified in Canada rather than doing petty jobs to survive … In a nutshell, Nigerians love to study and stayed informed. Hence, they would do whatever it takes to get them [to] the level they believe they deserve. (Key Informant Interview, Participant 5)

Another agreed: "We are beginning to witness the rise in the Nigerian professional migrants. Recently, there was an establishment of a Council of Nigerian Lawyers. This is a glimpse of the future we look forward to. We have other professionals, such as doctors, realtors, politicians, and so on" (Key Informant Interview, Participant 7).

When asked what changes they have noticed since Nigerian migrants came to Canada, some key informants stated that women migrants, in particular, had taken advantage of the opportunities for continuous education and training:

> Migration creates awareness and exposure to the most African community. Lots of things that are taken for granted are made accessible. Further, our people are exposed to the importance of education. Back in Africa, most mature women especially give up in education, while [a] great number of them now are back in school and having degrees in various fields like nursing and social work. To also add to this, the availability to go back to school is available, which encourages most adults relative to Africa, where there is no government assistance program [for continuing education]. (Key Informant Interview, Participant 5)

These results reflect those of James (2009), who also found that migrants understood education to be "one of the tickets" that would give Black people opportunity in Canada.

This section has analysed how migrants themselves perceive the influence of their own migration decisions, and the data reveals that migrants are identifying with the brain (re)train discourse. While loss is evident in the migration journey, other opportunities for migrants and their families are presented as positive outcomes, and key among them are the chances to gain new skills and educational experiences. There were, of course, some cases more closely associated with brain drain. For example, in one case, a respondent reflected on her experience of moving from being a physician in Nigeria to becoming a registered practical nurse in Canada:

> On getting here, I realized it's a good place to raise my family, so as I was in school I started processing my application as a doctor and very soon I realized it's not going to be as easy as I thought. After [a] few months, I changed my program to the RPN so I can graduate and provide for my family. (Registered Practical Nurse, 50–60 years old, Interview Participant 15)

This is a clear example of deskilling and arguably brain waste, but for the respondent the necessity was to find a way to provide a stable and promising future for her family.

Consequences of Migration

Respondents were asked about the consequences of migration as they had experienced it. Major themes were selected from the surveys and interviews. The major themes discussed in this section represent the key narratives that emerged: more education, increased independence, children, loneliness, and challenges.

Most survey respondents and interviewee participants said they felt more accomplished in their educational level post-migration:

> Today, I feel accomplished. I went to high school and finished as a registered nurse at York [University]. I am currently searching admissions to the master's program. I am grateful to Canada. Some of my friends in Nigeria spent eight years doing a 3–4 year degrees because of labour dispute strikes, but I didn't have to do that. (Registered Nurse, 30–40 years old, Interview Participant 9)

Some respondents selected nursing as an occupation after they had moved to Canada, and their experiences were similar to those of other Nigerians, now resident in Canada, who saw migration as a chance to remake their occupational identity through access to quality training and education:

> My life today is way better. Now that I think about it. It was at the beginning that was rough. I didn't know where to go to how to get things done. Now I am a registered nurse, and I am thankful to Canada for giving me that opportunity to further my studies, which wouldn't have been possible in Nigeria. (Registered Nurse, over 60 years old, Interview Participant 7)

Women were especially grateful for the opportunity to relaunch a career after rearing children: "I am more exposed, more educated, and feel more successful in my career and as a mother. I went back to school, now I have my master's and teach part time" (Registered Nurse, 30–40 years old, Interview Participant 4).

Another participant echoed the comments of several respondents about the quality and opportunity of continuing studies and education in Canada regardless of age: "Sky is the limit about education. While I was in school, I saw old mummies and ladies studying to get better and I am more committed to my education [now] rather than when I was back home" (Registered Practical Nurse, under 30 years old, Interview Participant 14).

Even if participants themselves were not interested in furthering their education, the educational orientation of Nigerians in Canada was commented on in terms of the opportunities it presents for the next generation: "Yes, Nigerians we love to go to school and study more; the ability to be thinking about more knowledge. In Nigeria, I used to say let my children read the rest" (Registered Practical Nurse, 40–50 years old, Interview Participant 13). Education and social change were also linked to broader forms of social and civic engagement in Canada. Another interviewee said that she now "knows there are different beliefs and rights and [I] take people as they are – LGBT, other religions, food, music; all these are because of migration" (Registered Nurse, 30–40 years old, Interview Participant 4). Another interviewee echoed the notion of having gained from the positive influence of her new country: "I am not just a nurse now, but I am the general manager of my husband's clinic where I am responsible for not just saving lives, but [also managing] staff. Canada gave us an opportunity" (Registered Nurse, 40–50 years old, Interview Participant 12).

Several participants also commented on their educational experiences. A survey respondent said: "Quality educational system played a great role in what I am today. Awareness of cultural diversity, which I would have gained back in Nigeria, but [have] gained [to a] greater extent upon migration" (Registered Nurse, 50–60 years old, Survey Respondent 2).

A key informant also mentioned the following:

> Many of my community members have been able to take advantage of the diverse nature of Canada. Many of them have navigated to another career when they could not find fulfilment in their original career. By so doing, they have overcome the barriers that clog the wheel of progress for many. Migrating to Canada makes many immigrants embrace some of Canadian culture, such as respect for law and order. Most of my community members now have [a] better understanding of their rights and obligations as citizens of a free state. (Key Informant Interview, Participant 3).

The educational benefits respondents gained from being in Canada were apparent in some of the interviews: "I have grown more mature now. I appreciate the good education that I got in Canada" (Registered Nurse, 30–40 years old, Survey Respondent 40).

About 80 per cent of those interviewed in Canada agreed they had gained a new skill or acquired new knowledge in Canada. The results also show that there was an increase in the level of education for Nigerian migrants once they arrived in Canada. These findings are in line with Lowell et al. (2004), who elaborated on how a diaspora creates new knowledge as well as transfers skills. It seems possible that these results are due to the fact that Nigerian's place immense value on education, regardless of where they find themselves. The argument that Nigerian immigrants are a relatively well-educated group is echoed in research from the US and Europe (Bakewell & de Haas, 2007; Joshua et al., 2014). Most of the reasons respondents gave for why they elected for migration were consistent with those cited by other authors (see Buchan et al., 2003; Cabanda, 2015; Clark et al., 2006; Crush & Pendleton, 2010; Freeman et al., 2012; Kingma, 2006; Kline, 2003; Oni, 2000; Walton-Roberts et al., 2017). Some scholars have suggested that emigration sometimes occurs because of "oversupply" or over-education of the population (Pellegrino, 2001), but this does not appear to be the case for this sample since the ratios of health workers to the population in Nigeria indicates there is still a need for more health workers. Rather, the issues are more related to the general standard of living in Nigeria and opportunities for further education and training, something the majority of the sample pursued once in Canada. In some cases, this was determined as the correct strategy even before participants had left Nigeria.

In this study, not all the participants were educated as nurses in Nigeria; rather, they were from several different occupational fields. After arriving in Canada, most of them were advised to join the health field by family members or friends, or they found they needed to go back to school to retrain in order to secure a job in their previous occupation. Bauder (2003) argues that Canada devalues existing immigrant credentials, but this research suggests some nuances to this process. First, an inability to secure a job in their field forces migrants to return to some kind of educational programming. After completing this training, those interviewed for this research indicated that, in their perception, they now possessed greater social capital than prior to their journey to Canada. Since their migration, most participants attributed positive outcomes in their lives to the impact of their move. Participants stated they saw positive changes in their careers, that they are more independent, and that their educational credentials have increased. Secondly, in this case of Nigerian health workers, it appears many used migration explicitly as a route to enhance their professional status. Admittedly, most of these migrants were originally educated in Nigeria and re-educated in Canada to fit Canada's labour market requirements, still their own interpretations of this process were mostly expressed in positive terms related to the overall improvement in their post-migration situation.

Conclusion

This chapter set out to understand the experience of internationally educated health care professionals on a micro level. The study has argued that the Nigerian participants appear to be having positive experiences in Canada, despite the challenges they faced in the process of settlement and integration, compared to the macro-scale analysis provided by the human capital approach. This finding is in line with Blain et al. (2017), who argue that human capital approaches are inadequate in explaining the variable trajectories and experiences of health care labour immigrants. Although some of sample had been economically successful in Nigeria, they migrated for various reasons, often with their families. While in Canada, they went through an adjustment period where they struggled with issues such as school, settlement, family dynamics, lack of a job, lack of Canadian work experience, and licensing issues. Most participants expressed the feeling that although the adjustment was challenging, the most important consequences of their migration were that they increased their education, improved their careers, and felt more independent.

The present study has been one of the first examinations of Nigerian health care workers in Canada. The findings of this study enhance our understanding of the experiences of Nigerian immigrants in the health care field in Canada. They add to the body of literature on Nigerians in Canada, an understudied population in Canada. Overall, this study strengthens the idea that micro-scale analysis of migrant experiences can yield valuable findings useful for policy-makers, health care practitioners, and academics. The findings from this study also differ from most work on the professional integration of immigrants, which tends to focus on the recognition of credentials acquired before migration.

While fewer researchers have examined the process of ongoing educational transition and career choice for newcomers, this study indicates that immigrants see their move as one that will facilitate further (re)training as part of longer-term plans. The implications of these findings suggest that research and resources need to be directed to innovative and flexible forms of education and training for newcomers, who may seek to redirect their careers post-migration.

Efforts at making the professions accessible to immigrants are already in place in Canada (see Jafri, chapter 11 in this volume), and this research suggest that these practices, and wider opportunities for retraining, are creating important opportunities for immigrants to achieve success.

REFERENCES

Aboderin, I. (2007). Contexts, motives and experiences of Nigerian overseas nurses: Understanding links to globalization. *Journal of Clinical Nursing*, *16*(12), 2237–45. https://doi.org/10.1111/j.1365-2702.2007.01999.x.

Adelakun, F.E. (2013). *The journey of foreign educated nurses integrating into the American health care workforce* [Unpublished doctoral dissertation]. Augsburg College.

Astor, A., Akhtar, T., Matallana, M.A., Muthuswamy, V., Olowu, F.A., Tallo, V., & Lie, R.K. (2005). Physician migration: Views from professionals in Colombia, Nigeria, India, Pakistan and the Philippines. *Social Science & Medicine*, *61*(12), 2492–500. https://doi.org/10.1016/j.socscimed.2005.05.003.

Bakewell, O., & de Haas, H. (2007). African migrations: Continuities, discontinuities and recent transformations. In L. de Haan, U. Engel, & P. Chabal (Eds.), *African Alternatives* (pp. 95–117). Brill. https://doi.org/10.1163/ej.9789004161139.i-185.38.

Bauder, H. (2003). "Brain abuse", or the devaluation of immigrant labour in Canada. *Antipode*, *35*(4), 699–717. https://doi.org/10.1046/j.1467-8330.2003.00346.x.

Beckles, H., Perry, A.M., & Whiteley, P. (2002). *The brain train: Quality higher education and Caribbean development*. University of West Indies Press.

Bhagwati, J., & Hamada, K. (1974). The brain drain, international integration of markets for professionals and unemployment: A theoretical analysis. *Journal of Development Economics*, *1*(1), 19–42. https://doi.org/10.1016/0304-3878(74)90020-0.

Blain, M., Fortin, S., & Alvarez, F. (2017). Professional journeys of international medical graduates in Québec: Recognition, uphill battles, or career change. *Journal of International Migration and Integration*, *18*(1), 223–47. https://doi.org/10.1007/s12134-016-0475-z.

Brush, B.L., Sochalski, J., & Berger, A.M. (2004). Imported care: Recruiting foreign nurses to U.S. health care facilities. *Health Affairs*, *23*(3), 78–87. https://doi.org/10.1377/hlthaff.23.3.78.

Buchan, J., Sochalski, J., & Parkin, T. (2003). *International nurse mobility: Trends and policy implications*. World Health Organization. http://apps.who.int/iris/bitstream/handle/10665/68061/WHO_EIP_OSD_2003.3.pdf.

Cabanda, E. (2015). Identifying the role of the sending state in the emigration of health professionals: A review of the empirical literature. *Migration and Development*, *6*(2), 215–31. https://doi.org/10.1080/21632324.2015.1123838.

Ciumasu, I. M. (2010). Turning brain drain into brain networking. *Science and Public Policy*, *37*(2), 135–46. https://doi.org/10.3152/030234210X489572.

Clark, P.F., Stewart, J.B., & Clark, D.A. (2006). The globalization of the labour market for health-care professionals. *International Labour Review*, *145*(1/2), 37–64. https://doi.org/10.1111/j.1564-913X.2006.tb00009.x.

Clemens, M.A., & Pettersson, G. (2008). New data on African health professionals abroad. *Human Resources for Health*, *6*(1), 1–11. https://doi.org/10.2139/ssrn.983116.

College of Nurses of Ontario. (2014). *New members of the general class 2013*. https://www.cno.org/globalassets/docs/general/43011_trendsnewmembers.pdf.

College of Nurses of Ontario. (2016). *Applicant statistics report*. http://www.cno.org/globalassets/2-howweprotectthepublic/statistical-reports/applicant-statistics-report-2016.pdf

Couton, P. (2013). The impact of communal organizational density on the labour market integration of immigrants in Canada. *International Migration, 51*(1), 92–114. https://doi.org/10.1111/j.1468-2435.2010.00673.x.

Creswell, J.W. (1994). *Research design: Qualitative and quantitative approaches.* Sage Publications.

Crush, J., & Pendleton, W. (2010). Brain flight: The exodus of health professionals from South Africa. *International Journal of Migration, Health and Social Care, 6*(3), 3–18. https://doi.org/10.5042/ijmhsc.2011.0059.

de Haas, H. (2010). Migration and development: A theoretical perspective. *International Migration Review, 44*(1), 227–64. https://doi.org/10.1111/j.1747-7379.2009.00804.x.

de Haas, H. (2012). The migration and development pendulum: A critical view on research and policy. *International Migration, 50*(3), 8–25. doi: 10.1111/j.1468-2435.2012.00755.x.

de Haas, H., & Vezzoli, S. (2013). Migration and development on the South–North frontier: A comparison of the Mexico–US and Morocco–EU cases. *Journal of Ethnic and Migration Studies, 39*(7), 1041–65. https://doi.org/10.1080/1369183X.2013.778019.

El-Khawas, M.A. (2004). Brain drain: Putting Africa between a rock and a hard place. *Mediterranean Quarterly, 15*(4), 37–56. https://doi.org/10.1215/10474552-15-4-37.

Ezeonwu, M.C. (2013). Nursing education and workforce development: Implications for maternal health in Anambra State, Nigeria. *International Journal of Nursing and Midwifery, 5*(3), 35–45. https://doi.org/10.5897/IJNM12.014.

Faist, T. (2008). Migrants as transnational development agents: An inquiry into the newest round of the migration–development nexus. *Population, Space and Place, 14*(1), 21–42. https://doi.org/10.1002/psp.471.

Fajobi, T.A., Olatujoye, O.O., Amusa, O.I., & Adedoyin, A. (2017). Challenges of apprenticeship development and youths unemployment in Nigeria. *Journal of Sociology and Criminology, 5*(2), 1–8. https://doi.org/10.4172/2375-4435.1000172.

Freeman, M., Baumann, A., Blythe, J., Fisher, A., & Akhtar-Danesh, N. (2012). Migration: A concept analysis from a nursing perspective. *Journal of Advanced Nursing, 68*(5), 1176–86. https://doi.org/10.1111/j.1365-2648.2011.05858.x.

Guilmoto, C.Z., & Sandron, F. (2001). The internal dynamics of migration networks in developing countries. *Population: An English Selection, 13*(2), 135–64. https://horizon.documentation.ird.fr/exl-doc/pleins_textes/divers17-02/010047710.pdf.

Harvey, W.S. (2008). Brain circulation? *Asian Population Studies, 4*(3), 293–309. https://doi.org/10.1080/17441730802496516.

International Organization for Migration. (2013). *World Migration Report 2013: Migrant Wellbeing and Development.* United Nations Publications.

James, C.E. (2009). African-Caribbean Canadians working "harder" to attain their immigrant dreams: Context, strategies, and consequences. *Wadabagei: A Journal of the Caribbean and Its Diaspora, 12*(1), 92–108.

Joshua, S., Olanrewaju, I.P., & Ebiri, O. (2014). Leadership, brain drain and human capacity building in Africa: The Nigerian experience. *Research Journal in Organizational Psychology and Educational Studies, 3*(4), 283–90.

Kalipeni, E., Semu, L.L., & Mbilizi, M.A. (2012). The brain drain of health care professionals from sub-Saharan Africa: A geographic perspective. *Progress in Development Studies, 12*(2–3), 153–71. https://doi.org/10.1177/146499341101200305.

Kingma, M. (2006). *Nurses on the move: Migration and the global health care economy.* Cornell University Press.

Kline, D. (2003). Push and pull factors in international nurse migration. *Journal of Nursing Scholarship, 35*(2), 107–11. https://doi.org/10.1111/j.1547-5069.2003.00107.x.

Koser, K., & Salt, J. (1997). The geography of highly skilled international migration. *International Journal of Population Geography, 3*(4), 285–303. https://doi.org/10.1002/(SICI)1099-1220(199712)3:4<285::AID-IJPG72>3.0.CO;2-W.

Ku, J., Bhuyan, R., Sakamoto, I., Jeyapal, D., & Fang, L. (2019). "Canadian Experience" discourse and anti-racialism in a "post-racial" society. *Ethnic and Racial Studies, 42*(2), 291–310. https://doi.org/10.1080/01419870.2018.1432872.

Liebig, T. (2003). Migration theory from a supply-side perspective. Discussion Paper No. 92. Research Institute for Labour Economics and Labour Law, University of St. Gallen, Switzerland.

Lowell, L., Findlay, A., & Steward, E. (2004). Brain strain: Optimizing highly-skilled labour from developing countries. *Asylum and Migration Working Paper 3.* Institute for Public Policy Research, London. https://www.ippr.org/files/images/media/files/publication/2011/05/brainstrain_1365.pdf.

Lucas, R.E. (1988). On the mechanics of economic development. *Journal of Monetary Economics, 22*(3), 3–42. https://doi.org/10.1016/0304-3932(88)90168-7.

Nursing and Midwifery Council of Nigeria. (n.d.). Nursing Professionals across Nigeria. Retrieved 12 July 2021, from https://portal.nmcn.gov.ng/.

Oni, B. (2000, February). *Capacity building effort and brain drain in Nigerian Universities* [Conference presentation]. ECA Regional Conference on Brain Drain and Capacity Building in Africa, Addis Ababa, Ethiopia.

Papademetriou, D.G. (1985). Illusions and reality in international migration: Migration and development in post–World War II Greece. *International Migration, 23*(2), 211–23. https://doi.org/10.1111/j.1468-2435.1985.tb00316.x.

Pellegrino, A. (2001). Trends in Latin American skilled migration: "Brain drain" or "brain exchange"? *International Migration, 39*(5), 111–32. https://doi.org/10.1111/1468-2435.00174.

Remennick, L. (2013). Transnational lifestyles among Russian Israelis: A follow up study. *Global Networks, 13*(4), 478–97. https://doi.org/10.1111/glob.12033.

Salami, B., Dada, F.O., & Adelakun, F.E. (2016). Human resources for health challenges in Nigeria and nurse migration. *Policy, Politics, & Nursing Practice, 17*(2), 76–84. https://doi.org/10.1177/1527154416656942.

Samers, M. (1998). "Structured coherence": Immigration, racism and production in the Paris car industry. *European Planning Studies*, *6*(1), 49–72. https://doi.org/10.1080/09654319808720445.

Saxenian, A. (2005). From brain drain to brain circulation: Transnational communities and regional upgrading in India and China. *Studies in Comparative International Development*, *40*(2), 35–61. https://doi.org/10.1007/BF02686293.

Spoonley, P., Bedford, R., & Macpherson, C. (2003). Divided loyalties and fractured sovereignty: Transnationalism and the nation-state in Aotearoa/New Zealand. *Journal of Ethnic and Migration Studies*, *29*(1), 27–46. https://doi.org/10.1080/1369183032000076704.

Statistics Canada. (2016). Census Profile, 2016 Census. http://www12.statcan.gc.ca/census-recensement/2016/dp-pd/prof/index.cfm?Lang=E.

Teal, F. (2016). Are apprenticeships beneficial in sub-Saharan Africa? *IZA World of Labor*, *268*. https://doi.org/10.15185/izawol.268.

Varma, R., & Kapur, D. (2013). Comparative analysis of brain drain, brain circulation and brain retain: A case study of Indian institutes of technology. *Journal of Comparative Policy Analysis: Research and Practice*, *15*(4), 315–30. https://doi.org/10.1080/13876988.2013.810376.

Vinokur, A. (2006). Brain migration revisited. *Globalisation Societies and Education*, *4*(1), 7–24. https://doi.org/10.1080/14767720600554957.

Vujicic, M., Zurn, P., Diallo, K., Adams, O., & Dal Poz, M.R. (2004). The role of wages in the migration of health care professionals from developing countries. *Human Resources for Health*, *2*(3). https://doi.org/10.1186/1478-4491-2-3.

Walton-Roberts, M., Runnels, V., Rajan, I.S., Sood, A., Nair, S., Thomas, P., Packer, C., Labonte, R., Bourgeault, I., Tomblin-Murphy, G., & Mackenzie, A. (2017). Causes, consequences and policy responses to the migration of health workers: Key findings from India. *Human Resources for Health*, *15*, 28. https://doi.org/10.1186/s12960-017-0199-y.

Williams, A.M., & Baláž, V. (2003). What human capital, which migrants? Returned skilled migration to Slovakia from the UK. *International Migration Review*, *39*(2), 439–68. https://doi.org/10.1111/j.1747-7379.2005.tb00273.x.

World Health Organization. (2017). Global Health Workforce statistics database. Geneva. Retrieved 27 September 2021, from https://www.who.int/data/gho/data/themes/topics/health-workforce.

16 Peripatetic Physicians: Rewriting the South African Brain Drain Narrative

JONATHAN CRUSH

Introduction

A recent review of studies of health professional migration noted that "the opportunities of health workers to seek employment abroad has led to a complex migration pattern, characterized by a flow of health professionals from low- to high-income countries" (Aluttis et al., 2014). As a result, "recruiting doctors and nurses from a LMIC [low- to middle-income country] to serve the demand in HICs [high-income countries] effectively creates a shortage in the country of origin, and hence contributes to worse health outcomes." Like most of the literature, the premise presented here is that health professional migration is a zero-sum game: those who migrate are a permanent, and costly, loss to the country of origin and a permanent, and valuable, gain for the destination country. This "brain drain" narrative has a powerful hold on the way in which the migration of doctors is conceptualized and its impacts understood. National and international policy responses are similarly premised on the notion that all forms of cross-border migration are advantageous to destination countries in the Global North and damaging to health outcomes in origin countries in the Global South.

South Africa is often seen in the migration literature as an archetypal African medical brain-drain story (Bhargava & Docquier, 2008; Joudrey & Robson, 2010; Kaplan & Höppli, 2017; Oberoi & Lin, 2006; Özden & Philips, 2015). Although the idea of a brain drain in specialties such as surgery is contested (Hutch et al., 2017; Liu et al., 2015), the country's brain drain narrative has several core elements. The story begins with statistics collected by destination countries and collated by organizations such as the Organisation for Economic Co-operation and Development (OECD) (see, for example, OECD, 2016). These show the large number of South Africa–trained doctors who have immigrated to countries such as Australia, Canada, the UK, and the US, as well as the savings these countries accrue by hiring doctors trained elsewhere (Mills et al., 2011).

Second, the major drivers of the brain drain are seen as push factors (Arnold & Lewinsohn, 2010; Bezuidenhout et al., 2009; Crush et al., 2012; Joudrey & Robson, 2010). The departure of new and established medical graduates of South Africa's medical schools has been labelled "brain flight," "brain haemorrhage," and a "crisis" for the country's health system (Crush & Pendleton, 2010; Labonté et al., 2015). Extreme dissatisfaction with workplace and living conditions in South Africa is seen as the primary reason for very high levels of emigration (Crush & Pendleton, 2012; Crush et al., 2014; de Vries et al., 2010; George & Reardon, 2013; George et al., 2013; Pendleton et al., 2007).

Third, surveys of physicians and medical students that ask questions about emigration potential and actions to facilitate departure generally assume that this will be permanent. For example, a 2013 survey of South African doctors found that 40 per cent had given emigration a great deal of thought (Crush et al., 2014). Some 29 per cent said they were likely to leave within two years and 50 per cent within five years, while 51 per cent had registered or written licensing examinations with an overseas professional body. Surveys of medical students tell a similar story. A survey of 800 final-year students at eight South African medical schools found that 55 per cent planned to work abroad after graduation (de Vries et al., 2010). Another survey of 260 students at three South African medical schools found that three quarters had given some or a great deal of consideration to leaving South Africa; moreover, a quarter said they would likely leave within two years of qualification and 58 per cent that it was likely within five years (George & Reardon, 2013). Finally, a 2011 survey of health sciences students at three South African universities found that just over half intended to work in another country (Naidu et al., 2013). Thus, a large number of doctors and medical students are primed to leave and the brain drain will continue.

The fourth element of the brain drain narrative is that policy measures and retention strategies to mitigate the brain drain have been unsuccessful (George & Rhodes, 2012). The adoption of the World Health Organization (WHO) Global Code of Practice on the International Recruitment of Health Personnel was meant to slow the brain drain but has had little obvious impact on the outflow of doctors from Africa (Tankwachi et al., 2015). As Crush and Chikanda (2018) conclude "the current suite of retention strategies are either failing and/or targeting the wrong factors" (p. 354). Government policies are actually seen as a major driver of the brain drain. A new National Health Insurance scheme will supposedly cause major changes in private and public health care delivery in South Africa (Breakfast, 2018; Hofman et al., 2015; Naidoo, 2012). The medical brain drain will therefore only accelerate with a scheme that is universally unpopular amongst doctors.

The final element of the brain-drain narrative is that South Africans who have emigrated are a "disengaged diaspora." Crush et al.'s (2012) study of the medical diaspora in Canada found that, despite the perpetuation of a strong South African cultural identity, South Africa is seen as a racial dystopia. There is minimal interest in supporting development initiatives, relatively low levels and volumes of remitting, and a declining propensity to remit over time. Surveys of South African doctors abroad have shown they have very little inclination to return to their country of origin and engage in classic forms of "return migration" (Crush et al., 2012).

In the new world of transnationalism, a global skills market, and greatly increased mobility by health professionals, it is unlikely that the traditional, uni-directional, permanent settlement model of the brain drain narrative adequately captures all forms of health professional migration (see in this volume: Chikanda, chapter 8; Connell and Negin, chapter 12; Adekola, chapter 15). This chapter re-examines the narrative and the evidence for more complex forms of mobility of South African–trained doctors. To do this, it is first important to examine the opportunities that exist within destination countries for temporary employment. The following sections turn to the South African case, beginning with a description of the data on which the chapter is based. Then the findings on temporary migration are discussed. The final section returns to the issue of the brain drain narrative by asking if there are differences between the emigration potential of South African doctors who have worked outside the country and those who have not.

Temporary Physician Migration: A Review

Most the research on the drivers of temporary migration focuses on lower-skilled workers or, in the health professions, on nurses (Castles & Ozkul, 2014; Howe & Owens, 2016; Hugo, 2009; Kingma, 2006; Prescott & Nichter, 2014). Global data on the temporary migration of doctors is non-existent, which makes it difficult to understand this neglected aspect of Aluttis et al.'s (2014) "complex migration pattern" and to contest the dominant narrative that all doctor migration is permanent and therefore, by definition, harmful to countries of origin. The reasons for the growth in temporary employment opportunities are also relatively unexplored. Certainly, a prime reason is the growing shortage of physicians in many Western countries as demands on public health care systems increase with aging populations. A 2014 survey of OECD countries found that thirty-three of thirty-four countries had issues with doctor supply and distribution (Ono et al., 2014). As Gauld and Horsburgh (2015) conclude, "a majority of these countries face a range of medical workforce challenges with questions over how they will meet increasing public demand for health care, maintain a sufficient number of general practitioners, fill shortages in particular hospital specialties and, very importantly, ensure an even distribution of doctors across the population."

In addition to the mismatch between local-trained supply and demand, remote and under-serviced areas face particular challenges in attracting and retaining local physicians. Physician shortages are also exacerbated by the emigration of locally trained doctors to other countries. A recent study of doctors in New Zealand, for example, found that 44 per cent of doctors were international medical graduates, and that half of those were trained in the UK (Gauld & Horsburgh, 2015). Over two thirds of the UK-trained doctors surveyed cited better quality of life, better working conditions, better career opportunities, and a desire to leave the UK's National Health Service (NHS) as the most important reasons for leaving the UK. Other studies of physician out-migration have been conducted in New Zealand (Sharma et al., 2012), as well as Germany (Ognyanova et al., 2014; Pantenburg et al., 2018), Ireland (Humphries et al., 2015a, 2017), and amongst potential emigres in the UK (Lambert et al., 2018). In the case of Ireland, there have been "substantial levels of emigration," and a survey of Irish-trained doctors in other countries likened working in Ireland to a "slow death" (Humphries et al., 2015a).

Temporary migration for employment is often contingent on more general immigration policies for skilled migrants and the requirements of licensing and registration bodies. Humphries et al. (2013, 2015b) developed a typology of temporary migration and identified seven categories of physician migrants in Ireland: (1) career-oriented migrants, who migrate to obtain postgraduate medical training and are most likely to leave; (2) livelihood migrants who are motivated by greater earning potential in Ireland; (3) backpacker migrants, who see short-term mobility as a chance to experience other countries and health systems; (4) commuter migrants, who return regularly for short periods or repeated time-limited stretches; (5) potential returner migrants, who are employed in Ireland but intend to return home at some point; (6) family migrants, who accompany spouses on a spousal visa; and (7) safety and security migrants, who are primarily refugees.

Research in South Africa suggests that there are examples of all seven types of temporary South African physician migrant. While there are few, if any, examples of South African physicians obtaining refugee status in other countries, safety and security concerns are a strong motivator of out-migration. Ireland itself has become an increasingly important destination for South African doctors leaving permanently and temporarily. However, the World Health Organization (n.d.) reports that 20 per cent of doctors who qualified as medical doctors in South Africa and are registered in Ireland only practise in that country. This raises the obvious question about where the other 80 per cent might be practising instead of or in addition to Ireland. Ireland is a stepping-stone to other destinations and experiences its own brain drain, which explains some of this discrepancy. But there are some doctors registered in Ireland who are now back in South Africa or are practising in both countries at different times.

Raghuram (2008, p. 191) argues that the UK's NHS has from its inception been embedded in transnational medical training and shows how opportunities for non-UK trained doctors to work in the UK radically changed in 2006 by moving doctors into the 5-tier points-based system and abolishing the permit-free training category that had been in place since the 1980s. In recent years, severe staff shortages have opened up new opportunities for temporary work in the NHS through "locumization" (General Medical Council, 2018). Locum doctors now constitute 31,500 or 18 per cent of all NHS physicians, a one-third increase since 2013. Around 20 per cent of international medical graduates work as locums, either via locum agencies or being employed directly by the NHS. There is considerable controversy over whether the growing reliance on locums saves money for the NHS (the consensus is that it does not). Murray (2017) suggests two reasons for locumization: short term flexibility and long term recruitment problems. "The answer to short term staffing problems is usually not more permanent staff, and the flexibility offered by locums and agencies can make financial and operational sense" (p. 525). Underlying this are "deep seated staff shortages," which have become more acute over time. Chronic physician shortages in the NHS led the UK government to lift its cap on the issuance of "tier 2" visas for doctors in 2018 (Rimmer, 2018). The OECD puts the total number of South African–trained doctors in the UK at less than 2,000, while the General Medical Council has over 7,000 South African medical graduates registered to practise. This discrepancy suggests that there are more South African doctors registered than are actually practising and that some are undoubtedly doing regular or occasional locums.

A recent report on doctor migration to Australia distinguishes permanent and temporary health professionals who came to the country to work between 2005/6 and 2009/10 (Hawthorne, 2012). As Hawthorne (2012) notes, the temporary pathway is highly attractive to governments and employers given the potential to prescribe foreign doctors' location as a condition of visa entry, allowing them to work for up to four years at undersupplied sites. As table 16.1 shows, temporary workers (primarily on 457 visas) outnumbered permanent immigrants (by 71 per cent to 29 per cent). Of the 34,870 temporary migrants, 44 per cent (or 15,342) were doctors. A further 2,420 temporary visas were awarded in 2010–11: 1,190 for general medical practitioners and 1,230 for resident (house) medical officers. Most of the permanent and temporary South African health professionals are probably doctors. In which case, 78 per cent were temporary migrants and 22 per cent were permanent immigrants. Some would undoubtedly use the 457 visas as a pathway to permanence but, equally, some would have returned to South Africa.

In Canada between 2010 and 2017, 4,161 specialists and 2,709 general practitioners entered the country on temporary work permits under the country's Temporary Foreign Work Program (TFWP) (table 16.2). Campbell-Page et al.

Table 16.1. Top source countries for permanent and temporary health professionals in Australia, 2005–2006 to 2009–2010

Permanent		Temporary	
Country	No.	Country	No.
UK	4,120	UK	9,350
India	1,510	India	6,420
Malaysia	1,300	Philippines	1,850
China	970	South Africa	1,770
Philippines	510	Malaysia	1,570
South Africa	500	Ireland	1,560
South Korea	480	China	1,380
Egypt	420	Zimbabwe	1,180
Singapore	390	Canada	950
Ireland	350	United States	830
Total all sources	13,880	Total all sources	34,870

Source: Hawthorne (2012).

(2013) note that instead of immediately immigrating, international medical graduates (IMGs) often start with a temporary work permit. However, "because it can frequently be acquired more quickly than permanent residency, it is often used as a bridge towards permanent residency, allowing IMGs to begin their Canadian employment" (p. 3). Their assumption is that IMGs are "cutting the immigration queue" by using temporary status to gain permanent residence. But this may not be the motive at all for career-oriented migrants who genuinely want work experience in Canada before returning home.

In Canada, Lu and Hou (2017) show that within five years of receiving their first work permits, only 9 per cent of all temporary foreign workers who arrived between 1995 and 1999 became permanent residents. The level increased to 13 per cent for arrivals between 2000 and 2004, and 21 per cent for arrivals between 2005 and 2009. The transition rate was not much higher for higher-skilled than lower-skilled temporary workers. This suggests that the majority of skilled temporary workers do not or cannot transition to permanent residence. Canada's TFWP has certainly been used by a number of doctors and their Canadian employers in recent years, although this seems to have tailed off (see table 16.2). Unfortunately, Lu and Hou (2017) do not show the rate of transition for particular countries of origin or professions.

Most Canadian provinces also have provisions for IMGs to work temporarily without being fully licensed. Provisional licenses allow IMGs to practise without passing Medical Council of Canada examinations or completing the requisite Canadian postgraduate medical training. These licenses are given different names in different provinces: "public service," "restricted," "defined," "conditional," or "temporary." They are usually given to IMGs willing to work

Table 16.2. Doctors working temporarily in Canada under TFWP

Year	Specialist physicians	General practitioners and family physicians
2010	1,870	440
2011	516	461
2012	531	470
2013	415	384
2014	237	408
2015	207	265
2016	198	187
2017	187	94
Total	4,161	2,709

Source: Data retrieved from Employment and Social Development Canada (2017).

in under-serviced communities (Audas et al., 2005). In some provinces, such as Newfoundland, there are more provisionally licensed than fully licensed doctors. Audas et al. (2005) found that South Africa was the major source country for provisionally licensed IMGs in at least five provinces (Alberta, British Columbia, Manitoba, Newfoundland, and Saskatchewan). Such provisional licenses were also issued to South African IMGs in Ontario and PEI. While provisional licensing may be a step on the road to permanent immigration for IMGs who can fulfil the terms and conditions for full licensing, many IMGs in Canada are not able to get residencies and fulfil the other criteria for licensing (Mathews et al., 2017a; Neiterman et al., 2017). For others, provisional licensing is an opportunity to acquire the benefits of temporary migration from countries like South Africa.

Another example of temporary doctor migration is to take up residencies and fellowships in other countries before returning home. The International Medical Graduate Training Initiative in Ireland is one example of an initiative in which qualified overseas postgraduate medical trainees undertake a fixed period of active training in clinical services in Ireland. Likewise, Canadian teaching hospitals have around 2,000 so-called visa trainees in residencies and fellowships. At least half are occupied by beneficiaries of the Saudi Postgraduate Medical Program, which has expanded dramatically in recent years as Canadian medical schools derive significant financial benefit from the arrangement (Mathews & Bourgeault, 2018). A recent diplomatic dispute between Canada and Saudi Arabia threatened the future of the program, but it now looks set to continue (Vogel, 2018). A recent study of where visa trainees were working after completing their training found that only 24 per cent were in Canada two years later (Mathews et al., 2017b). South Africa has no equivalent programs for its medical graduates with Canada or other countries, although agreements exist with Cuba at the medical degree level (Hammett, 2014) and for UK medical graduates to get training and experience in South Africa (Connor et al., 2014; Kong et al., 2015; Reardon et al., 2015).

Increasing numbers of skilled South Africans, including doctors, have also moved temporarily to the United Arab Emirates (UAE) and Saudi Arabia to work. Migration is motivated primarily by the opportunities for earning significantly more than in South Africa. A survey of skilled South Africans working in the UAE showed that few, if any, intend to remain there permanently (Fourie, 2006). They go home frequently for visits and most intend to return permanently after several years. Some intend to move on to other countries. The high-profile Karabus case – involving the airport transit arrest and trial in Abu Dhabi of South African paediatric oncologist Cyril Karabus on manslaughter charges from 2002 – may dampen doctor migration to the Gulf in the future (Davis, 2012). However, this population of circular migrants represents another migration stream that is not consistent with the medical brain drain narrative of permanent departure and disengagement.

Methodology

The remainder of this chapter is based on both quantitative and qualitative research. The Southern African Migration Programme (SAMP) has conducted two large-scale online surveys of South African–trained physicians working in South Africa in 2007 and 2014. The survey was completed by 745 doctors in 2007 and 860 doctors in 2014. Because this was not a random sample but a self-selected group of respondents, the responses may not be representative of the profession as a whole. Nevertheless, these are still the largest migration surveys of South African doctors in the country and provide valuable insights into the extent of temporary migration. The detailed survey findings are discussed in Pendleton et al. (2007) and Crush et al. (2014). The 2007 survey found that that more than one third (35 per cent) of the respondents had worked outside of South Africa. In the 2014 survey, this figure was closer to one half. The 2014 survey incorporated additional questions to probe further on the issue of temporary migration and return. These findings were supplemented with answers to open-ended questions at the end of the 2014 survey and key informant interviews with South African doctors in Canada and South Africa in 2017 and 2018.

Mobile Medical Migrants

Some 49 per cent of the South African doctors who responded to the 2014 online survey had work experience in at least one other country, 15 per cent had worked in at least two other countries, and one physician had worked in as many as seven. Table 16.3 shows the overwhelming importance of the UK as a site of employment outside of Africa, with 56 per cent of the South African

Table 16.3. Other countries of employment of South African doctors

	1st country no.	Total %	2nd country no.	Total %
United Kingdom	231	55.8	29	7.0
Canada	35	8.5	8	1.9
Ireland	28	6.8	9	2.2
Australia	15	3.6	10	2.4
New Zealand	13	3.1	12	2.9
United States	12	2.9	9	2.2
Gulf States	11	2.7	8	1.9
Asia	6	1.4	6	1.4
Other Europe	25	6.0	14	3.4
Other Africa	33	8.0	15	3.6
Other	5	1.2	6	1.4
Total	414	100.0	126	30.3

Table 16.4. Period spent working in another country

Period worked	1st country (%)	2nd country (%)
< 1 year	35.3	63.6
1–3 years	46.7	24.6
4–9 years	15.6	10.3
>10 years	2.6	1.6
Total	100.0	100.0

doctors having work experience there as place of employment. Canada was second at 10 per cent, followed by other European countries combined (including Germany, the Netherlands, and Belgium) at 9 per cent, Ireland (9 per cent), and Australia and New Zealand (both 6 per cent). Around 5 per cent of the South African doctors had worked in newer destinations such as the UAE and Saudi Arabia. Nearly 90 per cent had worked in their first overseas country for less than four years (see table 16.4). Those who had worked in a second country had been there for even less time (64 per cent less than one year versus 35 per cent of first country respondents). The vast majority of doctors who had gone overseas thus did so temporarily and for a limited time.

The temporary nature of migration is reflected in the doctors' responses to the question of why they came back to South Africa. Respondents were asked to rate the importance of various push and pull factors (table 16.5). As many as 61 per cent said they had returned because their job overseas was temporary in nature and 64 per cent because they had a permanent position to go to in South Africa. Other important factors (which help explain why they did not permanently stay abroad) included family who wanted to return and a desire to be closer to family still in South Africa (45 per cent push and 70 per cent pull), environmental

Table 16.5. Main reasons for return to South Africa (% important/very important)

Push factors in destination countries	
Job was only temporary	61.2
Family wanted to return	44.8
Inhospitable climate	41.3
Poor social life	38.7
No job satisfaction	35.3
Poor prospects for professional advancement	27.1
No job security	22.5
Insufficient remuneration	18.2
Inadequate benefits	17.9
Patients too demanding	16.5
Workload too heavy	11.0
Pull factors in South Africa	
Lifestyle preferable	77.1
South African culture	73.9
Closer to family	69.9
Permanent job in South Africa	64.0
Better social life	60.4
Greater job satisfaction	51.8
South African landscape/climate	51.3
Use skills to serve underprivileged	49.0
Poor prospects for professional advancement	39.7
Better education for children	37.0
Better access to medical care	34.7
Cost of living lower	34.3
Better job security	33.5
Better remuneration	29.7
Paid off medical school debts	26.4
Better benefits	26.3
Patients less demanding	18.1
Lighter workload	12.5

factors such as climate (41 per cent push and 53 per cent pull), and social life (39 per cent push and 60 per cent pull). Other non-economic pull factors included a preferable lifestyle (77 per cent) and South African culture (74 per cent). A desire to use their skills to serve the underprivileged was cited by nearly half of the respondents. In general, quality of life factors were rated more highly than work-related reasons such as remuneration, job satisfaction, benefits, and workload. Overall, pull factors were more important than push factors.

Complementary in-depth insights come from the doctors' unsolicited responses at the end of the survey. For example:

> I worked in England just after my house doctor year for six months in order to make some money to travel. At the time, we did not need much to be working there. Lately it is more difficult to just go over and work. The South African skills

> are very practically orientated and we are hard workers. That is why we are in demand all over the world. (Respondent 1)

> I trained here and went abroad to earn an income that would allow me to continue to work in the public sector on my return without being saddled with a huge mortgage. What was supposed to be a two-year commitment abroad was extended as house prices sky-rocketed and then the lure of obtaining a British passport. There had never been an intention to permanently emigrate. (Respondent 2)

> I was very happy living and working in the UK. I returned simply because I had met my future husband on a visit back home. I got an "equivalent" post in SA to the one I'd left behind in the UK, but can assure you that all similarities were on paper only. Working in the government health sector in this country is unpleasant. (Respondent 3)

> I was on a contract for two years and after extending my contract, I stayed for another nine months and returned home mainly to enrol my son at a South African school in the Western Cape. (Respondent 4)

What motivates South African doctors to work for periods of time outside the country? de Vries et al.'s (2010) survey of final-year students at South African medical schools found that of the 55 per cent who planned to work abroad, three quarters wanted to do so temporarily. One motivation is to get training and qualifications in a medical specialization through residencies and fellowships. In most countries to which young South African doctors are attracted, obtaining residencies is difficult since most positions are reserved for locally trained doctors or for sponsorship agreements with other countries. For example, in 2017–18, only 14 of the over 2,000 "visa trainee" residencies and fellowships in Canada for foreign-trained doctors were occupied by South Africans (Canadian Post-M.D. Education Registry, 2018). Access to UK training program posts is somewhat easier, particularly for South Africans with British connections who are able to avoid visa requirements and take out British citizenship. In de Vries et al.'s survey, a total of fifty-three doctors (or 10 per cent of the total who had worked overseas) had their highest qualification from an institution in another country (primarily the UK).

Another route for South African doctors wishing to work abroad is to take up temporary employment in another country, particularly via locums. In the UK, the NHS offers locums at £18–£25 per hour before tax, but hours are limited to fifty-six per week. A study by Rogerson and Crush (2008) on the recruiting of South African doctors located several South African–based locum agencies recruiting for the UK and elsewhere. In Canada, the locum route has traditionally been a popular option for newly qualified South African graduates. This is certainly not confined to new doctors as there are also locum opportunities for specialists. Locum recruiting companies in Canada typically advertise internationally as follows:

> Most Canadian physicians have time off during the year, whether it be for annual leave or vacation, and a very common practice in Canada is to Locum. If you are an International Medical Graduate (IMG), and would like increase your Medical experience in a Canadian environment this would be a financially and professionally rewarding route to take. You may even be considering a full-time job in Canada, and locums would be a good path to pursue in order to gauge if Canada is somewhere, you'd be comfortable in working. Whatever your specialty is, often times there is a demand. You may be a general surgeon, Internist, or Psychiatrist, it doesn't matter – the need is there. (Physician Canada, n.d.)

Data on locums is unavailable, but Saskatchewan and Alberta are major destinations. In these provinces, South African physicians made up 18 per cent and 8 per cent respectively of the total physician workforce in 2009 (Joudrey & Robson, 2010). These doctors provide locum opportunities for South African doctors within their networks. However, locum migration is not confined to these two provinces. One South African anaesthetist interviewed had done locums in Toronto and Vancouver. Some doctors had done locums in more than one country.

Responses from South African locum doctors show that they tend to be career-oriented, livelihood, or backpacker migrants. For example:

> I have worked in the UK, Brunei, Oman, the Emirates and New Zealand doing locums over the years and building up a nest egg abroad. (Respondent 5)

> I have been overseas (UK) to locum a number of times, to supplement income and to be able to holiday abroad. I have never had any intention of emigrating. (Respondent 6)

> I have lived almost for my entire life in South Africa, but have worked for short periods overseas. At this stage of my life I would love to work and travel in a different country for the purpose of not only treating medical conditions but also to experience different cultures and to delve into the history of that country. (Respondent 7)

> When I work overseas, I still work in Ireland for two months every eighteen months for money and a break. (Respondent 8)

During the course of the research, other kinds of temporary or short-term working arrangements also emerged. In one case, four South African general practitioners in practice in a major South African city also had a practice in the UK. Over the course of a year, each spends three months in London. Another doctor in Canada had dual citizenship and spent six months of the year in South Africa and six months working in a walk-in clinic in Alberta.

Other opportunities for work placements exist in the UK in private facilities. There are an estimated 548 private hospitals and between 500 and 600 private clinics offering a range of services (Mossialos et al., 2015, p. 52). Most are required to have a Resident Medical Officer (RMO) on site twenty-four hours a day to cover for cardiac arrests on behalf of consultants. The hospitals vary in size, medical complexity, and workload, but a typical private hospital will handle elective surgery. Only larger hospitals have intensive care units or intensive therapy units, and most of the load is ward work and managing post-operative complications. Cape Medical Services in Worcester is run by a South African doctor who organizes 6–12 month placements for South Africans in private facilities (see Cape Medical Services, n.d.). Most are contract positions as RMOs in acute care facilities. They usually work 168-hour shifts, and are required to remain on the hospital site at all times. The RMOs cover emergencies and general ward work, and some hospitals require RMOs to assist in theatre. As well as a salary, they receive free flights, onsite accommodation, General Medical Council registration, and advanced life support training courses. When off duty, they can undertake locums in the NHS.

South Africa's two largest private hospital companies have expanded overseas over the last decade, especially to the UK and the Gulf. Netcare, for example, owns fifty-seven acute care private hospitals in the UK with 2,788 registered beds and 188 operating theatres. In 2015, Mediclinic purchased a 29 per cent share in Spire, one of the UK's largest private health care providers. It is likely that expansion has opened up new opportunities for South African doctors who work for the two companies, and the survey showed that the UAE and Saudi Arabia have become desirable destinations for South African specialists.

Return and Remain?

The final question is how content temporary South African medical migrants are to continue working and living in South Africa on return. Does the opportunity to work in other countries make these medical migrants more or less satisfied with living and working conditions at home? The baseline for comparison are doctors who have never worked outside of the country (non-migrants). The major survey finding is that there is not much difference in the attitudes and perceptions of the two groups.

With regard to working conditions, migrants have higher satisfaction levels on twelve of the eighteen indicators but the differences are very small (4 per cent at most) and do not fit a discernible pattern that could be related to their experience abroad (see table 16.6). The only indicator with a higher spread was income, where migrants are more satisfied (by 7 per cent), possibly because of the opportunities to supplement local salaries with working abroad. Migrants

Table 16.6. Comparison of satisfaction with working conditions (% satisfied/very satisfied)

	Migrants	Non-migrants	% Difference
Appropriateness of training	83	80	+3
Relationship with colleagues	82	84	–2
Ability to find desirable job	68	67	+1
Workplace infrastructure	63	59	+4
Job security	61	63	–2
Workplace resources to do job	61	57	+4
Workplace morale	55	51	+4
Personal security in the workplace	49	46	+3
Prospects for professional advancement	43	47	–4
Workload	44	42	+2
Level of income	42	35	+7
Further educational/career opportunities	38	37	+1
Risk of contracting Hepatitis B	29	27	+2
Risk of contracting HIV/AIDS	28	27	+1
Relationship with management	29	30	–1
Risk of contracting MDR TB	23	23	0
Fringe benefits	20	16	+4

Table 16.7. Comparison of satisfaction with living conditions (% satisfied/very satisfied)

	Migrants	Non-migrants	% Difference
Medical services for family/children	68	64	+4
Desirable housing	66	55	+11
Good school for children	50	49	+1
Affordable quality products	31	27	+4
Cost of living	21	18	+3
Level of fair taxation	11	10	+1
Customer service	10	7	+3
Personal safety	7	8	–1
Children's future in South Africa	7	10	–3
Family safety	6	7	–1
Quality upkeep of public amenities	3	1	+2
HIV and AIDS situation	3	5	–2

are also less likely to be satisfied with the prospects for local professional advancement, which could be due to experience of other health care systems.

Migrants are more satisfied with living conditions in South Africa than non-migrants, with higher scores on nine out of twelve indicators. However, the differences are again not very significant (see table 16.7). On only one indicator – ability to obtain desirable housing – is there more than a 10 per cent spread. This could be related to higher earnings while abroad, facilitating access to higher-priced housing at home. Migrants are less enamoured with

Table 16.8. Satisfaction with government policies (% satisfied/very satisfied)

	Migrants	Non-migrants	% Difference
Import of foreign health professionals	6	8	–2
Government policy towards health sector	4	7	–3
Affirmative action	4	7	–3
Government economic policies	4	4	0
Black Economic Empowerment (BEE)	4	7	–3
Levels of corruption	0.2	0.5	–0.3

Table 16.9. Likelihood of permanent emigration by migration status (% very likely)

	Migrants	Non-migrants	% Difference
In the next six months	12	9	+3
In the next two years	33	26	+7
In the next five years	52	48	+4

South African government policies towards the health sector but the levels of dissatisfaction do not vary significantly and are extremely low (less than 10 per cent satisfied) amongst both groups (see table 16.8).

In general, levels of satisfaction amongst migrants are quite low across a range of living and working measures. Less than half of the migrants were satisfied with ten of the seventeen workplace indicators and ten of the thirteen living conditions indicators. Only 50 per cent of those who had worked in a foreign country said that return to South Africa was permanent, which suggests that continued temporary or permanent emigration is a strong possibility. Overall, those with experience working in other countries have higher emigration potential. For example, 46 per cent of migrants said they have given a great deal of consideration to emigration compared with only 36 per cent of non-migrants. Migrants also indicated that there was a greater likelihood of leaving than non-migrants. At each of three different time periods (within six months, two years, and five years) the proportion of migrants likely to leave for good was slightly higher (see table 16.9).

Conclusion

SAMP's 2007 and 2014 surveys of doctors in South Africa found deep dissatisfaction with jobs and life in the country and that many respondents were contemplating departure. As a result, it appeared that the brain drain to countries in the Global North was bound to continue, particularly as the government's retention strategies were also ineffectual. One of the unanticipated findings was that a significant minority already had experience working in other countries. The concept of "return migration" is usually associated in the migration literature with individuals who have settled in another country and later decide to re-emigrate to their "home" country. This conceptualization, however, does not adequately capture the complexity of doctor mobilities in the modern world. While some South African physicians fit the traditional picture, other research shows that the return intentions of emigres are very low. What we are dealing with, instead, is a complex category of physicians who have worked (and continue to work) abroad for varying lengths of time and for various reasons. Although they exhibit slightly more positive attitudes than non-migrants to working and living in South Africa, only half say they are committed to remain.

The phenomenon of temporary doctor migration does complicate the conventional brain drain narrative, which sees all departure as permanent and all effects as negative. In part, it is the hegemony of this narrative that explains why so little attention has been paid to peripatetic physicians and why, as a result, there is little concrete data and research on the subject. While this form of mobility is undoubtedly more common in other health professions, particularly nursing, the evidence presented in this chapter suggests that it is not absent amongst doctors. Certainly, it is a phenomenon that requires additional research, not just in South Africa but more broadly.

ACKNOWLEDGMENTS

I would like to thank the International Development Research Centre (IDRC) for research funding and Dr. Genevieve Crush for her assistance.

REFERENCES

Aluttis, C., Bishaw, T., & Franks, M. (2014). The workforce for health in a globalized context: Global shortages and international migration. *Global Health Action*, *7*(1), 23611. https://doi.org/10.3402/gha.v7.23611.

Arnold, P., & Lewinsohn, D. (2010). Motives for migration of South African doctors to Australia since 1948. *Medical Journal of Australia*, *192*(5), 288–90. https://doi.org/10.5694/j.1326-5377.2010.tb03511.x.

Audas, R., Ross, A., & Vardy, D. (2005). The use of provisionally licensed international medical Graduates in Canada. *Canadian Medical Association Journal*, *173*(11), 1315–16. https://doi.org/10.1503/cmaj.050675.

Bezuidenhout, M., Joubert, G., Hiemstra, J., & Struwig, M. (2009). Reasons for doctor migration from South Africa. *South African Family Practice*, *51*(3), 211–15. https://doi.org/10.1080/20786204.2009.10873850.

Bhargava, A., & Docquier, F. (2008). HIV pandemic, medical brain drain, and economic development in sub-Saharan Africa. *World Bank Economic Review*, *22*, 345–66. https://doi.org/10.1093/wber/lhn005.

Breakfast, S. (2018, June 26). National Health Insurance could see mass emigration of SA doctors. *The South African*. https://www.thesouthafrican.com/national-health-insurance-could-see-mass-emigration-of-sa-doctors/.

Campbell-Page, R., Tepper, J., Klei, A., Hodges, B., Alsuwaidan, M., Bayoumy, D., Page, J., & Cole, D. (2013). Foreign-trained medical professionals: Wanted or not? A case study of Canada. *Journal of Global Health*, *3*(2), 1–8. https://doi.org/10.7189/jogh.03.020304.

Canadian Post-M.D. Education Registry (CAPER). (2018). *Annual census of post-M.D. trainees 2017-18*. https://caper.ca/sites/default/files/pdf/annual-census/2017-18-CAPER_Census_en.pdf.

Cape Medical Services. (n.d.). About. https://www.capemed.com/about.html.

Castles, S., & Ozkul, D. (2014). Circular migration: Triple win, or a new label for temporary migration? In G. Battistella (Ed.), *Global and Asian perspectives on international migration* (pp. 27–49). Springer International.

Connor, K., Teasdale, E., & Boffard, K. (2014). Working in South Africa. *British Medical Journal*, *348*, g2918. https://doi.org/10.1136/bmj.g2918.

Crush, J., & Chikanda, A. (2018). Staunching the flow: The brain drain and health professional retention strategies in South Africa. In M. Czaika (Ed.), *High-skilled migration: Drivers and policies* (pp. 337–59). Oxford University Press.

Crush, J., Chikanda, A., Bourgeault, I., Labonté, R., & Tomblin Murphy, G. (2014). Brain drain and regain: The migration behaviour of South African medical professionals. Southern African Migration Programme.

Crush, J., Chikanda, A., & Pendleton, W. (2012). The disengagement of the South African medical diaspora in Canada. *Journal of Southern African Studies*, *38*(4), 927–49. https://doi.org/10.1080/03057070.2012.741811.

Crush, J., & Pendleton, W. (2010). Brain flight: The exodus of health professionals from South Africa. *International Journal of Migration, Health and Social Care*, *6*(3), 3–18. https://doi.org/10.5042/ijmhsc.2011.0059.

Crush, J., & Pendleton, W. (2012). The brain drain potential of students in the African health and nonhealth sectors. *International Journal of Population Research*, *2012*, Article ID 274305. https://doi.org/10.1155/2012/274305.

Davis, R. (2012, November 21). The Abu Dhabi nightmare continues for Dr. Karabus. *Daily Maverick*. https://www.dailymaverick.co.za/article/2012-11-21-the-abu-dhabi-nightmare-continues-for-dr-karabus/.

de Vries, E., Irlam, J., Couper, I., & Kornik, S. (2010). Career plans of final year medical students in South Africa. *South African Medical Journal*, *100*(4), 227–8. https://doi.org/10.7196/SAMJ.3856.

Employment and Social Development Canada. (2017). Temporary foreign worker program 2010–2017. Government of Canada. https://open.canada.ca/data/en/dataset/c65d2014-ef25-4781-b9b2-e13a7293b72d.

Fourie, A. (2006). *Brain drain and brain circulation: A study of South Africans in the United Arab Emirates* [Unpublished doctoral dissertation]. University of Stellenbosch.

Gauld, R., & Horsburgh, S. (2015). What motivates doctors to leave the UK NHS for a "life in the sun" in New Zealand; and, once there, why don't they stay? *Human Resources for Health*, *13*(1), 75. https://doi.org/10.1186/s12960-015-0069-4.

General Medical Council. (2018). What our data tells us about locum doctors. *Working Paper No. 5*. General Medical Council.

George, G., Atunja, M., & Gow, J. (2013). Migration of South African health workers: The extent to which financial considerations influence internal flows and external movements. *BMC Health Services Research*, *13*(1), 297. https://doi.org/10.1186/1472-6963-13-297.

George, G., & Reardon, C. (2013). Preparing for export? Medical and nursing student migration intentions post-qualification in South Africa. *African Journal of Primary Health Care and Family Medicine*, *5*(1), 1–9. https://doi.org/10.4102/phcfm.v5i1.483.

George, G., & Rhodes, B. (2012). Is there really a pot of gold at the end of the rainbow? Has the occupational specific dispensation, as mechanism to attract and retain health workers in South Africa, leveled the playing field. *BMC Public Health*, *12*(1), 613. https://doi.org/10.1186/1471-2458-12-613.

Hammett, D. (2014). Physician migration in the Global South between Cuba and South Africa. *International Migration*, *52*(4), 41–52. https://doi.org/10.1111/imig.12127.

Hawthorne, L. (2012). International medical migration: What is the future for Australia? *MJA Open*, *1*(S3), 18–21. https://doi.org/10.5694/mja12.10088.

Hofman, K., McGee, S., Chalkidou, K., Tantivess, S., & Culver, A. (2015). National Health Insurance in South Africa: Relevance of a national priority-setting agency. *South African Medical Journal*, *105*(9), 739–40. https://doi.org/10.7196/SAMJnew.8584.

Howe, J., & Owens, R. (Eds.). (2016). *Temporary labour migration in the global era: The regulatory challenges*. Bloomsbury.

Hugo, G. (2009). Best practice in temporary labour migration for development: A perspective from Asia and the Pacific. *International Migration*, *47*(5), 23–74. https://doi.org/10.1111/j.1468-2435.2009.00576.x.

Humphries, N., Crowe, S., McDermott, C., McAleese, S., & Brugha, R. (2017). The consequences of Ireland's culture of medical migration. *Human Resources for Health*, 15(1), 87. https://doi.org/10.1186/s12960-017-0263-7.

Humphries, N., McAleese, S., Matthews, A., & Brugha, R. (2015a). "Emigration is a matter of self-preservation. The working conditions ... are killing us slowly": Qualitative insights into health professional emigration from Ireland. *Human Resources for Health*, *13*(1), 35. https://doi.org/10.1186/s12960-015-0022-6.

Humphries, N., McAleese, S., Tyrrell, E., Thomas, S., Normand, C., & Brugha, R. (2015b). Applying a typology of health worker migration to non-EU migrant doctors in Ireland. *Human Resources for Health*, *13*, 52. https://doi.org/10.1186/s12960-015-0042-2.

Humphries, N., Tyrrell, E., McAleese, S., Bidwell, P., Thomas, S., Normand, C., & Brugha, R. (2013). A cycle of brain gain, waste and drain: A qualitative study of non-EU doctors in Ireland. *BMC Human Resources for Health*, *11*(1), 63. https://doi.org/10.1186/1478-4491-11-63.

Hutch, A., Bekele, A., O'Flynn, E., Ndonga, A., Tierney, S., Fualal, J., Samkange, C., & Erzingatsian, K. (2017). The brain drain myth: Retention of specialist surgical graduates in East, Central and Southern Africa, 1974–2013. *World Journal of Surgery, 41*(9980), 3046–53. https://doi.org/10.1007/s00268-017-4307-x.

Joudrey, R., & Robson, K. (2010). Practising medicine in two countries: South African physicians in Canada. *Sociology of Health and Illness, 32*(4), 528–44. https://doi.org/10.1111/j.1467-9566.2009.01231.x.

Kaplan, D., & Höppli, T. (2017). The South African brain drain: An empirical assessment. *Development Southern Africa, 34*(5), 497–514. https://doi.org/10.1080/0376835X.2017.1351870.

Kingma, M. (2006). *Nurses on the move: Migration and the global health care economy*. Cornell University Press.

Kong, V., Odendaal, J., Sartorius, B., & Clarke, D. (2015). International medical graduates in South Africa and the implications of addressing the current surgical workforce shortage. *South African Journal of Surgery, 53*(3/4), 11–14. https://www.researchgate.net/publication/281490879_International_medical_graduates_in_South_Africa_and_the_implications_in_addressing_the_current_surgical_workforce_shortage.

Labonté, R., Sanders, D., Mathole, T., Crush, J., Chikanda, A., Dambisya, Y., Runnels, V., Packer, C., MacKenzie, A., Tomblin Murphy, G., & Bourgeault, I. (2015). Health worker migration from South Africa: Causes, consequences and policy responses. *Human Resources for Health, 13*, 92. https://human-resources-health.biomedcentral.com/articles/10.1186/s12960-015-0093-4.

Lambert, T., Smith, F., & Goldacre, M. (2018). Why doctors consider leaving UK medicine: Qualitative analysis of comments from questionnaire surveys three years after graduation. *Journal of Royal Society of Medicine, 111*(1), 18–30. https://doi.org/10.1177/0141076817738502.

Liu, M., Williams, J., Panieri, E., & Kahn, D. (2015). Migration of surgeons ("brain drain"): The University of Cape Town experience. *South African Journal of Surgery, 53*(3/4), 20–2. PMID: 28240477.

Lu, Y., & Hou, F. (2017). *Transition from temporary foreign workers to permanent residents, 1990 to 2014*. Statistics Canada.

Mathews, M., & Bourgeault, I. (2018). Saudi visa trainees called home from Canada in diplomatic dispute. *The Lancet*, 392(10150), 815–16. https://doi.org/10.1016/S0140-6736(18)31950-0.

Mathews, M., Kandar, R., Slade, S., Yi, Y., Beardall, S., & Bourgeault, I. (2017a). Examination outcomes and work locations of international medical graduate family medicine residents in Canada. *Canadian Family Physician, 63*(10), 776–83. PMID: 29025807.

Mathews, M. Kandar, R., Slade, S., Yi, Y., Beardall, S., Bourgeault, I., & Buske, L. (2017b). Credentialing and retention of visa trainees in post-graduate medical education programs in Canada. *Human Resources for Health, 15*, 38. https://human-resources-health.biomedcentral.com/articles/10.1186/s12960-017-0211-6.

Mills, E., Kanters, S., Hagopian, A., Bansback, N., Nachega, J.B., Alberton, M., Au-Yeung, C.G., Mtambo, A., Bourgeault, I.L., Luboga, S., Hogg, R.S., & Ford, N. (2011). The financial cost of doctors emigrating from sub-Saharan Africa: Human capital analysis. *British Medical Journal, 343*, d7031. https://doi.org/10.1136/bmj.d7031.

Mossialos, E., Wenzl, M., Osborn, R., & Sarnak, D. (Eds.). (2015). *International profiles of health care systems*. Commonwealth Fund.

Murray, R. (2017). The trouble with locums: It's not all about the money. *British Medical Journal, 356*, j525. https://doi.org/10.1136/bmj.j525.

Naidoo, S. (2012). The South African National Health Insurance: A revolution in health-care delivery! *Journal of Public Health, 34*(1), 149–50. https://doi.org/10.1093/pubmed/fds008.

Naidu, C., Irlam, J., & Diab, P. (2013). Career and practice intentions of health science students at three South African health science faculties. *African Journal of Health Professions Education, 5*(2), 68–71. https://doi.org/10.7196/AJHPE.202.

Neiterman, E., Bourgeault, I., & Covell, C. (2017). What do we know and not know about the professional integration of international medical graduates (IMGs) in Canada? *Healthcare Policy, 12*(4), 18–32. https://doi.org/10.12927/hcpol.2017.25101.

Oberoi, S.S., & Lin, V. (2006). Brain drain of doctors from Southern Africa: Brain gain for Australia. *Australian Health Review, 30*(1), 25–33. https://doi.org/10.1071/AH060025.

Ognyanova, D., Young, R., Maier, C., & Busse, R. (2014). Why do health professionals leave Germany and what attracts foreigners? A qualitative study. In J. Buchan, M. Wismar, I. Glinos & J. Bremner (Eds.), *Health professional mobility in a changing Europe* (pp. 203–32). European Observatory on Health Systems and Policies and WHO.

Ono, T., Schoenstein, M., & Buchan, J. (2014). Geographic imbalances in doctor supply and policy responses. *OECD Health Working Papers No. 69*. Organization for Economic Co-operation and Development.

Organisation for Economic Co-operation and Development (OECD). (2016). *Health workforce migration: Foreign-trained doctors by country of origin*. https://stats.oecd.org/Index.aspx?DataSetCode=HEALTH_WFMI.

Özden, C., & Philips, D. (2015). What really is a brain drain: Location of birth, education and migration dynamics of African doctors. *KNOMAD Working Paper No. 4*. World Bank.

Pantenburg, B., Kitze, K., Luppa, M., König, H-H., & Riedel-Heller, S. (2018). Physician emigration from Germany: Insights from a survey in Saxony, Germany. *BMC Health Services Research*, 18, 341. https://bmchealthservres.biomedcentral.com/articles/10.1186/s12913-018-3142-6.

Pendleton, W., Crush, J., & Lefko-Everett, K. (2007). The haemorrhage of health professionals from South Africa: Medical opinions. Southern African Migration Programme.

Physician Canada. (n.d.). Locuming in Canada. Accesssed 29 September 2021, from, https://physicianlocumscanada.com/2020/05/15/locuming-in-canada/.

Prescott, M., & Nichter, M. (2014). Transnational nurse migration: Future directions for medical anthropological research. *Social Science & Medicine, 107*, 113–23. https://doi.org/10.1016/j.socscimed.2014.02.026.

Raghuram, P. (2008). Reconceptualising UK's transnational medical labour market. In J. Connell (Ed.), *The international migration of health workers* (pp. 182–98). Routledge.

Reardon, C., George, G., & Enigbokam, O. (2015). The benefits of working abroad for British General Practice trainee doctors: The London deanery dut of programme experience in South Africa. *BMC Medical Education, 15*, 174. https://doi.org/10.1186/s12909-015-0447-6.

Rimmer, A. (2018). Scrap the cap: Applying for a tier 2 visa. *British Medical Journal, 361*, k2370. https://doi.org/10.1136/bmj.k2370.

Rogerson, C., & Crush, J. (2008). The recruiting of South African health care professionals. In J. Connell (Ed.), *The international migration of health workers* (pp. 199–224). Routledge.

Sharma, A., Lambert, T., & Goldacre, M. (2012). Why UK-trained doctors leave the UK: Cross sectional survey of doctors in New Zealand. *Journal of Royal Society of Medicine, 105*(1), 25–34. https://doi.org/10.1258/jrsm.2011.110146.

Tankwachi, A., Vermund, S., & Perkins, D. (2015). Monitoring sub-Saharan African physician migration and recruitment post-adoption of the WHO Code of Practice: Temporal and geographic patterns in the United States. *PLoS One, 10*(4), e0124734. https://doi.org/10.1371/journal.pone.0124734.

Vogel, L. (2018). Saudi medical trainees may keep posts in Canada. *CMAJ, 190*(37), E1120. https://www.cmaj.ca/content/cmaj/190/37/E1120.full.pdf.

World Health Organization. (n.d.). *A dynamic understanding of health worker migration*. World Health Organization.

17 Recasting the "Brain" in "Brain Drain": A Case Study from Medical Migration

PARVATI RAGHURAM, JOANNA BORNAT, AND LEROI HENRY

Introduction

For decades medical migration has primarily been viewed through the lens of "brain drain" (Bach, 2004; Mejía, 1979). The transfer of the value of medical knowledge and practice from sending to receiving states impacts the former and their ability to deliver medical services. The loss of those who have been trained by one state to another, which then benefits from this migration, has been problematized through the analytical lens of brain drain and very often through the calculative processes associated with migration across national boundaries (Martineau et al., 2004; Moullan & Chojnicki, 2017). In particular, the effects of migration from the Global South for national development and for modernization have been the object of discussions as well as policy initiatives (Department of Health, UK, 2001; Gish, 1971). The questions asked concerned the effects of skilled emigration, or brain drain, on source countries, of "brain gain" on destination countries, and of "brain waste," or in other words, the loss of human capital due to the lack of recognition and utilization of skills, for both individuals and destination and source countries. Of course, some positive effects of such migration, such as "transfer of knowledge" and the flow of remittances, have come to be recognized as well (Docquier & Rapoport, 2004; Lethbridge, 2004; Levy, 2003).

Although there have been attempts to refine and nuance the notion of brain drain through concepts such as "brain circulation," these concepts have largely been applied to migrant workers in sectors such as information technology (IT) (Saxenian, 2000); they are not usually used for medical migrants (but see Levitt & Rajaram, 2013). The reasons for this are, first, that although the conditions of recruitment of migrant medical workers have led to the circulation of migrants, the longer time period of this form of migration is different from the short-term nature of mobility experienced by many IT workers and scientists. Systemic knowledge plays a greater part in medical migration so that complete and frequent transference between different health systems cannot be achieved at the same frequency as in the case of, say, IT workers, whose work is, by its nature, deterritorialized. Second, both the state and professional bodies have large investments in enabling and regulating the movement of health workers, unlike in the IT sector, so that mobility itself takes much longer to organize and arrange. As a result, the extent and nature of mobility of medical professionals is different to that of IT workers. Finally, the effects of the migration of health care workers are also much more asymmetrical (Marchal & Kegels, 2003; Pang et al., 2003). The negative effects of the migration of nurses and doctors not only on the skills base of the source country but also on the provision of health care in the source country (Kingma, 2006), alongside the fact that much of the movement of professionals is from the South to the North, has meant that such migration raises important ethical questions (Chikanda, 2004; Friedman, 2004). The ethics of migration are conceptualized within the terms of redistributive justice (Mackintosh et al., 2006; Runnels et al., 2011). It is argued that the erosion of human capital has a direct impact on the provision of welfare and can be measured in terms of falling health indicators (Stillwell & Adams, 2004).

Another more recent critique of brain drain is offered by Bradby (2014), who goes beyond a focus on migrants to argue that the commercialization of the health sector and the wider political economy of health and development within which medical mobility is set need much more attention. Following Bradby, this chapter also reverses the tendency of most research on mobile health professionals to emphasize the "drain" in brain drain, with its emphasis on spatial binaries such as North–South or urban–rural. Instead of focusing on the spatialities of the "drain" and the mobility of people it shifts the analytical lens to the spatialities of the brain (i.e., of medical knowledge). This is done through an investigation of the experiences of South Asian geriatricians in the UK's National Health Service (NHS). Section two discusses the context while section three explains the methods adopted. This is followed by sections four and five, where two aspects of the spatialities of knowledge are discussed. Section four draws attention to the importance of place and of encultured medical knowledges that migrants come face-to-face with and learn to imbibe once they move, while section five brings to light the ways in which medical knowledge is carried through the bodies of migrants who move from place to place; that is, how medical knowledge is mobilized. Thus, the chapter unsettles notions of migrants as seamless purveyors of knowledge across boundaries as is often assumed in the literature on brain circulation by emphasizing the "stickiness" of knowledge to places, and how passing through particular places shapes migrants' medical knowledge. However, it also shows that knowledge is mobilizable under particular circumstances. These two perspectives are important additions to the analysis of medical migration because together they offer a way of conceptualizing medical knowledge as "in motion," as having its own spatialities, embodied in the agency of actors and thus requiring the mobility of people, but also having place-based origins. Health mobilities, thus, no longer fixate only on the people who move.

South Asian Geriatricians

Starting in the nineteenth century, the migration of doctors was part of a long-standing tradition of movement between South Asia and the UK. Development of a medical career often involved experience of overseas work so that movement across the Commonwealth countries, and especially to and from the UK and its colonies, was part of colonial history. Moreover, the reach of Western medicine was only made possible by this mobility as its spatial claims rested on movement – learning medicine from these Western centres and reproduction of its practices in hubs around the world (Raghuram, 2009). Hence, UK-trained doctors moved to countries like India (Fisher, 2004; Forbes, 1994), while Indian doctors moved to the UK to learn and to be trained. This movement also included the "White" Commonwealth, with doctors from Australia and New Zealand, for example, seeking training in the UK. Circulation gave antipodean doctors opportunities to see medical conditions that would only rarely be presented amongst their small populations (Armstrong, 2014).

Canada, on the other hand, appeared to have been a branch of the metropole, with doctors from the UK, including migrant doctors in the UK, going to Canada for short placements to learn specific skills. However, in the postcolonial period, once in the UK, immigration regulations, professional accreditation rules, social networks, and race *together* played a significant role in how migrant doctors were treated.

Immigration regulations have changed over time but have had some common features, most notably a sharp distinction between other migrants and migrant doctors, and a continuous tightening of migration rules since the introduction of the first regulation affecting Commonwealth citizens in 1962. Extant shortages in the medical labour market meant that just as immigration rules to reduce migration were introduced (see table 17.1), new forms of exemption were also created. These exemptions were based on expanding the notion of training and by using professional accreditation as a draw to find new workers.

To obtain *professional accreditation* and to practise in the UK, doctors must register with the General Medical Council. Although doctors from most Commonwealth countries had their qualifications recognized through reciprocal arrangements between the UK and their countries of origin, most of these arrangements were dismantled in 1985 with the introduction of a limited registration scheme for all doctors. This four-year limited registration clearly linked migration status to training so that doctors could only shift to permanent registration on obtaining advanced training. Career blockages, however, prevented

Table 17.1. Immigration rules affecting Commonwealth migrant doctors

Year	Regulation
1962	Voucher system was introduced.
1971	Vouchers were abolished; work permit system was introduced.
1985	Four-year permit-free training scheme was introduced.
1997	Nature of training available to non-EU migrants was altered. Training typically takes five years, but all training posts had shorter stay periods.
2006	Non-EU migration was virtually closed. However, some non-EU recruitment has occurred through the Medical Training Initiative.

migrant doctors from achieving the advanced training required for permanent registration. Doctors who had limited registration were, on the other hand, excluded from many career-grade posts, including entry into general practice.

However, *social networks* operated to allow some people through to higher posts, while others were blocked. Non-migrant networks were crucial to recruitment and they operated in *racist* ways to exclude non-White migrants. Barriers based in traditions of assumed superiority or straightforward prejudice presented substantial impediments to mobility inside the UK. For instance, letters of reference written by Indian doctors were considered inadequate for progression (Raghuram et al., 2010). The exclusivity or selective inclusion into networks formed by UK-trained doctors prevented entry and progression of South Asian doctors.

In sum, the medical profession in the UK had always been notorious for its privileged culture and closed systems and means of entry and progression (Mavromaras & Scott, 2006; Webster, 2002). As a result, migrant doctors found that despite the internationalization of the education they had received in South Asia and the dependence of the UK's NHS on migrant doctors, this international professional community had a preference for local graduates built into it, which was to direct their careers in ways that they had not expected.

One alternative was to move to particular parts of the country that were considered less desirable, leading to ethnic clustering (Raghuram et al., 2009). Another was to shift sideways into less desirable parts of the profession that were facing chronic staff shortages – specialties such as geriatrics, psychiatry, and general practice.

In the case of geriatrics, the nature of its patient group – frail older people – meant that the specialty suffered from a marginality within medicine and, as a result, UK-trained medical students tended to find it unattractive at a time when, in the mid-twenieth century, the treatment and care of older patients had become a pressing issue for the new NHS (Bridgen, 2001; Denham, 2004; Jefferys, 2000; Martin, 1995; Pickard, 2013). A crisis of staffing throughout the

1960s meant that the dependence on overseas-trained doctors in the field of geratrics was great (Rivett, 1998; Smith, 1980). Since its inception, geriatric medicine had been a "Cinderella specialty," its image being affected by ageist attitudes towards the patient group, older people, and its appeal limited amongst medical practitioners by a lack of access to acute beds and thus to private practice. These are general characteristics shared by the specialty internationally, however, developments in the UK, which proved to be pioneering, owed much to the historical coincidence of two factors: (1) early recognition of the possibility that some conditions in old age were recuperable; and (2) the inception of a socialized medical service in 1948. This led to the growth of geriatrics in a sustained way in the UK, unlike in other parts of the world (Bornat et al., 2016).

The growth of this fledgling specialty led to high staffing demands and shortages, especially in cities beyond the acknowledged centres where the discipline had developed. Moreover, the pyramidal nature of medical staffing meant that there was a large demand for those at the lower rungs of the medical hierarchy – in training posts. This offered an opportunity for migrant doctors who were struggling to get into higher training, so that by 1974, 31 per cent of consultant geriatric posts and 60 per cent of registrar posts were filled by overseas-trained graduates, the figures having risen from 15 and 33 per cent respectively in 1967 (British Geriatrics Society [BGS], 1975). A survey found that 40 per cent of geriatricians who were appointed as consultants in England in 1981–2 were overseas graduates (Goldacre et al., 2004). Migrant doctors were over-represented in hospitals in poorer areas, away from the more desirable locations in southern England and the teaching hospitals preferred by non-migrant doctors (Raghuram et al., 2009).

The coming together of two marginalized groups, older patients and South Asian doctors, suggests an interesting set of research questions around career training and work satisfaction amongst South Asian doctors working in the geriatric specialty, strategies for negotiating racism, cultural stereotyping, and career hierarchies and intercultural treatment issues. A case study of one particular specialty also provides an opportunity to investigate how individual doctors transformed the structural limitations of movement into opportunity. The next section outlines the primary research method adopted for this study – oral history.

Method

Oral history interviewing was chosen because it leads to rich, greatly nuanced theorizing as well as adding directly to knowledge of particular experiences (Thompson & Bornat, 2017). As a method, oral history has contributed a great deal to understanding migration processes, with the opportunity to explore decision-making, networking, encounters, and reflection over time (Thomson,

1999). This project drew on two main empirical sources. The first was the group of seventy-two interviews conducted in 1990–1 by a team led by Professor Margot Jefferys with the pioneers of geriatric medicine in the UK (Jefferys, 2000). Those interviewed often mention the role of overseas doctors in the history of the specialty, however only one doctor of South Asian origin was included (Bornat et al., 2012). In order to fill that gap, a second dataset of sixty interviews was conducted by generating oral history interviews with retired and serving South Asian geriatricians (henceforth SAG interviewees). Both sets of interviews are archived at the British Library. Interviewees for this second project were recruited through networks of overseas doctors (for example, the British Association of Physicians of Indian Origin), the British Geriatrics Society, and through snowballing as the project progressed. The project adhered to the ethical guidelines of the British Sociological Association and the Oral History Society. The proposal was successfully reviewed by the Open University's Human Participants and Materials Research Ethics Committee and the NHS's National Research Ethics Service (NRES). In response to NRES requirements, the invitation letter clarified the participants' right to anonymity and their right to withdraw participation, procedures for guaranteeing participant security, questions relating to mental competence, and how researchers might deal with possible criminal disclosure should this occur.

The SAG interviewees include doctors trained in India, Bangladesh, Sri Lanka, Pakistan, and Myanmar, ranging in age between forty and ninety-one and arriving in the UK from the early 1950s onwards. Two thirds of the interviewees were retired or semi-retired and had arrived in the UK prior to 1976. All except one had worked as consultants, and some also held academic posts such as that of professor. All except five of the interviewees were male. The interviewees were geographically dispersed but with clusters in the North West, Wales, and the northern fringes of London, reflecting some of the main centres where South Asian geriatricians had contributed to the specialty (Raghuram et al., 2009).

Both interview schedules used a life-history approach, asking participants to talk about their childhood, upbringing, education at school and college, and subsequent training and careers. The South Asian doctors were also asked about their training in their home countries and after arrival in the UK, about their reasons for migration to the UK, and arrival and subsequent career progression in the UK with a focus on opportunities, barriers, and sources of support (for a discussion of the use of the two datasets, see Bornat et al., 2012). Both sets of interviews were analysed following a grounded theory approach, drawing out key themes after ten interviews had been completed, transcribed, and reviewed by the team. A common coding strategy was then developed and used iteratively with new themes being added as these emerged in later interviews. The project also drew on literature and archival searches of the institutional histories of the development of the NHS. In what follows, we draw on the data to discuss the spatialities of knowledge that constitute brain drain. We consider how knowledge was attached to place and how migrant doctors mobilized the knowledge they had acquired between places.

"Sticky" Knowledge and the Importance of Place

Opportunities for learning and promotion became available to migrant doctors, many from the Indian subcontinent, who were prepared to switch to geriatric medicine (for similar trends in psychiatry and general practice, see Esmail, 2007; Simpson et al., 2010; Smith, 1980). Yet even in geriatrics, medical knowledge was encultured in that it involved place-based learning that was dependent on cultural codes and ways of understanding (Williams, 2006; Williams & Balaz, 2008). Professor John Brocklehurst was interviewed by both Jefferys and again for the SAG project because he had a particularly good reputation amongst the interviewed South Asian doctors as a teacher, colleague and clinician and, of course, as the first professor of geriatric medicine in the UK. He was asked what he thought was necessary if South Asian doctors were to succeed:

> BROCKLEHURST: Well they certainly needed to integrate a bit with the rest of medical society which was not always easy to do. They had to be competent, and good and er [*pause*], I just really don't know otherwise.
>
> INTERVIEWER: When you say integrate, how do you see that process?
>
> BROCKLEHURST: It would depend on going to take part in meetings in the hospitals, running the general hospitals sort of staff meetings and that sort of thing. On the whole I think most of these doctors would take part in clinical sessions with everybody else, and I guess they probably found that even more difficult than did the home products. Although for young people coming up speaking on presenting patients and so on, to a crowd of hard boiled consultants and their next rank down is not easy I'm sure. ... But, er, I don't think there was any special criteria that didn't apply to all people on their way up in the ladder. (Professor Brocklehurst, male, retired professor of geriatric medicine, born 1924, UK, BL catalogue C1356/62)

Being mobile means finding ways to engage with knowledge and practices that are already present in particular contexts, in this case the hospital ward and also the hierarchy of the medical profession. As Professor Brocklehurst points out, this was a process that was challenging even to young doctors whose whole medical training had taken place within the UK. For those from elsewhere, it meant learning and being able to practise their skills within a context that might feel familiar on the surface (because of shared histories of medical training across the Commonwealth), but which had its own practices and protocols.

Mobile doctors became aware of this and described various ways in which they experienced formal and informal processes of induction into the life and working conditions of a hospital doctor. At times, this place-based knowledge was difficult to access, or somehow obscured by local dialect or by practice:

> You know our junior jobs in India, consultants, they were quite strict, but again here of course things were totally different. You have to be up in right time, you have to be in the ward in right time, you have to prepare the cases for the consultant or registrar even for the ward round. You have to discuss with the sister and nurses what is going on, what is the present situation. You have to see sometimes persons hourly about their present condition. If any [complain, you have to determine] what is going wrong so that we could feed [that information to] the registrars and consultant. So it was very busy. I was very busy when we were working, but at the same it was rewarding for me because I was always getting new experiences, you see, which were so different than what I did. (Dr. Das Gupta, retired consultant physician, born 1933, India, arrived UK 1965, BL catalogue C1356/06)

In their turn, South Asian doctors drew on this experience when they came to set up their own departments and learning opportunities. In their interviews, they describe how they produced the conditions under which access to encultured knowledge was facilitated:

> Because to develop a department one has to make it attractive, you know, one has to make it good enough for yourself as well as for the others who could come and work with you and for you. And to get that you have to have support from your colleagues which fortunately, you know, I had. But it may not have been the same with everybody. (Dr. Kumar Sinha, male, retired consultant physician in geriatric medicine, born 1937, India, arrived UK 1968, BL catalogue C1356/21)

Importantly, these medical workplace cultures were not simply national or given. They were created in particular places. Because of the nature of the specialty, geriatrics came to be established not in the most prestigious teaching hospitals in the south of England, but in district hospitals in the north of England, in Wales and in Scotland, in towns and cities such as Hull, Leeds, Manchester, and Cardiff. One place that became particularly influential was Sunderland, an industrial city in the northeast of England, where few non-migrant doctors wanted to work. It was not only a peripheral part of England but also seen as peripheral within the region, insignificant compared to the hospitals that were attached to Newcastle University, for instance:

> And there were two hospitals here. I think one was the Royal Infirmary, [the] other was Sunderland General. So when you mention earlier in your discussion about

> the workhouse, Sunderland General used to be workhouse. Infirmary was the elite type of hospital so you will see more local graduates working in Royal Infirmary than Sunderland General. (Dr. Bansal, male, consultant physician in geriatric medicine, born 1947, India, arrived UK 1973, BL catalogue number C1356/04)

Moreover, medical knowledge did not simply consist of habits of working but involved considered styles of arranging care for patients. In the case of geriatrics this involved rearranging the spatial practices involved in geriatric care. One such rearrangement was the development and adoption of an age-related admission policy. All patients who were above a particular age – sixty-five when this was first established in Sunderland by pioneers like Dr. Oscar Olbrich and Dr. Eluned Woodford Williams, but later seventy and seventy-five in other centres – were admitted into a single ward irrespective of their reason for admission. This policy recognized that geriatric patients often did not have a single condition; rather, they were likely to have multiple pathologies requiring special skills and services related to the process of treatment (access to psycho-geriatricians, physiotherapists, etc.) as well as to discharge (e.g., involvement of a rehabilitation team and/or social services) (Kafetz et al., 1995). Geriatricians working with an age-related admissions policy thus developed and deployed a set of composite team-based skills that were developed in particular places that had staked a claim to innovativeness and thus became renowned centres of learning. Those who worked in Sunderland often ascribed their career success to having been trained in a setting where the new specialty of geriatrics had a particularly good reputation and to the learning environment in dealing with these multiple pathologies that this policy afforded.

An age-related admissions policy was also adopted and adapted by many of those who passed through Sunderland, increasing Sunderland's fame as a centre of innovation:

> I wanted to take geriatrics into what I used to do in Sunderland. To get fully functional, full-scale outpatient and patient facilities for [the] elderly. And as I said, within five years I was admitting all admissions over the age of seventy. (Dr. Hajela, retired consultant physician in geriatric medicine, born 1933, India, arrived UK 1956, BL catalogue number C1356/18)

From integrating care to subspecializing, there were local trends established in what constituted good geriatric care. Geriatric knowledge as a medical practice was thus being developed within particular spaces, wards, hospitals, and areas of the country, and in relation to the social and generational characteristics of a particular patient group. However, those who were encultured into these particular practices took their knowledge out to other centres, as discussed in the next section.

Mobilizing Knowledge

Despite the limits and challenges set by the contexts in which various forms of medical knowledge were exercised and developed, migrant doctors were able to mobilize knowledge. This mobilization became important not only for the migrants but also for the cementing of certain practices developed in centres of knowledge as the preferred models of care. It institutionalized knowledge developed in particular places by showing its more universal relevance.

Centres of good practices such as Sunderland and also Manchester were not only claiming good practices, but were becoming reified as centres of learning through which those in the learning stages of the medical career hierarchy (especially as registrars and senior registrars) had to pass to claim knowledge (Bornat et al., 2016). It was by passing through these centres that one claimed to be knowledgeable. As one SAG interviewee recalled, having been through a centre of learning clearly gave him an edge and shaped his career decisions:

> [I]f I were to make a career in [the] UK, geriatric medicine was perhaps a better career for me, especially being trained in Sunderland. But again, as I said, the career progression was so rapid in Sunderland that, you know, I just rode with it.
>
> INTERVIEWER: And how did you feel about going into geriatric medicine then?
>
> No problem because what I was seeing [was that] geriatric medicine there was [a] very appealing branch because ... funnily enough we also had a first special dedicated six bed ward for MI [myocardial infarction] care in Sunderland. (Dr. Hajela, male, Retired Consultant Physician, b 1933 India, arrived UK 1956, BL catalogue number C1356/18)

A career in geriatrics became appealing because of the spatial practices of the ward (having dedicated beds for subspecialties) and admissions (as we saw in Sunderland) of the eminence that such spatial practices, spearheaded by individuals like Woodford Williams and Oscar Olbrich, had given to centres adopting such practices. Passing through these centres gave geriatric trainees pride in their field but also a form of capital that they could use to develop their careers elsewhere. Association with such centres of learning (and their pioneers) thus shaped career trajectories.

However, it is not only people's careers that benefited from association with Sunderland – both the centre and its policies were also dependent on mobile bodies. Thus, the success of centres such as Sunderland was made possible when doctors who passed through them extended its reach by following practices that they had learned there. These place-based practices of innovation spread out to influence how others thought good geriatric medicine should look like and what might be the ideal solutions to the problems of caring for the growing numbers of older people. For instance, Sunderland's age-related admissions policy was adopted and adapted by many of those who passed through Sunderland, increasing Sunderland's fame as a centre of innovation and helping to establish age-related admission as a preferred style of care.

Geriatricians who passed through such centres also considered where they might have the most impact and where they might have the best chance to adopt and adapt these innovations when making decisions about their career. One SAG interviewee went for a post in Mansfield, deciding to develop the practice he was familiar with from his time at St James's University Hospital in Leeds:

> It's a mining town. Small town. Very little Asian population there. Mostly indigenous. The geriatric provisions in those days were very poor. There were something like 260 long-stay beds in three hospitals and there was one geriatrician there already. And very little support to develop the services from the management side, mainly because of cost. And it was convenient for everybody else, like the general physicians, to dump their patients into geriatric medicine and forget about them. So it was a kind of dead end that you go into a place where the service is mostly long-stay service, nobody goes home, dies there. There wasn't any nursing home in those days.
>
> People had the conception then, they believed very firmly, that they have worked all their life, paid taxes, they have got the right to stay in hospital for whatever length of time is needed, you see.
>
> And the facilities for rehabilitation, treatment, et cetera were also poor. Staffing level was very limited … It was very convenient for the GPs, you know. In between their times they come and write out a few prescriptions and go away, you see. And so in that way they didn't really contribute much towards treating and rehabilitating elderly ill patients at the hospital. (Dr. Rahman, male, retired consultant physician and geriatrician, born 1935, Bangladesh, arrived UK 1967, BL catalogue number C1356/10)

This doctor recognized the potential that the blank sheet gave him to practice the learning he had acquired in Leeds. It would allow him to set his own mark. As such, professional growth for these migrant doctors was intimately tied to practising innovation in new territories and spaces, thus extending the reach of these forms of innovation. It offered a way of forwarding their field.

Conclusion

This chapter takes a different tack from previous studies of brain drain. It focuses on the mobility of knowledge rather than the mobility of groups of individuals, and in doing so highlights how medical knowledge is more than what is learned as physical and organic, as it is also social and spatial, and makes five specific contributions to the literature on brain drain.

First, the chapter goes beyond the nation as the preferred spatial analytical scale at which medical migration is discussed by showing how migrants benefited from and contributed to medical knowledge and innovation at other levels – in this case both in small towns and cities in certain parts of the UK and in particular specialties. In doing so, it changes the scale at which the transfer of health workers is studied. It explores how the loss of value through emigration and as a consequence of race is partially rescued by moving into less desirable parts of the medical profession. However, these newer, less well-regarded areas of the profession also offer scope for advancing overseas doctors' professional careers and the opportunity to gain satisfaction and make one's name.

Second, medical knowledge is not simply removed from one country to another but comes up against encultured knowledges that operate locally. Migrants learn these knowledges in certain places but also take them beyond those places to new sites, spreading forms of good practice and medical innovation. The spatialities of knowledge and those of migration are therefore intertwined, and in this process, the "brain" in brain drain does not stand still. As seen in the case of centres such as Sunderland, places can become crucial to claims to become a model for medical practice. Thus, medical knowledge is based on enculturations of practice in place. Moreover, such knowledge requires mobilizing (circulation) in order to become established as a preferred way of practising medicine. Location and mobility are two aspects of the complex nature of the knowledge that is called forth in discussions of medical migration.

Third, by focusing on geriatrics, a specialty that did not exist in the countries of origin of the migrants we interviewed but was becoming instituted in the British medical system, we have been able to complicate the story of medical migration by showing how the structures migrants encountered offered opportunity for new knowledge to be developed and disseminated as well as opportunity for career development. In sum, the multiple actors, institutions, places, and people involved in the recognition, validation, and circulation of knowledge shows that the migrant is only one part (albeit an important one) of the terrain of mobile medical knowledges. The transfer of value across borders leads not only to deskilling but also to the acquisition of new skills not available in the sending countries. Moreover, the skills that are acquired are not simply transferred to migrants, but rather they are co-developed by migrants in their new contexts. Hence, knowledge and skills should not be considered as a fixed-sum game. Focusing on skills rather than migrants shows how knowledge and innovation has been driven by migrants. The implications of our research, therefore, are that migrant mobility should be analysed as more than the movement of migrants; the spatialities of medical knowledge should also be considered as mobilizing health requires and involves more than mobile bodies.

Fourth, our case makes the question of intersectionality in care more complex by drawing on an example of racialised migrants who were highly skilled and whose educational success in their countries of origin placed them high in the occupational order. Arriving in the UK to further their training, they moved down within the hierarchy of medical specialties in order to move up into consultant roles rather than inhabit the multiple forms of non-consultant career-grade roles that were mushrooming in order to fill the service requirements of the NHS. They got opportunities to train themselves and eventually to train others into the version of geriatrics that they had shaped in their local contexts. Much of the literature on intersectionality, often led by the experiences of Black women, has focused on those who have been multiply disadvantaged – by class, race, and gender. This chapter has attempted to complicate that picture by exploring the experiences of South Asian men, many of whom came from middle-class backgrounds but faced discrimination at the intersection of hierarchizing processes due to colour, citizenship, and immigration status and country of qualification.

Finally, we want to end by pointing to the continuing relevance of these issues today. It has become commonplace to argue that a large degree of care mobility is driven by the care needs of aging populations in the West. Although much of this literature has focused on the justifiable shortage of social care, medical care requirements also remain extant. Moreover, the shortage of social care has effects on medical care. Between 2013 and 2015, delays in transferring patients from the hospital rose 31 per cent in the UK and in 2015 accounted for 1.15 million bed days, with 85 per cent of the patients occupying those beds aged over sixty-five (Oliver, 2016). The demand for hospital care in the UK is weighted towards older age groups with around 7.6 million (41 per cent) of the 18.7 million adults admitted to hospital in 2014 being sixty-five years or above (NHS Benchmarking Network, 2015) even though in 2014 people sixty-five or older formed only 17.7 per cent of the UK's population (Office for National Statistics, 2016). Of course, not all the health care requirements of this population needs to be met by geriatricians, but the demand for geriatricians will nevertheless grow. In the US, it is estimated that 17,000 old age specialists are required to care for the 12 million geriatrics in the population (Olivero, 2015). However, in 2010, only 75 people entered geriatric training programs in the US (Mangipudi, 2017). Similarly, in Canada, there were only 242 certified specialists, of whom 35 per cent were aged fifty-five or over (Heckman et al., 2013); estimates of the numbers vary, but it has been suggested that around 700 certified specialists are required to adequately treat Canada's geriatric population. Although efforts to address the shortfall have been instituted in all these countries with the introduction of special programs (e.g., the Geriatrics Workforce Enhancement Program and the Medical Student Training in Aging Research Program in the US; the Resident Geriatrics Interest Group and the National Geriatrics Interest Group in Canada), geriatrics continues to remain a shortage specialty. Moreover between 2001–2 and 2017–18, the geriatric specialty declined in actual and population-adjusted (–58 positions, –23.3%) filled positions when hospice/palliative care was omitted (Petriceks et al., 2018). For instance, in the US, over 20 per cent of all training posts in geriatrics remain unfilled (Fisher et al., 2014). In comparison, in the UK in 2017 only 4.2 per cent of the training posts were vacant (British Geriatrics Society [BGS], 2017); this comes after real concerns about the shortage of geriatricians. Thus in 2010, while the number of acute and general physicians increased by 23.3 per cent, there was concern because not only had the number of consultant geriatricians in the UK fallen by 1.6 per cent from 1,129 to 1,111, but there was also concern that almost 10 per cent of these consultants were over sixty years old (BGS, 2010). These concerns are particularly acute given the lack of development of geriatrics as a specialty in most parts of the world so that direct overseas recruitment into the specialty is only a long-term option; it involves a degree of local training (BGS, 2017).

To conclude, our chapter offers a distinctive take on the question of transnational value and transfers. It highlights how race and the lack of transferability of some credentials has led some South Asian migrants to take a step back in their careers. However, it also relates how these migrants found new kinds of learning and added value not only to the labour market but to the knowledge and training involved within the health service. The chapter thus argues for going beyond narrating medical migration through the language of "drain" to instead focus on the places and spaces through which medical knowledge (the "brain") is acquired, adapted, and performed.

REFERENCES

Armstrong, J. (2014). Doctors from "the end of the world": Oral history and New Zealand medical migrants 1945–1975. *Oral History*, *42*(2), 41–9. https://www.jstor.org/stable/24343432.

Bach, S. (2004). Migration patterns of physicians and nurses: Still the same story? *Bulletin of the World Health Organisation*, *82*(8), 624–5. https://apps.who.int/iris/handle/10665/72850.

Bornat, J., Raghuram, P., & Henry, L. (2012). Revisiting the archives: A case study from the history of geriatric medicine. *Sociological Research Online*, *17*(2), 1–12. https://doi.org/10.5153/sro.2590.

Bornat, J., Raghuram, P., & Henry, L. (2016). "Without racism there would be no geriatrics": South Asian overseas-trained doctors and the development of geriatric medicine in the United Kingdom, 1950–2000. In L. Monnais & D. Wright (Eds.), *Doctors beyond borders: The transnational migration of physicians in the twentieth century* (pp. 185–207). University of Toronto Press.

Bradby, H. (2014). International medical migration: A critical conceptual review of the global movements of doctors and nurses. *Health*, *18*(6), 530–96. https://doi.org/10.1177/1363459314524803.

Bridgen, P. (2001). Hospitals, geriatric medicine, and the long-term care of elderly people 1946–1976. *Social History of Medicine, 14*(3), 507–23. https://doi.org/10.1093/shm/14.3.507.

British Geriatrics Society. (1975). Minutes of the Meeting of the Working Party on Geriatric Medicine, Aspects of Recruitment and Relationship to General Medicine, St. Pancras Hospital, 14 November.

British Geriatrics Society. (2010). *Consultant geriatricians workforce figures.* http://www.bgs.org.uk/factsnfigures/resources/intelligence/consultant-geriatricians-workforce-figures.

British Geriatrics Society. (2017). BGS *Workforce data: 2017 Summary.* https://www.bgs.org.uk/resources/bgs-workforce-data-2017-summary.

Chikanda, A. (2004). Skilled health professionals' migration and its impact on health delivery in Zimbabwe. *COMPAS Working Paper No. 4.* University of Oxford.

Denham, M.J. (2004). *The history of geriatric medicine and hospital care of the elderly in England between 1929 and the 1970s* [Unpublished doctoral dissertation]. University College London.

Department of Health, UK. (2001). *Code of practice for NHS employers involved in the international recruitment of NHS professionals.* Department of Health.

Docquier, F., & Rapoport, H. (2004). *Skilled migration: The perspective of developing countries.* World Bank. https://doi.org/10.1596/1813-9450-3382.

Esmail, A. (2007). Asian doctors in the NHS: Service and betrayal. *British Journal of General Practice, 57*(543), 827–34. https://www.ncbi.nlm.nih.gov/pmc/articles/PMC2151817/.

Fisher, J.M., Garside, M., Hunt, K., & Lo, N. (2014). Geriatric medicine workforce planning: A giant geriatric problem or has the tide turned? *Clinical Medicine, 14*(2), 102–6. https://doi.org/10.7861/clinmedicine.14-2-102.

Fisher, M.H. (2004). *Counterflows to colonialism: Indian travellers and settlers in Britain 1600–1857.* Permanent Black.

Forbes, G. (1994). Medical careers and health care for Indian women: Patterns of control. *Women's History Review, 3*(4), 515–30. https://doi.org/10.1080/09612029400200067.

Friedman, E. (2004). *An action plan to prevent brain drain: Building equitable health systems in Africa.* Physicians for Human Rights.

Gish, O. (1971). *Doctor migration and world health: The impact of the international demand for doctors on health services in developing countries.* G. Bell & Sons.

Goldacre, M., Davidson, J., & Lambert, T. (2004). Country of qualification, ethnic origin of UK doctors: Database and survey results. *British Medical Journal, 329,* 538–4. https://doi.org/10.1136/bmj.38202.364271.BE.

Heckman, G., Molnar, F., & Lee, L. (2013). Geriatric medicine leadership of health care transformation: To be or not to be? *Canadian Geriatrics Journal, 16*(4), 192–5. https://doi.org/10.5770/cgj.16.89.

Jefferys, M. (2000). Recollections of the pioneers of the geriatric specialty. In J. Bornat, R. Perks, P. Thompson, & J. Walmsley (Eds.), *Oral history, health and welfare* (pp. 75–97). Routledge.

Kafetz, K., O'Farrell, J., Parry, A., Wijesuriya, V., McElligott, G., Rossiter, B., & Lugon, M. (1995). Age-related geriatric medicine: Relevance of special skills of geriatric medicine to elderly people admitted to hospital as medical emergencies. *Journal of the Royal Society of Medicine*, *88*(11), 629–33. https://www.ncbi.nlm.nih.gov/pmc/articles/PMC1295386/pdf/jrsocmed00064-0029.pdf.

Kingma, M. (2006). *Nurses on the move: Migration and the global health care economy*. Cornell University Press.

Lethbridge, J. (2004). Brain drain: Rethinking allocation. *Bulletin of the World Health Organisation*, *82*(8), 623. https://apps.who.int/iris/handle/10665/72849.

Levitt, P., & Rajaram, N. (2013). Moving toward reform? Mobility, health, and development in the context of neoliberalism. *Migration Studies*, *1*(3), 338–62. https://doi.org/10.1093/migration/mnt026.

Levy, L. (2003). The first world's role in the third world brain drain. *British Medical Journal*, *327*, 170. https://doi.org/https://doi.org/10.1136/bmj.327.7407.170.

Mackintosh, M., Mensah, K., Henry, L., & Rowson, M. (2006). Aid, restitution and international fiscal redistribution in health care: Implications of health professionals' migration. *Journal of International Development*, *18*(6), 757–70. https://doi.org/10.1002/jid.1312.

Mangipudi, S. (2017, January 25). Who wants to be a geriatrician? A looming shortage in our time of need. *The Oxford Institute of Population Ageing*. http://www.ageing.ox.ac.uk/blog/2017-Who%20wants%20to%20be%20a%20Geriatrician#_ednref9.

Marchal, B., & Kegels, G. (2003). Health workforce imbalances in times of globalization: Brain drain or professional mobility. *International Journal of Health Planning and Management*, *18*(1), S89–101. https://doi.org/10.1002/hpm.720.

Martin, M. (1995). Medical knowledge and medical practice: Geriatric medicine in the 1950s. *Social History of Medicine*, *8*(3), 443–61. https://doi.org/10.1093/shm/8.3.443.

Martineau, T., Decker, K., & Bundred, P. (2004). "Brain drain" of health professionals: From rhetoric to responsible action. *Health Policy*, *70*(1), 1–10. https://doi.org/10.1016/j.healthpol.2004.01.006.

Mavromaras, K., & Scott, A. (2006). Promotion to hospital consultant: Regression analysis using NHS administrative data. *British Medical Journal*, *332*(7534), 148–51. https://doi.org/10.1136/bmj.38628.738935.3A.

Mejía, A. (1978). Migration of physicians and nurses. *International Journal of Epidemiology*, *7*(3), 207–15. https://doi.org/10.1093/ije/7.3.207.

Moullan, Y., & Chojnicki, X. (2017). Is there a "pig cycle" in the labour supply of doctors? How training and immigration policies respond to physician shortages. *IMI Working Paper No. 132*. International Migration Institute.

NHS Benchmarking Network. (2015). *Hospital episode statistics, admitted patient care, England – 2013–14*. https://digital.nhs.uk/data-and-information/publications

/statistical/hospital-admitted-patient-care-activity/hospital-episode-statistics -admitted-patient-care-england-2013-14.

Office for National Statistics. (2016). *Overview of the UK population: February 2016.* https://www.ons.gov.uk/peoplepopulationandcommunity/populationandmigration /populationestimates/articles/overviewoftheukpopulation/february2016.

Oliver, D. (2016, May 26). Why is it more difficult than ever for older people to leave hospital? *The Kings Fund.* https://www.kingsfund.org.uk/blog/2016/05/older-people -leave-hospital.

Olivero, M. (2015, April 21). Doctor shortage: Who will take care of the elderly? *U.S. News.* https://health.usnews.com/health-news/patient-advice/articles/2015/04/21 /doctor-shortage-who-will-take-care-of-the-elderly.

Pang, T., Lansang, M.A., & Haines, A. (2003). Brain drain and health professionals. *British Medical Journal, 324*(7336), 499–500. https://doi.org/10.1136/bmj.324.7336.499.

Petriceks. A.H., Olivas, J.C., & Srivastava, S. (2018). Trends in geriatrics graduate medical education programs and positions, 2001 to 2018. *Gerontology and Geriatric Medicine, 4*(1–4). doi:10.1177/2333721418777659.

Pickard, S. (2013). Frail bodies: Geriatric medicine and the constitution of the fourth age. *Sociology of Health & Illness, 36*(4), 549–63. https://doi.org/10.1111/1467 -9566.12084.

Raghuram, P. (2009). Caring about the "brain drain" in a postcolonial world. *Geoforum, 40*(1), 25–33. https://doi.org/10.1016/j.geoforum.2008.03.005.

Raghuram, P., Bornat, J., & Henry, L. (2009). Ethnic clustering among South Asian geriatricians in the UK: An oral history study. *Diversity in Health and Care*, 6, 287–96. https://diversityhealthcare.imedpub.com/ethnic-clustering-among-south -asian-geriatricians-in-the-uk-an-oral-history-study.php?aid=2050.

Raghuram, P., Henry, L., & Bornat, J. (2010). Difference and distinction? Non-migrant and migrant networks. *Sociology, 44*(4), 623–41. https://doi.org/10.1177 /0038038510369360.

Rivett, G. (1998). *From cradle to grave: Fifty years of the NHS.* King's Fund.

Runnels, V., Labonté, R., & Packer, C. (2011). Reflections on the ethics of recruiting foreign-trained human resources for health. *Human Resources for Health*, 9(2), e1–e11. https://doi.org/10.1186/1478-4491-9-2.

Saxenian, A. (2000). Silicon Valley's new immigrant entrepreneurs. *CCIS Working Paper No. 15.* University of California.

Simpson, J.M., Esmail, A., Kalra, V.S., & Snow, S.J. (2010). Writing migrants back into NHS history: Addressing a "collective amnesia" and its policy implications. *Journal of the Royal Society of Medicine, 103*(10), 392–6. https://doi.org/10.1258 /jrsm.2010.100222.

Smith, D.J. (1980). *Overseas doctors in the National Health Service.* Policy Studies Institute.

Stillwell, B., & Adams, O. (2004). Health professionals and migration. *Bulletin of the World Health Organisation, 82*(8), 560. https://apps.who.int/iris/handle/10665 /269212.

Thompson, P., with Bornat, J. (2017). *The voice of the past* (4th ed). Oxford University Press.

Thomson, A. (1999). Moving stories: Oral history and migration studies. *Oral History, 27*(1), 24–37. https://www.jstor.org/stable/40179591.

Webster, C. (2002). *The National Health Service: A political history* (2nd ed.). Oxford University Press.

Williams, A. (2006). Lost in translation? International migration, learning and knowledge. *Progress in Human Geography, 30*(5), 588–607. https://doi.org/10.1177/0309132506070169.

Williams, A., & Balaz, V. (2008). *International migration and knowledge*. Routledge.

Contributors

Sheri Adekola is a socio-cultural geographer and an interdisciplinary professional. She completed her PhD in Geography at Wilfrid Laurier University. Her research is concerned with questions of social inequality and the experiences of marginalized groups in Toronto. Her research and teaching interests address diversity, transnationalism, urbanization, globalization, community development, citizenship, and research methods.

Valentina Antonipillai is a postdoctoral fellow in Global Health, Faculty of Health Sciences at McMaster University. Her research writings focus on moving towards inclusive migration and health policies by examining the effects of inequitable access to health and prescription drug insurance in Canada and across other high-income countries.

Jelena Atanackovic, PhD, is a senior research associate in the School of Sociological and Anthropological Studies at the University of Ottawa. She obtained her PhD in Sociology from McMaster University. Her research interests lie in the areas of gender, immigration, and integration as well as health policy and health care delivery. She has been involved in several research projects on internationally educated health care professionals in Canada. Her dissertation explores experiences of immigrant live-in caregivers in Ontario.

Andrea Baumann RN, PhD, FAAN, CM, is the Associate Vice-President of Global Health, Faculty of Health Sciences, McMaster University, and Director of the World Health Organization Collaborating Centre in Primary Care and Nursing Health Human Resources. She has made substantial contributions as a health services researcher, writer, and award-winning academic. She has received numerous awards, including the Order of Canada in July 2018.

Joanna Bornat is Emeritus Professor at the Open University and has been an editor of the *Journal of Oral History* for over forty years. Through oral history interviews she has researched family, race, and class, publishing on topics of age, aging, intergenerational relationships, and the development of the geriatric specialty in the UK. She has a particular interest in the reuse of archived interviews.

Ivy Lynn Bourgeault, PhD, is a professor in the School of Sociological and Anthropological Studies at the University of Ottawa and the University Research Chair in Gender, Diversity, and the Professions. She leads the Canadian Health Workforce Network and the Empowering Women Leaders in Health initiative. She has garnered an international reputation for her research on the health workforce, particularly from a gender lens.

Abel Chikanda is an associate professor in the School of Earth, Environment and Society at McMaster University. He has held academic and research positions at the University of Kansas, the Balsillie School of International Affairs, Queen's University, and the University of Zimbabwe. He has conducted research on issues such as health worker migration, migration and development, immigrant integration, and food security in Southern Africa and North America.

John Connell is a professor of Human Geography at the University of Sydney and author of various papers and books on the migration of health workers, including *The Global Health Care Chain* (2009) and *Migration and the Globalisation of Health Care* (2010). Recent books include *Islands at Risk* (2013), *Change and Continuity in the Pacific* (with Helen Lee, 2018), *The Ends of Empire* (with Robert Aldrich, 2020), and *COVID-19 and Islands* (with Yonique Campbell, Palgrave Macmillan, in press, 2021).

Mary Crea-Arsenio is a senior research analyst in Global Health, Faculty of Health Sciences and a doctoral candidate in the School of Earth, Environment, and Society at McMaster University. Her areas of research include immigrant employment, workforce integration, and the migration of highly skilled health professionals.

Jonathan Crush is a professor at the Balsillie School of International Affairs at Wilfrid Laurier University, and an honorary professor at the University of the Western Cape. He is the director of the Southern African Migration Program (SAMP) (www.samponline.net) and the Hungry Cities Partnership (www.hungrycities.net) and has researched and published extensively on issues of migration and development in Africa and more generally. His latest co-edited book is the *Handbook on Urban Food Security in the Global South* (Edward Elgar, 2020).

Crystal Ennis is a scholar of global political economy and a lecturer at Leiden University. Her research examines labour and migration governance, and the political economy of dependency on hydrocarbon revenue and foreign labour in Gulf economies. Her publications have appeared in *New Political Economy*, *Global Social Policy*, *Third World Quarterly*, *International Journal of Middle East Studies*, and *Cambridge Review of International Affairs*, among others.

Héctor Goldar Perrote holds an MA in Human Geography from Wilfrid Laurier University. His master's thesis explores the intricacies of Japan's immigration policy. Most recently, he has been employed as a trainee at the European Commission.

Caitlin Henry is a lecturer in Human Geography at the University of Manchester. She researches the impacts of health care restructuring on the geographies of care, labour, and health in the US and the UK. Her recent work highlights the connections between spatial restructuring of institutional care provisioning and broader geographies of social reproduction.

Leroi Henry is a senior lecturer in Human Resource Management at the University of Greenwich. His recent research has explored the impact of COVID-19 on minority ethnic workers in the UK, career progression for minority ethnic workers in the UK civil service and higher education, racialized masculinity, and support for professional football trainees and career progression for black managers in professional football in the UK.

Nuzhat Jafri is the principal of Fair Practices Consulting, a management-consulting firm specializing in equity, inclusion, diversity, human rights, and accessibility. Nuzhat served as the first and only executive director of the Office of the Fairness Commissioner (OFC) from September 2007 to August 2017. She has wide leadership experience in the public, private, and not-for-profit sectors.

Heidi Kaspar is a social and health geographer at the Department of Health Professions, Bern University of Applied Sciences. Her research focuses on therapeutic mobilities, the global and intimate relations of care work and participatory health research. She co-leads the Competence Centre for Participatory Health Care.

Joel Negin has been the head of the School of Public Health at the University of Sydney since November 2015. He holds grants from the Australian NHMRC and the Department of Foreign Affairs and Trade for his research that focuses on strengthening health systems in low- and middle-income countries. He maintains collaborations in Uganda, Vietnam, Indonesia, and Fiji, and maintains a passion for capacity-building in the Asia-Pacific region.

Elena Neiterman, PhD, is a continuing lecturer and a teaching fellow in the School of Public Health Sciences at the University of Waterloo. Her research focuses on gender, immigration, and health, as well as health policy and health promotion. Her work concerns women's work, women's reproductive health, and mental health and work. She is actively involved in developing pedagogies and teaching strategies that are centred on mental well-being and promotion of equity, diversity, and inclusion.

Christine Noelle Peralta is an assistant professor in the Departments of History and Sexuality, Women's and Gender Studies at Amherst College. She is currently working on a book entitled *Insurgent Care: Reimagining the Health Work of Filipina Women, 1870–1948*, which documents Filipina women's medical knowledge, migration, and labour across the US empire.

Parvati Raghuram is a professor in Geography and Migration at the Open University. She has published widely on gender, migration and development, and on postcolonial theory. She is co-editor of *South Asian Diaspora, The Geographical Journal*, and the Palgrave Pivot series *Mobility and Politics*.

John Ravenhill is the chair of the Political Science Department at the University of Waterloo. He was previously Director of the Balsillie School of International Affairs, Head of the School of Politics and International Relations in the Research School of Social Sciences at the Australian National University, and Chair of the Department of Politics at the University of Edinburgh.

Arthur Sweetman is a professor in the Department of Economics at McMaster University where he holds the Ontario Research Chair in Health Human Resources. He is also a member of McMaster's Centre for Health Economics and Policy Analysis (CHEPA). Previously, he was the Director of the School of Policy Studies at Queen's University where he held the Stauffer-Dunning Chair in Policy Studies. He is currently a co-editor of the *Canadian Journal of Economics*, and undertakes quantitative research on health workforce issues as well as related issues in health economics and social policy.

Maddy Thompson is a human geographer with interests in health. She has examined the role of migration in global health provision, focusing on the migration of Filipino nurses, and is currently exploring the rise of digital health technologies in local and global contexts. She was awarded her PhD at Newcastle University and is now a Leverhulme Trust Early Career Fellow at Keele University.

Micheline van Riemsdijk is an associate professor of Geography at Uppsala University. She studies highly skilled migrants and the governance of international migration. She has edited the book *Rethinking International Skilled Migration* (with Qingfang Wang, 2017), as well as special issues of the *Journal of Ethnic and Migration Studies*, *Third World Quarterly*, and *International Migration*. Her research has been funded by the Swedish Research Council for Health, Working Life and Welfare, the US National Science Foundation, and the Research Council of Norway.

Margaret Walton-Roberts is a professor in the Department of Geography and Environmental Studies at Wilfrid Laurier University and is affiliated to the Balsillie School of International Affairs. She has published widely on issues related to gender and migration, immigrant settlement in mid-sized cities, and global health professional migration. Her current research focuses on the international migration of health care professionals within Asia and from Asia to North America and Europe.

Index